Springer Biographies

The books published in the Springer Biographies tell of the life and work of scholars, innovators, and pioneers in all fields of learning and throughout the ages. Prominent scientists and philosophers will feature, but so too will lesser known personalities whose significant contributions deserve greater recognition and whose remarkable life stories will stir and motivate readers. Authored by historians and other academic writers, the volumes describe and analyse the main achievements of their subjects in manner accessible to nonspecialists, interweaving these with salient aspects of the protagonists' personal lives. Autobiographies and memoirs also fall into the scope of the series.

More information about this series at http://link.springer.com/series/13617

John H. Cummings

Denis Burkitt

A Cancer, the Virus, and the Prevention
of Man-Made Diseases

 Springer

John H. Cummings
School of Medicine
University of Dundee
Scotland, UK

ISSN 2365-0613 ISSN 2365-0621 (electronic)
Springer Biographies
ISBN 978-3-030-88565-6 ISBN 978-3-030-88563-2 (eBook)
https://doi.org/10.1007/978-3-030-88563-2

© The Editor(s) (if applicable) and The Author(s), under exclusive license to Springer Nature
Switzerland AG 2022
This work is subject to copyright. All rights are solely and exclusively licensed by the Publisher, whether
the whole or part of the material is concerned, specifically the rights of translation, reprinting, reuse of
illustrations, recitation, broadcasting, reproduction on microfilms or in any other physical way, and
transmission or information storage and retrieval, electronic adaptation, computer software, or by similar
or dissimilar methodology now known or hereafter developed.
The use of general descriptive names, registered names, trademarks, service marks, etc. in this publication
does not imply, even in the absence of a specific statement, that such names are exempt from the relevant
protective laws and regulations and therefore free for general use.
The publisher, the authors and the editors are safe to assume that the advice and information in this book
are believed to be true and accurate at the date of publication. Neither the publisher nor the authors or the
editors give a warranty, expressed or implied, with respect to the material contained herein or for any
errors or omissions that may have been made. The publisher remains neutral with regard to jurisdictional
claims in published maps and institutional affiliations.

This Springer imprint is published by the registered company Springer Nature Switzerland AG
The registered company address is: Gewerbestrasse 11, 6330 Cham, Switzerland

For Ireland – its sons and daughters

Preface

Why write a biography of Denis Burkitt? Because he was the first to describe a cancer of children in Africa that would eventually be called 'Burkitt's lymphoma'. As a result of his legendary search for its geographical distribution, he started the quest for viruses that might cause human cancers. He then pioneered a particular form of chemotherapy, resulting in successful treatment for these children. Continuing his unique approach to the cause of disease once he had returned to England, he identified the key role of dietary fibre for health. Beyond fibre, but now fundamental to present-day public health nutrition policy, he developed with others the concept of 'Western diseases'. He considered these to be 'man-made diseases', which led him to propose key strategies for their prevention. We owe him a lot.

But there are already three books about him, in addition to many papers in learned journals and more popular publications, describing his life and work. So why more? Is it because I was asked to? That's not sufficient reason. Or because I knew him well having first met him in 1972 and worked in an area of medical and nutritional science, which he popularised, namely dietary fibre? We were always John and Denis from then onwards. Maybe being born into a family with strongly held religious beliefs, as was Denis, gives me an additional insight. But having already written an appraisal of some of his fibre work there seemed little point in writing more.

The person who realised that I might be persuaded to take on this project was Olive Burkitt, Denis' longsuffering but devoted wife, whom I had known since 1984 when we were on a trip to a meeting in Athens 'with wives'. I was interviewing her at home in Bisley, Gloucestershire whilst gathering information about Denis and dietary fibre when she interjected that no-one had written about Denis the scientist and what his real contribution to medicine and health had been. It was her way of saying that she was not overly enamoured by the existing accounts of Denis' life although in fairness to their authors each had a different story to tell.

The Long Safari by Bernard Glemser, published in 1970, brought to the attention of the public the heroic journey Denis made through East Africa with two colleagues to define the geographical limits of the childhood cancer, which he had discovered. The geography of disease was a concept that would define his career and life. In

Glemser's book is a significant part of the written record Denis made at the time of the journey, a valuable resource for this early adventure. Secondly, Kellock's *The Fibre Man*,1985, arose out of the difficulty Denis had in finding a publisher for his autobiography, which he was pleased to hand over to Brian Kellock. It is a faithful rewriting in a more readable form but is very much what Denis wanted his public to know about himself and his work. Thirdly, Dr Ethel Nelson's *Burkitt, Cancer, Fibre*, completed in 1992 but not published until 1998, was written specifically at the request of Denis to give credit to his family and his faith. Olive most definitely did not like it.

What carried me over the line of commitment to this project was the offer by Denis' and Olive's family, particularly his eldest daughter Judy, to give me access to the archive of papers, photographs, press cuttings, his personal diaries and other family memorabilia, which are in their safe keeping. There are 72 annual volumes of diaries starting in 1923, when he was aged almost 12, and ending in 1993 shortly before his death. Some of the diaries are small whilst other years extend into two substantial volumes. Three years of diaries are missing: 1932, 1933 and 1934. His writing is challenging to read but the reward is great in that it is a very personal account of his daily life, the people he met and his views on what was going on in the world around him or, importantly, his failure to comment on them. These complement other substantial archives at Trinity College Dublin and at the Wellcome in London. The first six chapters of this book describe the personal journey of Denis to find a meaning in life, before encountering in Africa the childhood cancer that would make his name.

He was brought up and educated in an Ireland that was undergoing historic changes, including a war of independence that led to a treaty with the English in 1921. The country was divided into a northern 6 counties, which remained part of the United Kingdom, and a southern 26 counties, which over the next 16 years evolved to become the Republic of Ireland. The significance for Denis who grew up during these years, attended Trinity College Dublin and started work as a doctor at its Adelaide Hospital in 1936 must have been great. His family home in County Fermanagh, where his father was the senior surveyor, was one of the six counties to remain 'British'. Did Denis feel a loss of his Irishness? Was this ever an issue for him? Reading the diaries or any of his other writings does not help with this.

Denis arrived in Africa in 1941 for military service when there was a world war in progress. He would return, to Uganda, in 1946 and would remain until 1966, a witness to the independence movement that was sweeping across the continent. He expresses regret at the conflicts that emerged and the death or displacement of many of his missionary friends. Denis was adamant he was never a missionary, but any analysis of the rights and wrongs of how the 'occupying powers', which included the UK, Belgium, France, Portugal and Germany, acted towards the peoples of these countries is not made.

I must apologise for lives described and incidents related, which smack of colonialism, and attitudes that today many readers will not find acceptable. Moreover, I have used the colonial names of some of the countries of East and Central Africa, such as Tanganyika for Tanzania, Northern and Southern Rhodesia which became

Zambia and Zimbabwe, and Nyasaland for Malawi until such time in this story they became independent. This is part of the essential context in which Denis, a child of his times, made his most important discoveries some 60–70 years ago. His various accounts of daily life do give some unique insights into a time of great change in Africa, even if Denis makes only the briefest of notes and comments on what was going on around him. He was very focussed on his research.

This book is not a detailed nor critical review of the present state of the art with regard to Burkitt's Lymphoma nor similarly the evolution of thinking about dietary fibre and its place in current recommendations for healthy diets. There are other more learned articles and papers in the scientific literature that deal with this. But in recognition of the importance to health of these subjects, I have attempted to set down the present understanding of these still-evolving concepts because they are part of the Burkitt legacy. They are of importance to all of us. Moreover, whilst these discoveries and concepts came to being in the new molecular age following the publication of the structure of DNA in 1953 and genetic code by Crick in 1966, this is not a story permeated with descriptions of high technology and complex nomenclature. In fact, Denis, aware of what was going on because he very quickly became a keynote speaker at major scientific meetings, took pride in telling everyone that he never worked in a laboratory, was not a scientist in a white coat and was never in receipt of large sums of money for his research. Reading his papers, you will find no biochemistry, no statistics except an occasional mean value for a series of observations he had made, little regard for the science of epidemiology but always painting a picture with a broad brush as he saw life around him. When the world's scientists started to take an interest in Denis' work, especially the possibility that cancer could be caused by viruses, he realised that it was time to move on. Back in the UK in 1966 when asked about the latest news of his lymphoma, he replied: 'It's all out of my hands now. All the really clever chaps in epidemiology, virology, immunology and biochemistry have left me in the dust'.

There is another aspect of Denis' life that needs to be aired before embarking on any journey with him, his faith in God. This account contains a lot about the Christian religion, God, the Bible or 'scripture' as Denis calls it, prayer and personal devotion. Brought up in a home where life was suffused with the Protestant ethic and observance of a religion defined by the Church of Ireland, Denis followed in the family tradition until, when at Medical School in Dublin, he underwent a transforming experience at a meeting of Christian Union students one evening. From that day onwards he became an evangelical Christian with a mission to 'save souls for God' pursued in such a way that almost certainly impacted his career. He was reprimanded by his consultant whilst a junior doctor for handing out religious tracts to patients in the ward. His initial failure to be accepted for a post with the Colonial Medical Service and to see active service or be given a substantial command whilst in the army must in part have been as a result of his proselytising zeal. But his faith gave him an inner strength and confidence in both the present and the life hereafter that served him well and brought with it a network of friends around the world in whose company he felt secure.

All this is part of the story of the life of Denis Parsons Burkitt and essential background to understanding the man and his motivation. But do not underestimate him. His successes have impinged on all our lives. Very few people have a disease named after them, still fewer a suggestion for its cause, a virus, which bought a new dimension to our understanding of cancer and its possible prevention by vaccination. The move by Denis into nutrition came about by his investigating why, as a surgeon, he never saw large bowel cancer in Uganda when it was so common in North America. This important observation grew into the worldwide recognition that a part of our diet, fibre, hitherto largely ignored, was the key to a new understanding of a collection of diseases that are major causes of death now all over the world. Initially called 'Western diseases' by Denis and Hugh Trowell, his co-investigator in later years, they include coronary heart disease, hypertension, diabetes, obesity, gallstones, other bowel disorders like diverticular disease, and venous afflictions such as varicose veins and haemorrhoids. Not daunted by the immensity of the problem of Western or 'man-made diseases', in the final part of his career he came to believe that the future lay in prevention rather than treatment.

The reason for my taking on this task at first seemed motivated by a sense of duty and loyalty to Denis and his family but it has proved to be an engrossing and rewarding journey with him through a life that had its frustrations but also achievements of historic moment. It is the first biography to be researched and written after his death, almost 30 years ago, a suitable time for reflection and the judgement of history. The first to have unfettered access to such a large and personal archive. The first by a medical scientist to look at Denis from this viewpoint and the first that hopefully Olive would have enjoyed.

Scotland, UK   John H. Cummings

Acknowledgements and Sources[1]

Without the help of Denis' family, this biography would not have been possible. Olive Burkitt,[2] whom I first met in May 1984, was a unique source of oral history. Together with her diaries, autobiographical notes and candid views on some of the people she met on her travels with Denis, she was able to give a balancing account of their family and social life. Their three daughters, Judith, Carolyn and Rachel have recalled their personal memories and given helpful and unquestioned access to their substantial archive of Burkitt material, which included the diaries of both Olive and Denis, a large collection of photograph albums, endless boxes of 35 mm slides, audio and video tape recordings of Denis lecturing or recounting events, movies he made, his personal papers, letters and mementoes of his world travels. Much else of the family history resides in this archive without access to which more of the personal side of Denis' life could not be told and the influence of his distinguished family on him. Other members of the family and friends were also helpful including, especially, Philip Howard, Judy's husband and generous host during my numerous stays with them; Elizabeth Dawson; Robin (Paddy) Burkitt,[3] son of Robert (Robin) Burkitt, Denis' brother; David Maurice, Rachel's husband, their son James Maurice; and John Cookson, friend of Olive and ex Dean Close School. Except where specifically mentioned, all of the photographs in this volume were taken by Denis, or at his request, with his camera and reside in the family archive.

Conscious towards the end of his career that he had achieved special status having had a cancer named after him and found, what he believed to be, the way to prevent many of the diseases of modern civilisation, he started in 1988, aged 77, to give much of the record of his scientific work to the Wellcome Tropical Institute, which subsequently became part of the Wellcome Library[4] and additionally to the

---

[1] Here and throughout the book, I have omitted people's titles.
[2] Olive Lettice Mary Burkitt (Rogers) SRN 1920–2017.
[3] Robin Patrick Parsons Burkitt OBE 1941–2019.
[4] Wellcome Library Papers of Denis Burkitt WTI/DPB/.

library of his alma mater, Trinity College Dublin.[5] There are thus two major reposi-
tories of Denis' papers.

My thanks are due especially to Dr Jane Maxwell, Principal Curator of the
Manuscripts & Archives Research Library at Trinity and to Dáire Rooney, Aisling
Lockhart and other staff of the library and Reading Room. The library has records
of Denis' time as an undergraduate, papers and research records relating to his work
on the lymphoma in Uganda and some of his early maps plotting disease incidence,
including his very first of hydrocele. There are patient notes, many clinical photographs of the lymphoma and, donated in 2011 on the occasion of a meeting to celebrate the 100th anniversary of Denis' birth, a much-annotated Bible apparently
used when he was in the Army serving mainly in Africa. In addition, there are
hugely packed ring binders of press cuttings and meeting programmes, which Olive
dutifully, perhaps also with pride, collected documenting his subsequent years
of travel.

The Wellcome Library has a much bigger archive physically with both early
material of Denis' relating to his lymphoma work, but also a comprehensive collection of his research on dietary fibre, diet and disease as well as his unpublished
autobiography. Other valuable related archives at the Wellcome include papers and
books relating to Hugh Trowell, Neil Painter, Richard Doll, Peter Cleave and many
other key individuals whom Denis met. My thanks for help in accessing these
archives are due particularly to Jenny Shaw, Collections Development Manager,
Amanda Engineer, Christopher Hilton, Edward Bishop, Peter Judge, Amelia Walker,
Arike Oke, Crestina Forcina, Holly Peel, Miriam Keaveny and other staff of
Archives and Manuscripts and those working from time to time in the Rare Materials
Room. Pride of place at the Wellcome should probably be given to Denis' account,
recorded daily, of the long safari of October to December 1961 along with accounts,
sometimes extensive, of his subsequent journeys.[6]

Other libraries with invaluable resources that I have consulted include that of the
Royal Society of Medicine in London, The British Library and, of a different scale
but key to understanding the life of James Burkitt, that of the British Trust for
Ornithology in Thetford, Norfolk, where the help of the librarian Carole Showell
saved me much time.[7] For assistance in defining the ancestry of the main players in
the family story, I thank the Civil Registration Service of Ireland, the General
Register Office for Northern Ireland and the General Register Office of the UK.

Denis made many audio and video recordings at the request of various organisations, some of which are with the family archives. Available online are four video/

[5] Trinity College Library, Manuscripts & Archives Research Library IE TCD MS 11268 Papers of
Denis Parsons Burkitt and MS 11387 Papers of Denis and Olive Burkitt.

[6] Wellcome Library, Papers of Denis Parsons Burkitt WTI/DPB/D/1-45.

[7] Library of British Trust for Ornithology; Burkitt, J P – Personal papers (Robin territories and
manuscript on survival) R 5-1-1.

audio tapes of interviews with Dr Max Blythe done on behalf of Oxford Brookes University in 1990 and 1991.[8]

My grateful thanks are also due to a number of individuals who knew Denis or worked at Makerere and have given me both first-hand information about Denis and other players in this story. They include Keith McAdam for insights into the development of the Medical School at Makerere, the influence of his father Ian McAdam as Professor of Surgery and family life there with the Burkitts; Dick Drown,[9] lifelong friend and mentor of Denis; Jack Darling,[10] whose notes written in 1991 for Ethel Nelson give a first-hand account of Denis' early years both in Ireland and Uganda; Tony Leeds, who first met Denis in Ilesha Uganda in 1970 and subsequently had extensive correspondence with Hugh Trowell and Alec Walker; Martin Eastwood, who invited Denis to the first meeting on dietary fibre in Edinburgh in 1973, bringing a new dimension to his thinking about fibre, for helpful discussions at an early stage of this project; Norman Temple, the last person to collaborate with Denis and who knew him well in his final years; Jim Mann for personal recollections of Denis and Hugh Trowell and GD Campbell in South Africa, who together added much to ideas about fibre and diabetes; Este Vorster,[11] who worked closely with Alec Walker over many years; John Church for his vivid recollections of safari in Rwanda with Denis; and Joan Church for her stories[12] of life in Africa.

My mentors in matters literary and biographical include Adam Sisman, who, at an early stage, introduced me to the realities of biography writing; George Misiewicz to whom, in 1970, I gave my first attempt to write a scientific paper only to have it returned covered in red ink, and he has been a most valued wise council over the years; Alison Lennox (Stephen), who sent me a number of biographies of other famous scientists and gave me the benefit of her knowledge of biography writing; Alison Ball of Springer New York, my thanks for practical help in turning my amateur text into a book and for her constant optimism; Richard Booth in Dublin, who helped me with some Irish history and reminded me that I was not doing my readership a favour but rather needed a compelling account with which they would engage enthusiastically; and John Gillespie for useful information about the history of Dublin and Richard Price, longstanding friend in Cambridge, who spent 7 years in East Africa as a teacher and who has helped me to relate with more accuracy matters relating to the geography of the region and culture of its people. In England, Margaret Barham enabled me to better understand the mind of Denis and shed some light on the principal influences that determined Denis' behaviour.

In all matters concerning children's health, cancer and viruses, I have to thank Owen Smith in Dublin who tolerated most diplomatically my ignorance in these

---

[8] https://radar.brookes.ac.uk/radar/file/c25588ea-b30a-46b6-9c74-4b9461787bad/1/Burkitt%2C%20Denis%201%201990.12.10%20MSVA053.pdf
[9] Rev. Richard Drown 1919–2018.
[10] John Singleton Darling OBE, FRCS 1911–2002.
[11] Hester Hendrina Vorster (Martins) DSc 1943–2020.
[12] Joan Church (2020) Life's an Adventure. Cambridge Red Graphic.

matters and spent time correcting and upgrading chapters on these subjects. Owen has done more than anyone to keep alive the memory of Denis in the country of his birth; he has documented and interpreted the importance of his work for the present and has established with the help of Denis' family the Burkitt lecture at Trinity College.[13]

My thanks are also due to Margaret Bray, Simon Akam, Joan Trowell, and the family of Hugh Trowell for help with archives about Uganda and particularly the unpublished biography of Trowell by Elizabeth Bray.[14] I would especially like to thank Thorold Masefield for introducing me to Dick Drown,[15] whom I was able to meet at Thorold's home at Brockenhurst in Hampshire where I enjoyed Thorold's hospitality and many stories of life in Uganda and especially about that of his wife Jenny's[16] father, Hugh Trowell.

To Mike Wilkinson for discussions on the early studies of fibre in animal nutrition, Prakash Shetty[17] for persuading me that this was a worthwhile project, and Susannah Wilson and family for valuable suggestions about the text. To Gillian McClure and John Maney for their knowledge of the world of publishing, and finally to Philip and Jean James for stimulating conversation and a bed for the night on my frequent visits to London.

Biblical references are all from the Revised Version with Revised Marginal References. Oxford University Press. First published in 1898.

Finally, to Wikipedia that much-maligned but endlessly informative soul of the Internet.

Illustrations The majority are from the photo albums, boxes of 35 mm transparencies and extensive records of his travels, which Denis, with a lifelong interest in photography, left for posterity. Wherever he went, he carried a camera. These are now in the family archive and are used here with their kind permission. Equally valuable are those from the archives of Trinity College Dublin Library and the Wellcome Library in London. These are acknowledged where they appear. Others have been provided for publication by individuals, journals or the press and their permission to publish is included in the figure legends. The quality of certain illustrations is not up to today's standards of digital imaging, this is because they are either old or remastering has not been possible because of lockdown during the pandemic.

---

[13] Smith O (2012) Denis Parsons Burkitt CMG, MD, DSc, FRS, FRCS, FTCD (1911–1993). Brit J Haematol 156;770–776.

[14] Wellcome Library Papers of Hugh Trowell, PP/HCT/A.5.

[15] Rev.Richard Drown 1919–2018.

[16] Jennifer (Jenny) Masefield nè Trowell 1940–2019.

[17] Prakash Sarvotham Shetty MD, PhD 1943–2018.

Contents

About the Author

Professor John Hedley Cummings
OBE, MA, MSc, PhD (Hon), FRCP (London and Edinburgh), RNutr.

John Cummings is Emeritus Professor of Gastroenterology and Nutrition at the University of Dundee.

Born in Lincolnshire, he was educated at a Methodist boys boarding school, Woodhouse Grove, in Yorkshire. He graduated in medicine from Leeds in 1964, and after undertaking specialist clinical training joined the staff of the Medical Research Council where he worked for 5 years in London at the Gastroenterology Unit, then for 23 years at the MRC Dunn Clinical Nutrition Centre and at Addenbrookes Hospital in Cambridge. There he gathered together a unique research group working on diet, the gut microflora and the role of the large bowel in human nutrition. He has published over 300 papers and articles on this subject and is known for his pioneering work on the importance of colonic bacteria to health; the role of carbohydrates, especially resistant starch and dietary fibre as substrates for fermentation; and the metabolism of short chain fatty acids. In 1998, he moved to Dundee, Scotland, at a time when the importance of large bowel bacteria, or the gut microbiome, was being recognised.

Professor Cummings has served on many national, European and international bodies concerned with diet and health, including the UK Government's Committee on Medical Aspects of Food and Nutrition Policy (COMA) and Scientific Advisory Committee on Nutrition (SACN). He was chairman of the IUNS Diet and Cancer Committee and a member of the EC concerted action PASSCLAIM, where he chaired the Working Group on Gut Health and Immunity and was a member of the Consensus Group, which drew up criteria for substantiation of health claims on food in 2005. He has served WHO in various roles including as a member of the FAO/WHO Expert Consultation on Carbohydrates in Human Nutrition (1997), the Scientific Update on Carbohydrates in Human Nutrition (2006), the WHO Nutrition Guidelines Advisory Group (NUGAG) from 2009–2019 and the Expert Advisory Panel on Nutrition (2011–2021).

In 1993, Professor Cummings was awarded the Caroline Walker Science Prize and the Cleave Trophy of the McCarrison Society in recognition of his work on public health aspects of diet. In June 2008, he was made an OBE in the Queen's birthday honours for his services to medicine and nutrition and awarded the British Nutrition Foundation Prize in the same year. In May 2009, the North West University in South Africa made him an honorary PhD for the work he had done with them in helping to establish their department of nutrition over a period of 25 years. He was awarded the Rank Prize Funds Nutrition Prize in 2012.

Abbreviations

AFIP	Armed Forces Institute of Pathology (USA)
AIM	Africa Inland Mission
AMFA	Affordable Medicines for Africa
BMA	British Medical Association
BSG	British Society of Gastroenterology
CHIP	Coronary Health Improvement Project
CMS	Church Missionary Society
CMF	Christian Medical fellowship – CMS in the USA
CMO	Chief Medical Officer
DIY	Do it yourself
DRC	Democratic Republic of the Congo
DMO	District Medical Officer
DMS	Director of Medical Services
EBV	Epstein-Barr Virus
ENT	Ear, Nose and Throat
ENSA	Entertainments National Service Association
FRCS	Fellowship of the Royal Colleges of Surgeons (London, Edinburgh, Glasgow or Dublin)
GOS	Great Ormond Street Hospital, London
HPV	Human Papilloma Virus
IARC	International Agency for Research on Cancer (Lyon, France)
ICRF	Imperial Cancer Research Fund, London
IM	Infectious Mononucleosis (Glandular fever)
MAP	Medical Assistance Programme
MRC	Medical Research Council of the UK
MS	Master of Surgery
NCD	Non-Communicable Diseases
NCI	National Cancers Institute of the USA
NHS	National Health Service of the UK
NIH	National Institutes of Health (USA)
PA	Personal Assistant

RAF	Royal Air Force (of the UK)
RAMC	Royal Army Medical Corps
RC	Roman Catholic
RSM	Royal Society of Medicine (London, UK)
RMO	Resident Medical Officer
RSO	Resident Surgical Officer
SAIMR	South African Institute of Medical Research
SCM	Student Christian Movement
SDA	Seventh Day Adventists
SKI	Sloan Kettering Institute, New York
TCD	Trinity College, Dublin
UCI	Uganda Cancer Institute, which in 2015 became the Uganda Cancer Institute/Fred Hutchinson Cancer Research Center Clinic and Training Institute
UCSF	University of California San Francisco
UICC	Union Internationale Contra le Cancer. Union for International Cancer Control
UK	United Kingdom of Great Britain and Northern Ireland
VIP	Very Important Person
YMCA	Young Men's Christian Association

Chapter 1
For God, the World and the Robin

Denis was born into a family with strong ecclesiastical roots where service abroad, in whatever profession, was an accepted norm. He would acquire a faith that would at times motivate his whole being and be important in his decision to work in Africa. But what about the robin? Is the early study of this familiar garden bird the reason why wholegrain bread is an essential part of dietary recommendations for many countries today? How did the name Burkitt become associated with the robin and its behaviour, and did Denis learn anything from his father's study of it?

Denis' father, James Parsons Burkitt an engineer by profession, was one of the greatest amateur ornithologists of his time. In a landmark series of studies, he defined the territories of the robins in his 15-acre garden in Ireland by using rings, which he made himself, attached to their legs. The maps he drew of these territories have passed into history as one of the defining contributions to the understanding of bird behaviour. Some of this interest and knowledge was transmitted to Denis because he became a keen collector of the eggs of wild birds.[1] The mapping of bird territories must also have been noticed by Denis as it was clearly reprised by him in his own pioneering studies of the geography of human health but gets scant acknowledgement by him. What then inspired Denis to draw maps of the prevalence of many diseases and become the most famous descendant of this talented family?

David Lack's classic book *The life of the Robin*, published in 1943 [1], gives an important clue as to the significance of the robin to the Burkitts. The result of 4 years of Lack's observations in the woodland, orchards and fields around Dartington Hall School in South Devon, it is a very readable book, which ran to five editions and many printings, including as a Penguin paperback. Whilst scholarly in its meticulous study of the robin, the book entertains and entices you into it by asking you 'to consider a bird whose life is devoted almost exclusively to fighting'. In the index of people mentioned in the text, Charles Darwin is there, as are Konrad Lorenz, Julian

---

[1] Collecting the eggs of wild birds is no longer legal in either Ireland or the UK.

© The Author(s), under exclusive license to Springer Nature Switzerland AG 2022
J. H. Cummings, *Denis Burkitt*, Springer Biographies,
https://doi.org/10.1007/978-3-030-88563-2_1

Huxley and Eliot Howard, a noted amateur ornithologist of the time. But one person receives more mentions than any other, namely, J.P. Burkitt, whom Lack credits with elucidating the 'main outlines of the life-history of the robin' some 10 years before he had started his own studies. He notes that James Burkitt was the first observer of birds to use coloured rings and says he should be given full credit for his original work. At the back of the book, there are 16 pages of references, but Lack says that the work of Burkitt is referred to by him so often that he will simply cite his papers once at the beginning.

James Burkitt (1870–1959) was not a professional ornithologist, just a very able and intelligent observer of all that went on around him. According to Lack he did not start his observations of birds until he was 37, at which time he had no bird books or ornithological friends, had never joined the British Ornithologists' Union (BOU) nor did he have much contact with other ornithologists. Yet, at a time in the early part of the twentieth century, when Britain led the world in this field, Lack named James Burkitt as one of the seven pioneers of research into the life of birds [2]. And what James Burkitt did, in mapping the territories of birds to learn of their behaviour, was probably important not only for the world of natural history but also for his son's future career and therefore for all of us.

If Denis' interest in birds was confined to egg collecting, did he gain anything from his other talented forbears? They were Irish and adventurous, had careers abroad and were imbued with a strong belief in God.[2] Denis would continue much of the family's lifestyle and philosophy. James' father, Denis' grandfather, was the Reverend Thomas Henry Burkitt (1835–1922) a Presbyterian minister. Thomas was born in Wexford, a coastal town to the southeast of Ireland, where his father was a successful solicitor who had earlier been Commandant of Police in British Honduras. The Burkitt family had emigrated from Northamptonshire in England as farmers some 200 years earlier. Thomas went to Trinity College Dublin and then travelled to Toronto University in Canada to finish his education, which was not unusual at that time. But his experience in Toronto was to change his life completely, and possibly that of his descendants, because there he joined a group of very committed young Presbyterians whose faith and influence led him to leave the Church of Ireland, the church of his family, and eventually to become ordained as a Presbyterian minister. On his return home, his relations were not at all pleased by this change in allegiance, but such was the strength of his belief that he turned his back on them and took a posting as minister in a small and spread-out Presbyterian community along the coast of Donegal, the focus of which was at Killybegs, far from where he had been brought up. The life of a Presbyterian minister at this time was one of hardship, which meant that his family would grow up to know the rigours of faith with penury. His parishioners were spread widely across the countryside, and many were visited

---

[2] Pedigree of the Family of Burkitt of Northamptonshire and later of Sudbury, Suffolk. Compiled by Major H.R.E. Rudkin in 1934 and revised in 1996 by David E Burkitt. Family archives; notes of Robert Townsend Burkitt of Norfolk; Burkitt, RT. Our side of the Burkitt family from 1835, the date of my Grandfather's birth. Unpublished manuscript. Burkitt family archives. Updated by the author. With thanks to Tim Cummings for the construction of Fig. 1.1.

by him on horseback, carrying a revolver in case he was threatened. In 1897 the family moved to Athenry, near Galway, where they had far fewer parishioners and had to lead an even more frugal lifestyle (Fig. 1.1).

In 1867,[3] Thomas had married Emma Eliza Parsons (1838–1931) the daughter of the Reverend James Parsons of York and Emma Wilks from where both Denis' and his father James' middle name of Parsons originates. The Parsons and the Wilks were well-known non-conformist families in the north of England. Emma Wilks was the daughter of James Wilks, MP for Boston (1776–1852), whilst James Parsons had started a career as a solicitor but was called to the Ministry and was for 50 years in charge of the main Congregationalist Church in York. He was close friends with other famous preachers of that time including George Whitfield the evangelist who, with John Wesley, helped to found Methodism. Denis' lineage in the Christian faith was therefore strong and extended back at least three generations. Apparently undaunted by the privations that his paternal grandfather, Thomas Burkitt, had endured for his faith, Denis would become committed to an evangelical form of Christianity.

The Rev Thomas and Emma Burkitt had seven children born at Killybegs. The five boys all had distinguished careers, whilst the two girls were bound by the social mores of the Victorian era that did not expect girls to earn a living. Neither Norah (1874–1965) nor Mary (May) (1878–1966) married. Norah did not have any higher education and instead spent her life caring for her parents and then other family members ending her days at Ballycastle, a coastal town in Northern Ireland where she would frequently play host to her siblings and their children. May became a schoolteacher and was a governess for some years before being appointed Head of Leeson House, a small private girls school at Langton Matravers, a village in Dorset close to the Jurassic Coast, to which a number of the girls in the Burkitt clan would be sent. May also spent her last years in Ballycastle. Denis recalls time with Aunt May with a fondness born out of the security of family life.

Aside from James and the two girls, there were four other boys in the family, Robert, Roland, Harold and Francis, whose careers took each of them away from Ireland. The life of Robert (1869–1945) the eldest child is something of an enigma. Perhaps the truth about him has been buried with him. He was sent out at age 14 to Nova Scotia to finish his schooling and went on to Halifax University, graduating in 1890 with honours in mathematics before proceeding to Harvard where he studied mining. At Harvard he formed a strong friendship with George Byron Gordon (1870–1927) who was of a similar age and with whom he shared common interests. Robert joined Gordon in Central America in 1894 and worked as his assistant at Copan, learning the basics of archaeology. When Gordon left for Cambridge at the end of the excavations, Robert decided to stay on having become fascinated by the language and culture of the Mayan people. He spent the rest of his life in Guatemala, becoming Professor of Archaeology at the University of Guatemala and an authority on Mayan folklore, ritual, crafts and language. He attained recognition for his

---

[3] or the beginning of 1868.

Family tree of Denis Parsons Burkitt

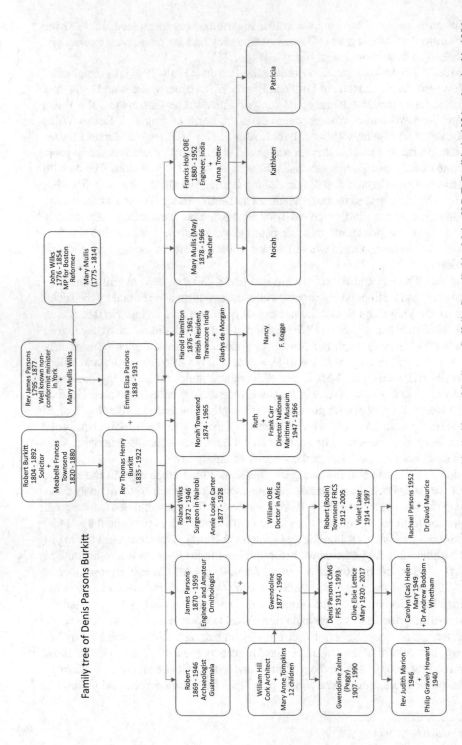

Fig. 1.1 Pedigree of the Family of Burkitt of Northamptonshire and later of Sudbury, Suffolk. Compiled by Major H.R.E. Rudkin in 1934 and revised in 1996 by David E Burkitt. With amendments from the family archives and notes of Robert Townsend Burkitt of Norfolk, 'Our side of the Burkitt family from 1835, the date of my Grandfather's birth'. (Unpublished manuscript.) Updated by the author. With thanks to Tim Cummings for the construction of this family tree

knowledge of the Kekchi language and was working on a grammar and dictionary of the language when he had a stroke in 1945. He is buried, alone, in the British Cemetery in Guatemala City [3].

Robert never married and was not a family man. In the early years of his time in Guatemala, he wrote regularly to his father and brothers sending back money, but then quite suddenly all correspondence stopped, from which time he never again communicated with any of them. A reason suggested for this is that very early in his time in Guatemala, he became a Roman Catholic, and he knew how intensely this would offend his rigidly Protestant parents and family. Certainly, his father the Reverend Thomas had strong views about the Catholic Church, enough to make him move to a new house, but were there other aspects of Robert's lifestyle that might have offended the family much more? Denis never met him and never considered himself to be an academic.

Harold and Francis, the fourth and fifth sons, both worked for most of their lives in India. Harold (1876–1961) went to Trinity College Dublin and joined the Indian Civil Service in 1900 starting as district officer and becoming British resident in Travancore, a province at the south-west tip of India. A road was named after him, Burkit Road, in the Thyagaraya Nagar district of Chennai (Madras)[4] where he was President and Mayor of the Corporation from 1917 to 1918. Francis Holy Burkitt (1880–1952) was the youngest of the seven children. He also spent his career in India where he became Chief Engineer and Secretary to the Governor of Northwest Province and was awarded the OBE for his services. Denis was aware of his uncles from India but was not close to either. Instead of India he would work in Africa and have a ward at the Tallaght University Hospital in Dublin named after him.

More thought needs to be given to Roland Wilks Burkitt (1872–1946) who qualified as a surgeon in Dublin and spent most of his working life in Kenya and as Uncle Roland featured frequently in the life of Denis. His career deserves more recounting because he was almost certainly one factor in putting Denis, and his brother Robert, known as Robin, off from starting to study medicine 30 years later at Trinity College.

Roland qualified in 1906 and then worked for Sir Miles Thomas at Richmond Hospital where he learnt his surgical skills and how to give anaesthetics. After a brief encounter with London society medicine, which suited him not at all, he joined a thriving practice in Sussex where there were plenty of cases requiring surgery including removal of tonsils and adenoids, appendicectomy, hernia repairs and various lumps and bumps. It was an ideal job for him, but soon the patients were refusing to see the young doctor for fear of what he might remove next. He was politely asked to leave and, given his limited experience of medicine, applied for an appointment on a tea plantation in Assam. The climate, however, proved to be far from healthy, and at the invitation of Captain Robert Hill, who was Secretary of the Nairobi Club and connected to Roland by marriage, he went in 1911 to Nairobi where he stayed until 1938 [4] (Fig. 1.2).

---

[4] Burkit Road. Appears on Google Maps today as a tree-lined boulevard of mainly commercial properties. Note spelling.

Fig. 1.2 Dr Rowland
Wilks Burkitt (1872–1946)

Roland became a competent and experienced surgeon who took every opportunity to learn new techniques on trips back to England, but it is as a physician that he acquired his reputation as 'Kill or cure Burkitt'. He is best known for his approach to fever. He carried around with him an oil sheet which he would put under a febrile patient on their bed, remove all the bedclothes, strip them naked and then either pour cold water over them or arrange for a cold bath, until such time as their fever abated. Babies were placed in a cool draught and sprayed with water from a watering can. So drastic were his measures that he would be sent for only as a last resort, but it was acknowledged that he saved lives and allegedly did not lose a single patient during the Spanish 'flu epidemic of 1918'. He published a booklet entitled *Treatment of a fever for those remote from a Doctor*, which became widely used throughout East Africa where medical help was not always available. There were other less than orthodox treatments such as drawing off a pint of blood to relieve blood pressure, injecting 400–600 mg of phenobarbitone ('Luminal') intramuscularly for the cure of blackwater fever [5][5] or tapping spinal fluid to relieve headache. It was an encounter of the 'what will he remove next?' style with Denis and Robin in their home in 1918 that was to put them off doing medicine.

---

[5]A serious complication of malaria in which the products of broken-down red blood cells are excreted in urine.

Roland married Annie Louisa Carter (1877–1928), and they had one child, William, who studied medicine at Cambridge and then St Bartholomew's Hospital and had an interesting career in Africa. After the army and various hospital postings, he was asked by the Kenya Society for the Blind if he would consider starting a mobile eye surgery unit. The scheme, financed out of William's own resources, was a great success for which he was appointed OBE. He had spent his childhood holidays in Ireland and came to know Denis well when he was in Uganda. With his wife Elizabeth, they would be frequent visitors to the Burkitt household, who took them under its wing during the post-African part of all their careers.

Roland died in December 22 aged 74 following a serious car accident in Kenya and was buried in Dublin. He had left Africa long before Denis arrived there but was well remembered by all who met him. He would encounter Denis in Ireland in 1918.

Denis' father James [6] was altogether more conventional but a man of outstanding ability and achievement. Born in Killybegs he was educated at home along with his other siblings until 1885 when the family moved south to Athenry. There the boys were sent to Galway Grammar School from which James moved on to Queen's College Galway in October 1888 to study engineering and mathematics. He gained a scholarship in each of the 4 years that he was a student and graduated Bachelor of Arts in 1891 and Bachelor of Engineering in 1892 both with first class honours. He was clearly very able and although small in stature (see Chap. 8, Fig. 8.6) was not reticent about his abilities. After graduation he had a visiting card printed which noted that he was 'First of all Competitors in each Professional Examination' and 'First Honorman and Exhibitioner in Mathematical Science' and announcing that he had become 'Senior Scholar and Lecturer in Mathematics, Queens College, Galway'. He did not remain as a lecturer for long soon moving into the world of civil engineering. After working on bridge construction and repair, the railways and waterworks, towards the end of 1898, he applied for the post of County Surveyor for Fermanagh. To obtain this post, he had to sit competitive exams, which took place over 5 days in November. There were nine subjects. James was first in all of them except hydraulic engineering and was top of the Civil Service Commission exams with 739 marks, the next in class having 638. He was appointed, at the unusually young age of 28, as County Surveyor for Fermanagh, where he would remain for the rest of his working life [7, 8] (Fig. 1.3).

James took over the post of County Surveyor at a time when there were few cars and the roads were made primarily of broken stones. He is credited with introducing the use of steamrollers, which used together with the newly discovered Tarmacadam, patented in 1902 by Edgar Purnell Hooley, allowed him to oversee the making of the first road in Fermanagh, the Sligo Road near Enniskillen, of 'water bound macadam'. Apparently, a great joy to the local cyclists [9]. With the creation of two distinct jurisdictions in Ireland in 1921 came the need for roads that avoided crossing the new border. This presented several challenges for the county surveyor's department since the border wrapped around the north-western part of Lower Lough

Fig. 1.3 James Parsons
Burkitt (1870–1959)

Erne.[6] To allow travellers to avoid having to cross into what would become the
Republic of Ireland, a series of bridges were built under the direction of James link-
ing Boa Island with the mainland, which, with the whole of Lough Erne, now lay
entirely in the UK. The bridges were officially opened by the Premier[7] on February
12, 1927 (Fig. 1.4).

James moved to 3 Alexandra Terrace, Enniskillen, on the south-east shores of the
Lower Lough around 1900, with his younger sister May to keep house for him. On
February 10, 1903, he married Gwendoline Hill of Audley House in Cork, daughter
of William Henry Hill, an architect of Cork County, and his wife MaryAnn
Tompkins.

Fewer records exist of the ancestry and achievements of Gwendoline and the Hill
family. Her father was clearly an architect of some distinction but a man without
pretention. He was responsible for the design of the courthouse in Cork and was
subsequently offered a knighthood, which he declined saying that 'he preferred to

---

[6] The nomenclature for the parts of Lough Erne is slightly confusing. The more northerly part of
the Lough is called 'lower' and the southerly extension 'upper'. Both are in present-day 'Northern
Ireland'.

[7] Premier. It is not clear who this is. Note is from Gwen's 'diary', which had a page for every day
of the year and in which she recorded important events in the life of the family. Since the bridge
was on the border, the Premier might well have been the Premier of Ireland at that time WT
Cosgrave or his equivalent in the north James Craig.

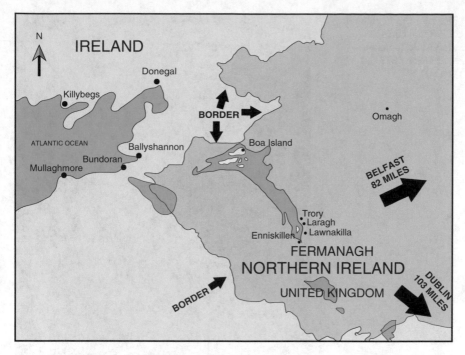

Fig. 1.4 Map showing Lower Lough Erne, Boa Island and the 1921 border between Northern Ireland and the Republic of Ireland

live and die as plain Mr Hill and wanted nothing else' [10, 11]. He married MaryAnn Tompkins (1877–1960), and Gwendoline was the 9th of their 12 children another of whom, Uncle Louis, would have a profound influence on Denis during his years as an undergraduate in Dublin (Fig. 1.5).

The Hill family lived at Audley House in Cork, whilst Gwendoline's sister Muriel married Percy Worth Newenham, owner of a nearby grand country house named Coolmore, which would provide a welcome and familiar refuge for the Burkitt family at holiday times in future years. Gwen was brought up as a member of the established church the Church of Ireland and is said to have 'needled her rather more intellectual husband, because he retained a degree of loyalty to his non-conformist, Presbyterian, upbringing'. Frugal by nature she mended the family's clothes, knitted sweaters, darned socks and wasted nothing (Fig. 1.6).

James and Gwendoline had three children, Gwendoline Zelma (Peggy) (1907–1990), Denis Parsons (1911–1993) and Robert (Robin) Townsend (1912–2005). James and Gwendoline's lives were in many ways very simple, in that James, always referred to as 'Jim' by his wife, stayed in the same job throughout his career. They travelled little, except to visit their boys at school, and moved house only twice, from Alexandra Terrace in Enniskillen to Lawnakilla a couple of miles north in April 1911 and then in March 1926 to a house named Laragh, which they

Fig. 1.5 The 12 children
of William Hill and
MaryAnne Tompkins.
Note Gwen 1877 and
Louis 1880

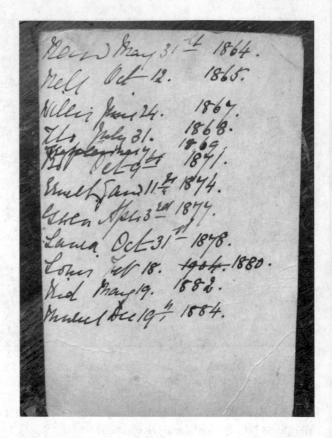

purchased, 3 miles from Ballinamallard a little further north again but still close to
Lough Erne, where they remained (Fig. 1.7).

It was a household where the Christian faith determined the rules by which they
lived. Each morning they had prayers after breakfast, much as both James' and
Gwendoline's parents had done and where the word of God, as made evident
through the Bible, provided a code of practice to which they could always refer.
They were regular churchgoers attending the nearby Trory Parish Church with
James, a teetotaller, as churchwarden and Sunday school teacher. Life at home was
conventional in its order with Gwendoline managing domestic affairs and James
responsible for the garden and the fabric of the building. He had a well-equipped
workshop as one might expect of an engineer. They were comfortably off financially. The 1911 Census of Ireland shows that in March, just before they moved to
Lawnakilla, they employed two servants, Sabina Henry aged 27 as a nanny and for
domestic duties and Rose Scollan as a cook. Once at Lawnakilla they added a gardener. At Laragh they provided a home and secure base for over 30 years to which
the family could always return. Following his retirement as county surveyor in
1940, James continued his bird-watching, reading his Bible and supporting the
church. He was the longest-serving member of the Synod of the Church of Ireland.

Fig. 1.6 Gwendoline Hill.
Mother of Denis
(1877–1960)

Fig. 1.7 Lawnakilla north
of Enniskillen where James
did his most notable work
with the robin

Gwendoline kept, in a small leather-bound volume entitled 'The Soul's Enquiries Answered' given to her by her mother in 1903, brief factual records of life events in the family for over 50 years.

What about the birds? They were a lifelong interest of James. He had 'learnt as a boy to distinguish by sight and sound all the sea birds of that wild Atlantic coast (of County Donegal) and knew all the nesting sites of the rare as well as the common birds' [7]. He had a considerable collection of birds' eggs, and his notebook records his observations, the earliest of which is dated June 13, 1906. Not only did James, and his friends, collect eggs in their early years but also went shooting duck and snipe. But it was his study of the robins in his garden at Lawnakilla that established his reputation as one of the British pioneers of ornithological research.

James studied bird song, and what it meant in relation to the retention of territory and breeding [12, 13]. In October 1922 he started ringing adult robins with a view to defining their territory and behaviour. He is credited by David Lack with introducing the use of coloured rings for bird identification, but Lack subsequently retracted this because he discovered that James was colour blind, a trait he did not pass on to either of his sons (Fig. 1.8).

James was not the first person to ring birds, credit for which is given to a Danish ornithologist Hans Christian Cornelius Mortensen (1856–1921) who successfully ringed starlings in 1899 [14]. But James, as Lack says, had 'initiated a technical revolution in the field study of the individual bird' by attaching metal rings of different widths to the legs of the birds in his garden. It was being able to identify individual birds that enabled James to define their territories [15] (Fig. 1.9).

Fig. 1.8 Rings for bird identification used by James Burkitt following on from the ones he initially made himself

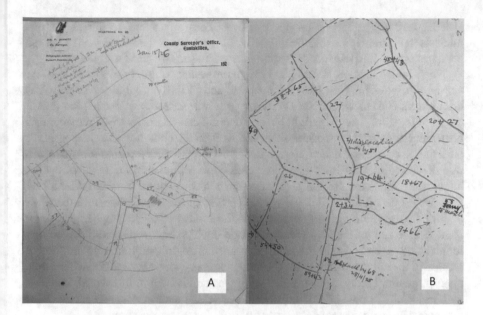

Fig. 1.9 James Burkitt's maps of the territories of robins at Lawnakilla
A. January 15, 1926, map shows overview of garden and its internal divisions with robins by number and their location
B. Detail from map of nesting sites dated 1925 showing territories of pairs clearly defined (By kind permission of the British Trust for Ornithology, Thetford, UK)

James wrote up his observations in a series of five papers published in *British Birds* between 1924 and 1926 [16–19]. There is also a sixth paper, 'Longevity of Birds in the Natural State' which was never published and which resides in the archives of the British Trust for Ornithology in Thetford, Norfolk.[8] Having ringed 81 birds, James used his evident talents as a mathematician to draw up equations to come up with a figure of 2.8 years for their average life span although admits that this may be too high.

The move to Laragh did not end James' interest in birds. He notes that 'I have just acquired a rookery with a residence attached, and a barn owl thrown in' [20]. In 1938 when he saw a female who had been ringed in 1927, he wrote a note for the journal entitled 'Eleven-year-old robin' in which he remarks 'I think this case is a record'. His ideas were not recognised by the British ornithological community until Lack acknowledged his original work in his book in 1943 and classed him amongst the pioneers of British ornithology in 1959. Perhaps the lack of recognition was because James did not seek publicity for his work, never joined the British Ornithologists' Union and rarely spoke at meetings about his interest. He did have companions with whom he would go on birding trips one of whom must have been

---

[8] British Trust for Ornithology Burkitt JP – Personal papers (Robin territories and MS on survival) (Burkitt-JP) C023323.

H E Howard who travelled to the north-west coast of Ireland to pursue his own interests in bird territories. Little did either of them know that many years later one of James' granddaughters, Judy, would marry into the Howard family.

When James died on March 30, 1959, there was an obituary in *The Irish Times* and later tributes in other Irish newspapers[9] and in the journal *British Birds* [21], brief biographies in the *Oxford Dictionary of National Biography* and later a mention in Wikipedia. They highlight his engineering career, brilliance as a mathematician, pioneering studies in ornithology and work in the church where 'he became an outstanding figure and had conferred upon him every office open to a layman'. He was described as a 'A man of tireless energy, he would think nothing of cycling home many miles for breakfast, (after an early birdwatching trip) and after a brief meal would be off to a full day's work. He was a lover of solitude, and there was nothing that gave him more pleasure than the chance of a "birding" expedition to some remote spot of the wild Western seaboard which he loved so well. He brought to the study of birds a mind as sharp as a needle. As in everything he did, he was ruthlessly thorough – ruthless to himself.' '…a very shy young man, with a quiet manner, gentle, kindly, forbearing…', 'He was on his own in this world, in a class by himself.', '…a man of wonderful patience…', 'If ever there was a man of God, that man was James Parsons Burkitt' [6].

Not one of James' papers contains any drawings or photographs, which is particularly remarkable given his interest in the territories of birds. But there exist nine maps drawn by James on which he has marked the territories of the robins he identified at Lawnakilla and the dates of observation. It is perhaps not surprising that with his training as a surveyor, he would quite naturally understand the value of maps in defining a problem and its solution. Mapping of bird territories had rarely been done before this [22], so these maps would have enhanced considerably the impact of James' papers if he had added them to his manuscripts. Could they, nevertheless, have importance? Denis would have been 15 in 1926 and old enough to take an interest in his father's activities. He and Robin are known to have accompanied James on his bird-watching trips in the school holidays. Despite not following in his father's footsteps into ornithology, Denis was to become one of the pioneers in using the mapping of disease occurrence to help identify its cause, and to this end many of his early papers and lectures included maps. He must surely have seen the maps that his father was drawing but would make only a passing reference to them in his own writings and mapping of human diseases.

How and in what way was the life of Denis Parsons Burkitt influenced by this pageant of family life? There was a clear thread of Christian belief coming down to Denis through the generations, of ambition and achievement, of love of the countryside and of nature especially the birds, of a frugal lifestyle despite adequate financial resources, of travel and work in many different parts of the world, of loyalty and of hard work. What about James' writing papers about his birds for good-quality

[9]W.E.T. Jas. Parsons Burkitt. An Appreciation. The Irish Times April 2, 1959. Trinity College Dublin, Manuscripts & Archives Research Library IE TCD MS 11387 – Papers of Denis and Olive Burkitt 11387/2/6).

scientific journals, despite the subject being well outside his normal professional duties as an engineer? Denis, likewise, and without any training or instruction in writing for journals, had an enormous output of papers and articles, about 400, some in the most prestigious publications. He also had to fight for recognition in the early years of his career and would work very much on his own. With this background he went on to become the family's most famous descendent.

References

1. Lack D. The life of the robin. London: H.F. & G. Witherby Ltd; 1943.
2. Lack D. Some pioneers in ornithological research 1859–1939. The Ibis. 1959;82:299–324.
3. http://snaccooperative.org/ark:/99166/w64k4pvx Guide to the Burkitt, Robert James, 1869–1945 Papers, 1892–1896, inclusive, (Peabody Museum Archives, Harvard University): Danien, E. Send me Mr Burkitt…some whisky and wine: early archaeology in Central America Expedition: The University Museum Magazine of Archaeology/Anthropology, University of Pennsylvania. Expedition.1985; 27(3) 26–33: Danien EC. Treasure in the stable. Expedition. 2008;250(2):40–41.
4. Gregory JR. Under the sun. A memoir of Dr R.W.Burkitt, of Kenya. Printed in Kenya by The English Press Limited (Undated). Family Archives.
5. Burkitt RW. Treatment of Blackwater Fever. BMJ. 1943;4300(1):737.
6. Nelson EC, Haffer J. The ornithological observations of James Parsons Burkitt in County Fermanagh, Northern Ireland. Arch Nat Hist. 2009;36:107–28.
7. Nelson EC. Burkitt, James Parsons (1870–1959), Oxford Dictionary of National Biography, Oxford University Press, Sept 2013 https://doi.org/10.1093/ref:odnb/101124;
8. Wikipedia. https://en.wikipedia.org/wiki/James_Parsons_Burkitt
9. Trimble WE. Jas. Parsons Burkitt. An appreciation. The Impartial Reporter and Farmers' Journal. 1959; April 2nd.
10. Glemser B. The Long Safari. London: the Bodley Head; 1970. Published in the United States as Mr Burkitt and Cancer.
11. Kellock B. The Fibre Man. Tring Herts: Lion Publishing; 1985.
12. Burkitt JP. Relation of song to the nesting of birds. Ir Nat. 1919;28:97–101.
13. Burkitt JP. The relation of song to the nesting of birds. Ir Nat. 1921;30:113–24.
14. Preuss NO. Hans Christian Cornelius Mortensen: aspects of his life and of the history of bird ringing. Ardea. 2001;89:1–6.
15. Burkitt JP. A study of the robin by means of marked birds. Br Birds. 1924;17:294–303.
16. Burkitt JP. A study of the robin by means of marked birds (Second paper). Br Birds. 1924;18:97–103.
17. Burkitt JP. A study of the robin by means of marked birds (Third paper). Br Birds. 1925;18:250–7.
18. Burkitt JP. A study of the robin by means of marked birds (Fourth paper). Br Birds. 1925;19:120–4.
19. Burkitt JP. A study of the robin by means of marked birds (Fifth paper). Br Birds. 1926;20:91–101.
20. Burkitt JP. Rooks and "territory". Ir Nat J. 1927;1:228.
21. Ruttledge FR. James Parsons Burkitt (1870–1959). Br Birds. 1959;52:308–9.
22. Howard HE. Territory in bird life. London; 1920.

Chapter 2
Formative Years

The first entry in the personal diaries of Denis Parsons Burkitt is for January 1, 1923, when Denis was approaching age 12, and starts with:

> morning played with peggy. After dinner went to Orchard terrace for tea. Walked in and drove in went to town and got 7 pieces of chocolate for 2^d. then did a lot of tracking with Robin & Peggy. Made a card house 4 stories high

Peggy and Robin were his elder sister and younger brother. In Orchard House lived 'Grannie', most likely James' mother Emma who would have been aged 85 and whose husband Thomas had died the year before. Tracking was a game in which one of the group would go ahead and lay a trail for the others to follow. Chocolate was a recurring delight for Denis throughout his life.

The diaries give a more or less continuous daily record from 1923 until a few days before his death in March 1993 leaving a, mostly brief, written account of 70 years of his life. They consist of a summary of what he did, whom he met, his travels, family life, the weather and occasionally his thoughts. They are factual rather than philosophical. In addition to the diaries, he wrote or typed extensive descriptions of his travels especially in Africa, and in 1980, whilst in his 70s, he wrote a 92,000-word autobiography, which has never been published[1] (Fig. 2.1).

Why did he start, at a very young age, to write about himself? Why does anyone keep a diary? Possibly the initial impetus was that of being given one for Christmas in 1922 by a cousin. The flyleaf is inscribed 'Laddie Burkitt from Marty Hill', who was one of his mother's family from Cork. It may simply have reflected the fashion of the day for diary keeping. His father kept a diary. But why did Denis, in addition to his many scientific papers, books, long letters to Olive and extensive descriptions of his travels, keep this record? From time to time, he writes very personal notes to himself about his behaviour, his perceived failures, his obedience to God and about

---

[1] Burkitt DP circa 1985. Autobiography. Wellcome Library, London. Papers of Denis Burkitt WPI/DPB/F/1.

© The Author(s), under exclusive license to Springer Nature Switzerland AG 2022
J. H. Cummings, *Denis Burkitt*, Springer Biographies,
https://doi.org/10.1007/978-3-030-88563-2_2

Fig. 2.1 Early diaries of Denis Burkitt, 1923–1931. The open diary is from 1923. It recounts a week at Trearddur House School on the Tuesday of which he 'Played hockey. got a bad knock on my knee'

some of the people he met, which provide unique insights not otherwise available. Did he expect that one day people might read these? Rarely is there any comment on the politics of the day or the great issues of the time.

Despite all his writing, there is no proper record of the first 12 years of Denis' life, which centred mainly around the family, early school days and his growing independence. Enough oral history and family papers exist to tell us that by the age of 12, he had attended two schools, lost an eye and suffered an assault on his tonsils by 'kill or cure' Uncle Roland.

Denis was born on February 28, 1911, at the family home on Alexandra Terrace, Enniskillen, Ireland. Five weeks later the family moved to Lawnakilla another rented house just north of Enniskillen where they were to reside for 15 years and where James would carry out his studies of the robin. James, as already related, was very able with a first class brain and many practical skills as an engineer. A loyal family man, he is described by Denis as being shy in company, having great humility and given the nature of his bird-watching interests, perhaps something of a loner. The family never appreciated the outstanding achievements of James until the public recognition given to him at the time of his death, causing Denis to regret in later years that he never followed in his father's footsteps to explore the wonders of the natural world. More likely to have influenced Denis' early outlook on life is his father's Christian faith. James' knowledge of the Bible was detailed, and his ability to express his faith verbally and cogently probably started Denis on his path to belief.

Mother, Gwendoline (Gwen), was an altogether different person. She acted out the part of a women born in the later Victorian period by devoting herself to her husband and family. Denis describes her as being 'a lady to her fingertips, with impeccable manners and with the knack of putting everyone at their ease'. She didn't have her husband's intellect and singleness of purpose, and when he, frequently, disagreed with her, she would not be drawn into arguments. She was a peacemaker who was the source of love within the family and had a wide circle of friends in the surrounding countryside. Gwen was musical and would lead the family hymn singing on Sunday evenings by playing the piano, which talent was not shared by James who was tone-deaf, a trait passed on to Denis who confessed to not understanding music. Unlike James, Gwen was born into a relatively wealthy family, that of an architect who designed public buildings. Nevertheless, she was quite frugal by nature and is recorded by Denis as reusing Christmas cards by removing the names and written greetings and resending them, carefully avoiding returning them to the original sender.

There is not a lot of emotion in Denis' brief description of his parents in his autobiography, albeit many years after they had died. But there was a loving and caring environment at home, and Denis laid great store by the support he would get from his family. He was known by them to be warm and affectionate.

> When at the age of 84 my mother passed on she left a £10 note in an envelope for each of us her children for any special needs we may have had. It was somehow particularly characteristic of her…but so typically a final gesture of love. Tears streamed from my eyes when I opened my envelope.

Early family life for Denis very much revolved around sister Peggy and brother Robin, but from the age of 7, and for the succeeding 20 years, he shared his education with Robin as they proceeded through school, on to Trinity College Dublin and into medical practice. Robin is the person most frequently mentioned in the first 20 years of diaries. The boys got on well and experienced many things together including the visit in 1918 by Uncle Roland, when they were aged 7 and 5, in which Roland decided, without much consultation with parents or patients, to remove their tonsils. Denis recalls how Uncle Roland asked the boys to open their mouths, put a finger into their throats and declared that they needed their tonsils out. Roland asked Gwen to cover the drawing room furniture with sheets, and the next morning the boys were carried, in turn by their father, downstairs where Uncle Roland gave them an anaesthetic and removed their tonsils. Denis, much later, remarks that this was 'a terrible judgement in the light of current knowledge of the immunological importance of these structures'. And that 'having removed numerous tonsils myself I shudder at the risks incurred, but in the past this operation was commonly performed by general practitioners on kitchen tables'. Robin, who was also a victim of this misplaced strategy in preventive medicine, makes no mention of it in his later writings about the family; neither does the event feature in Gwen's lifetime diary. Three years after the tonsil episode, Denis was to commence writing his diary.

Other accounts of his life and work especially in the scientific press, other media, in his correspondence and his comprehensive accounts of his travels provide many additional sources from which to piece together the characteristics of this exceptional man (Fig. 2.2).

With regard to schooling, initially they were educated at home by a governess and then at a local school. In 1920 the boys were sent to the preparatory branch of Portora Royal School in Enniskillen, a leading Irish public school, which claimed both Oscar Wilde and Samuel Beckett as former pupils. In the summer of 1922, an accident occurred on the driveway leading to the school that was to change both their education and Denis' life. For the exact dates of what happened over the next few weeks, we have to thank Gwen and her 'Souls Inquiries' diary entries, which generally note only major events in the family. In June 15 at school, two groups of boys were playing outside, throwing stones at each other. Denis was a bystander but had the misfortune to be hit by a stone that shattered his glasses, and the broken pieces entered his right eye. The boys cycled home, and the family doctor was called who immediately arranged for Denis to be taken to Belfast, a distance of over 80 miles, to see the leading eye surgeon in Ireland, Mr Craig. A day or 2 later, Mr Craig operated to remove the broken glass, which had lacerated Denis' cornea and showed

Fig. 2.2 Burkitt children with their mother circa 1920. Back row, Denis, Gwen; front row, Robin, Peggy

no signs of healing. The surgeon advised that Mrs Burkitt be sent for to sit with Denis. Four weeks later, in July 15 Denis had a second operation in an attempt to save the eye, but it became infected, and James was now sent in order for Mr Craig to explain the likely course of events. He told Gwen and James that a condition known as sympathetic ophthalmia might develop[2] and that it might be necessary to remove the eye. Gwen records for July 24, 1922, 'Denis eye out at quarter to nine pm'.

Denis made very little of this disaster. In his diary he notes that he went to collect 'glass', presumably his eye implant, on his birthday 8 months after the accident – on a visit to Belfast the following January 'got 2 eyes for me' and in March 1925 'Dropped my glass eye and broke it'. Otherwise the problem is rarely mentioned in conversation or in his writings. But for a surgeon to have only one eye would have been quite a handicap. Those who met him in later life were often not aware that his right eye was artificial, it being a very good likeness.

His parents were naturally much more upset by and concerned about the effect this trauma would have in their son. As a form of compensation, James bought him a model steam engine that really worked and cost £1 5s (equivalent to £80 today), a generous gift, perhaps typical of an engineer and indicative of the concern they had for Denis at this time. In a more practical vein, a further gift from his parents was a Meccano set with a wind-up motor, which must have been the envy of his friends, the Meccano brand having been established only 14 years earlier. But a gift from Denis' wealthy uncle Sir Alfred Sherlock, married to his mother's youngest sister, Muriel, was to give him skills that he would use throughout his life. The gift was a 1A Kodak folding camera that allowed Denis to set both shutter speed and distance. He was already the owner of a Kodak Brownie box camera, but now he could take a much greater variety of pictures, and from this time on, photography was to be another way that he could record places and events that featured in his life and that of his family. He would move to doing colour photography as soon as the technology became available in the 1940s and in the 1960s acquire a cine camera. Later he would illustrate his lectures with slides that he had made himself. They provide an outstanding record of his life and work.[3]

As a result of the accident, Denis' parents decided that returning to Portora School was not in the boys' best interests and arranged to send them away to Trearddur House School, just south of Holyhead in North Wales, in September 1922. It is not clear why this school was chosen, but Denis and Robin spent an enjoyable 3 and 4 years there, respectively, commuting back and forth to Ireland accompanied initially by Aunt May (Mary). It was a new school at this time comprising only 22 pupils, a headmaster and 1 other teacher. Life comprised daily prayers, lessons in the morning Monday to Saturday and sport or similar recreation in the afternoons with further lessons later in the day. Reading books was

---

[2] Sympathetic ophthalmia in which, following an injury to one eye, inflammation occurs in the other necessitating removal of the injured eye to prevent complete blindness.

[3] The photograph albums of which there are about 60 reside in the family archives, whilst his lecture slides are mostly in the Wellcome Library DTI/WPB.

encouraged and there was a school library. Subjects that Denis learnt included English language and literature, French, Latin, algebra, geometry, scripture, history and natural history. Academically Denis was not top of the class in the end of term exams in any subject, and in some, such as Latin, he confesses to doing rather badly. Robin always did better. Sundays were for letter writing and attendance at church morning and sometimes evening also. There was some free time for walks, to spend time on the beach and swim or to visit Holyhead in the school pony and trap accompanied by Mrs Williams and Miss Potter of the school staff. The main sports in the winter were football (soccer), hockey and rugby and in the summer tennis and cricket, which was played four or five times a week when it was not raining. Gym and skipping seem to have been daily activities with running most weeks and boxing for the more senior pupils. Stamp collecting was popular at the school and could also be continued at home, which Denis took to enthusiastically. Letters were written to Stanley Gibbons, the leading stamp dealer of the day, and stamps were ordered by post accompanied by postal orders (e.g. 2s 6d). Irish stamps in 1922 and onwards would have been collectable as the country implemented partition by overprinting the stamps of Great Britain with numerous varieties of ink and text. Denis also had an album for collecting cigarette cards. He developed practical skills especially carpentry and developed his own photographs in the school darkroom, talents that were to prove very useful to him in his professional life. He was already quite inventive and developed a cable device for taking selfies. He would go on to experiment with trick photography, always keeping abreast of the latest technology (Fig. 2.3).

Carpentry was clearly a great interest of Denis, who made a cage in which to keep mice and jerboa, which they were allowed to have at school as pets. He would go on to develop this skill as he grew older initially from his father but later by attending carpentry classes, even when he was in Africa. Each boy had their own garden at the school where Denis grew parsnips, radishes, violets, cowslips and other flowers some of which he brought from home. Drawing and painting was something that Denis also enjoyed although there is no record of this possible talent being carried through into his adult years. The illustrations for his lectures were always drawn by his eldest daughter Judy supplemented by photographs from his increasingly large collection. Other social skills and recreations included dancing but without the benefit of girls at the school, playing cards, chess, charades, singing, learning to fish, sail a boat and shoot, which latter was practised two or three times a week. Denis' scores for shooting were usually in the 40–60/100 range with Robin frequently doing better. But in order to shoot, Denis had to learn to use his left eye, which may not have been his natural choice.

During holidays from school, Denis enjoyed family life with outings in the 'motor' to local beauty spots such as Boa Island and to the mountains where James was constructing new roads. There were visits to the homes of Aunt May, Aunt Norah, Grannie in Orchard Cottage and to meet local friends. Play and adventure was with Peggy and Robin. Bicycle rides, tennis, tracking and cowboys with pistols loaded with caps, which produced a loud bang and smell of gunpowder on pulling the trigger, were favourite games. A donkey, called Nelly, was purchased on Tuesday April 10, 1923, and housed in its own stable; then the next day a puppy, Bumble,

Fig. 2.3 Robin and Denis.
An early 'selfie' done with
a device made by the
budding photographer,
Denis (right of picture)

arrived. There was also a pony, which they learned to ride, and there were hens, which produced eggs for the family, and cats. They built a model railway in the attic of the house. Frequent trips into Enniskillen were made, either walking or on bicycles, to run errands for 'Mammy'.

A notable asset was the acquisition of a 'Woodhouse' in the garden at Enniskillen, positioned sufficiently far from the main house, past the rockery to afford the children some privacy and sense of adventure. The structure had probably been in the garden for some years because their first job was to clean it by sweeping it out after which they tried to plaster the inside walls and failed. They may have been more successful thatching the roof with whins and ruches.[4] They then made two benches, one of which was from an old trunk they found in the back of the pantry in the main house and the other from an apple box. A shelf made at Trearddur was put up and a

---

[4]Whins or furze being colloquial names for gorse, a member of the pea family.

path created around the house. A bell followed and a nameplate. Some cups and plates were borrowed and clandestine meals were enjoyed away from the critical eyes of parents. A Woodhouse Magazine was started and Peggy wrote two poems for it, one about the house and the other about Roland. Perhaps surprisingly, rarely if ever are other children mentioned as visiting the Woodhouse although at the back of his 1923 diary is a list of nine 'My Friends' first amongst which is Robin. Of the others only one is given as living in Enniskillen. In all aspects for which we have any insight, their lives at home were close to being idyllic in these early years.

The weather determined their daily lives to some extent, described as 'horrible' or 'a very nasty wet morning', 'awfully hot in the afternoon'. A church service might be 'very long', whilst the consumption of cake is usually accompanied by descriptions such as 'lovely chocolate cake'. Only once Denis remarks 'Felt rather homesick' on arriving back at Trearddur. The diary for 1924 ends abruptly on Thursday May 29, without explanation leaving a complete blank for 7 months until January 1925 when the daily record of Denis' activities resumes. Enlisting in the army, he was already in the OTC (Officers' Training Corps); he enjoyed wearing the uniform, parades and learning about guns. Events of note during these early months of 1924 include 'Daddy let us drive the car' – this was either en route to or in Enniskillen. Pretty remarkable of James when Denis would have been aged 13 and Robin 11.

Something of the daily life at school can be gleaned from this entry for Sunday March 16, 1924:

> I did not have any cold bath because of my cold. Boiled eggs for breakfast. After breakfast I got quinine. Wrote nearly three pages of home letter. I was not allowed out at all. Morning walk was to windmill with overcoats. I swapped six alpine flowers to Barton for a character from Dickens. Scripture after dinner till 2.30. Walk was loop the loop. Silence read was at 3.30. til 4.30. Mr Piper read to us instead of Shakespeare. Did not write any letters. Read my bible in bed. Nice day.

The cold lasted a week, a not infrequent occurrence for Denis, and meant he was confined indoors and given quinine every day, but the upside was that he did not have to have a cold bath or go on walks, which pleased him, and Robin looked after his jerboa.

The boys spent 3 years at Trearddur moving on to Dean Close School in Cheltenham in 1925 and 1926. During the final term at Trearddur, Denis played a lot of cricket and continued lessons, carpentry and singing. He did not distinguish himself either academically or in sport and regarded these school years as inconspicuous, but he was clearly very good at photography receiving an A+ in June 21 for a photograph taken, developed and printed at school and in July 22 was awarded the annual photo prize, a colour filter for his camera. The school year ended without much fanfare in July 27 and the boys set off for the boat to take them back to Ireland.

   The summer holidays of 1925, which marked the transition between the two schools, started with the whole family driving to Mullaghmore[5] for a camping holiday. This small town is situated on the north-west coast of Ireland in the County of Sligo, an area of exceptional beauty but fully exposed to the Atlantic Ocean with its powerful seas. For Denis and Robin, it provided an ideal place to bathe, walk on the shore, trawl for shrimps in the shallows and with the others drive to the mountains, pick mushrooms and walk. They borrowed a friend's boat and went out, with James in charge, catching lots of pollock. The men played golf; they all visited friends and after 11 seemingly ideal days returned home taking with them 28 of the fish they had caught. Back at Lawnakilla a notable visitor again was Uncle Alfred (Sherlock) who gave Robin and Denis 10/- each. In August 12 they went to Ballinamallard, a small village a couple of miles north of where they were living, where they watched a march of over 600 Orangemen. A rare reference by Denis to the existence of a long established but polarising Protestant Order and perhaps a nod to the situation which they, as a Protestant family, faced in Ireland. At the time of Irish independence from the UK in 1922, the country had been divided into a predominantly Protestant group of northern counties and much larger area in the south that was very much Catholic. Apart from this diary entry, at the age of 14, rarely if ever does Denis comment on the tension that existed between the followers of the two faiths.
   The rest of the holidays passed without incident with the boys enjoying the freedoms of the countryside, a comfortable and stable home environment and attendance at church and without any major concerns despite Denis having to leave for a new school in September 21. Robin returned for another year to Trearddur during which he was head boy and awarded a scholarship to Dean Close. The journey from Lawnakilla and later from Laragh to Dean Close was considerable, which they did on their own. It comprised a train journey to Belfast, then an 8-hour overnight sea crossing to Liverpool, breakfast at Lime Street Station and a further train with the final leg to the school in the centre of Cheltenham by taxi. Dean Close[6] is a public school that had been established in 1886 and had a strong evangelical Protestant Christian foundation, which served to reinforce the beliefs and traditions that Denis had been brought up to respect. This was probably important in James' and Gwen's choice of school, along with its relative proximity to the Irish Sea. The school had about 250 pupils and a new headmaster in 1924, Percy Bolton ex Oundle School.[7] On arrival Bolton described Dean Close as being 'drab, dismal or dull' and the chapel as 'of distressing gauntness without and grim austerity within' with a prevailing pattern of 'institutional severity'. But he was a reforming head and life must have

---

[5] Mullaghmore in County Sligo is known as the place where Earl Mountbatten was killed by a bomb placed in his boat during the night by the IRA on August 27, 1979. See Fig. 1.4.

[6] Whitney CE. At Close Quarters [1]. This is a history of the school. Special mention of Denis' achievements in later life is given on p. 329.

[7] Bolton, born in 1890, was a scholar at Queen Elizabeth's Grammar School, Blackburn; read mathematics at King's College, Cambridge; and then, after a year teaching at Cheltenham College, went to Oundle, aged 24.

improved whilst the boys were there although Denis at no time complained about it (Fig. 2.4).

Arriving at Dean Close at 1:30 pm on Tuesday September 22, 1925, Denis quickly slipped into the school's routine – 'Chaple', lessons, prep, walks, which could be up to 10 miles, swimming, visits to the tuck shop and letters received and sent home. There was the expected emphasis on physical education, team games and personal fitness, very much as had been the case at Trearddur. He immediately joined the OTC and clearly enjoyed being on parade in uniform and shown how to handle a rifle. There was a nearby shooting range. James visited at half term and took Denis to lunch and tea in Cheltenham, and on Sunday they caught a bus to the nearby town of Tewkesbury where they went to a service in the fine Norman abbey. Term ended with exams but apparently little in the way of Christmas festivities, and he was home at Lawnakilla on Friday December 18 to be joined by Robin and Peggy at the end of their school terms 4 days later. There was snow and frost but this did not stop James and now Denis going out shooting. Denis managed to get a pigeon, his skills now improving only 3 years after losing his eye. Christmas at home was traditional with church, presents and a turkey for dinner. Robin developed

Fig. 2.4 'Dean Close School 1929 July'. Early photograph and entry in album by Denis

a sore throat, was confined to bed for 2 days and visited by the local doctor. Denis stayed with him much of the time reading to him and doubtless talking about what awaited Robin at Dean Close. The two brothers remained very close throughout their lives.

In March 1925, when Denis was 14, his height measured at school was 5'$^3/_8$" and a year later was 5' 1½". His eventual height, according to data in his passports, was 5'10½", so the 4 years at Dean Close included his growth to maturity, which is not a matter for comment by Denis at any stage. He did get a new glass eye in April 1927, which might indicate that his was growing, but there is never any remark about needing a bigger bicycle, or clothes, that his athletic prowess had changed in any way or of the thoughts and emotions of a teenage boy. His handwriting remained that of a young person throughout 1925 and 1926 but then gradually changed so that by 1929, when he was 18, the style, spelling and legibility were that which he would use throughout his life.

1926 brought an important change for the whole family. Early in March Lawnakilla was sold. Presumably James and Gwen had been planning this because within 2 weeks they were moving into a more substantial property, which they had purchased, situated between the village of Ballinamallard and the Lough about 2 miles north of where they had been living for the past 15 years. Laragh was a five-bedroomed mid-Victorian house standing in its own grounds, which extended to 25 acres, and afforded easy access to the Lough where the boys would have their own small boat and could swim, boat and fish in safety. By March 26 James and Gwen had moved to Laragh although continued packing up at Lawnakilla for a further week. Term ended at Dean Close for Denis in March 25 but had not yet finished for either Peggy or Robin, so on the Sunday he went to Trory Church for the service with James and spent the rest of the day exploring the new house and area. Trory Church, overlooking Lough Erne, the Parish Church of St Michaels is Church of Ireland. Despite his Presbyterian upbringing, James would become one of its most active and valued members as he pursued a parallel career in the service of God. Laragh remained the family home for the rest of the lives of James and Gwen, and they would be buried at Trory Church in 1959 and 1960. A plaque in the church commemorates their lives (Fig. 2.5).

After the move the boys' main focus of activity was the river flowing west to the Lough where they fished for trout and rowed their boat. Cycling around the countryside was an almost daily activity to visit friends or find a spot for lunchtime picnic. They collected eggs from the nests of ducks, plovers, snipe and ring plover and went shooting with James and to church on Sundays. School continued to be inconspicuous with no mention of academic or sporting achievements for Denis, except a second place in a maths exam. Robin, always more able, visited Dean Close to take the scholarship exam in late June and was informed a week later that he had been successful and would receive £30 annually whilst at the school. Denis continued to be an enthusiastic member of the OTC enjoying parades, drill and anything where he could wear his uniform. Cricket was replaced largely by tennis, perhaps because cricket is less forgiving of errors in technique, for those with only one eye, than tennis. By the end of July, everyone was back at Laragh for the summer holidays. They

Fig. 2.5 Laragh, to which the family moved in 1926

all set off in the car almost immediately for a 3-week trip to Cork to see Granny Hill and various uncles and aunts. James and the boys had brought their guns and had several successful shooting expeditions. A family outing to Blarney Castle allowed Denis to kiss the Blarney Stone[8] although mother and father may not have been prepared for the risk to life and limb that was required to touch the stone. Back at Laragh a short camping holiday on Long Island on the Lough, followed with James joining in but boating around the islands, was limited because of bad weather. The final part of the holiday included a 3-day bicycle tour to Mullaghmore on the coast, a round trip of about 80–90 miles and a favourite destination for the family.

The 7-week summer holiday was soon over, and now Robin accompanied Denis back to Dean Close where they would be together for 3 years. Life slipped quickly into routine with lessons, prep, workshops, chapel, walks, games, mostly soccer, running, letters sent and received and hot baths. OTC activities were very much looked forwards to by Denis, including parade, cleaning rifles, route marches and full dress parades for special events such as Commemoration Day (November 1) and Field Operations where they were given ten blank cartridges each to add a touch of reality. Term ended with exams and the boys were home by Friday December 17, immediately going out on a shoot where Denis got a pheasant, the first of four shoots over the next 2 weeks. Academically the year had been unremarkable for Denis although he came first in mechanics and did better in maths than in other

---

[8] Blarney Stone. Part of the battlements of Blarney Castle. When kissed the stone was believed to endow people with the gifts of eloquence and flattery.

subjects. Practical skills were more to Denis' liking. He was given a fretwork kit for Christmas and was becoming known amongst the family for his art as a conjuror.

Cycling had become a major activity especially during the holidays when Denis was out most days, always recording the miles he had travelled. Daily distances could be as much as 60 miles, for example, to Belleek a busy market town on the border with the Republic and famous for its Belleek china, but more usually were less than 10. Such detailed recordkeeping clearly appealed to Denis perhaps not so much as a measure of his physical prowess but as a way of defining his relationship with the world around him. Recordkeeping was to be core to his discovery of the cause of the childhood cancer that would bear his name. Robin was a constant companion, and both he and Denis were confirmed as members of the Church of England by the Bishop of Bristol on Friday March 25, 1927. Academic distinction continued to elude Denis although he remarks on his ranking in the form maths exams, occasionally topping the class, and in French where he usually did well. More animated accounts are given of shooting, fishing trips, boating on the river and during term time the OTC. He continued to enjoy making things, which included an alarm clock, a pinhole camera and an electric light for his bicycle. Both boys remained healthy although they had mumps when at school in June 1927, which kept Denis below par for 6 weeks.

As evidence that Denis was growing up, on January 2, 1928, approaching his 17th birthday, Joyce Hill, a cousin, appeared in his life. Back at school later in the month, he wrote to Joyce, an unprecedented epistle since he had hitherto written only to close family. A couple of weeks later, Joyce sent him a cake, always a favourite food of Denis prompting him to write again. Further parcels arrived from her, and when back at home for the Easter holidays, he immediately took Peggy and Joyce on a river trip, and the following day he was out with Joyce alone in the boat. He took her for a drive in the car and to tea with an aunt. He then notes that 'Joyce went at 1.30' after which she is never mentioned again. Denis records the facts of this short liaison, as he sets down the everyday events in his life, without revealing his emotions. But being alone with Joyce was clearly a moment that put all other events of those few months into the background for him.

Attendance at church is routinely mentioned on Sundays as a dutiful exercise but seemingly not anything that might challenge the events of the week that followed. He was by now a successful shot and bagged 12 and 11 grouse on 2 days in August. Tennis was his main sporting activity, whilst back at school for the autumn term, he was learning semaphore, wireless communication, lifesaving, reading at least one book a week, a practice he would continue throughout his life, and writing articles for the House Magazine. He was made assistant librarian and a lance corporal in the OTC. Life seemed predictable, manageable and even enjoyable but 1929 was to be a year of change.

With his 18th birthday approaching, Denis was maturing fast. He now owned his own car and had started to keep detailed accounts of his monthly spending. Presumably James was giving the boys an allowance. He bought his first razor and attended the local hospital dance on New Year's Day, staying up until 4 am, and the League of Pity dance on the third. Back at school he was elected to the Finance

Committee and promoted to Commander of No 4 Section of the OTC, which specialised in signalling. Denis was clearly interested in communication because he set out to build a battery-powered portable wireless. A list of almost 30 parts was required, which cost him nearly £6, and included condensers, valve holders for Dario and Marconi valves, transformers, an accumulator with acid and charging facility together and many small items. He made a wooden cabinet for it, the whole project taking him about a month. He then took it up the school tower where there would have been a good signal and after only a few days and with help from one of his fellow pupils, Brookes, was able to show that it worked. The wireless was to become his companion both at school and home where he would listen daily, but in addition it was portable and could be taken out on picnics. Quite an achievement in 1929. He recalls listening to Prime Minister Ramsay McDonald's speech from New York in October 4, relayed via the Daventry Transmitter.

Back at school for his final term at the end of April, Denis was elected to the Senior Room Finance Committee but not the Science Society, despite his skills with the wireless. Academic achievement seemed to be beyond his ability and there were no prizes for him at his final speech day. But he had developed practical skills that would bring a lifetime of benefits to those around him, and, despite this apparent lack of academic ability, he would go on to be feted for his breakthroughs in medical science and receive many honours and prizes. Perhaps as a result of his achievements at school, he would never describe himself as clever (Fig. 2.6).

Fig. 2.6 In Officers'
Training Corps uniform
1929/1930

Before an end of term OTC camp at Strensall near York in England, where there is an army firing range and where he was promoted sergeant, there was an important visit to Dublin. Taking the boat from Liverpool to Dublin rather than Belfast in July 3, he was met by James who, as proud father, supported Denis. In July 5 he took the entrance exams for the Trinity College Dublin Engineering School. The topics were maths and mechanics. Back at school a week later, Denis learned that he had been successful and given a place to study engineering. Why engineering? It had always been assumed in the family that Denis would follow his father into this field, and he had been better at maths at school than any other subject. Denis was also very practical, loved making things and had no great sense of vocation at this time. Moreover, he and Robin had spent some time discussing their futures whilst at Dean Close, and both were certain of one thing, that they would not do medicine or dentistry, perhaps because of distant memories of Uncle Roland and his attack on their tonsils and for Denis' frequent visits to the dentist for 'stoppings' (fillings), known today as conservation work. And so, it was that Denis went up to Trinity to study engineering, whilst Robin would follow a year later to study modern literature. How was it that both boys would change to medicine and eventually become medical doctors?

Reference

1. Whitney CE. At close quarters. Herefordshire: Logaston Press; 2009.

Chapter 3
Trinity College Dublin; A New Direction

The first year at Trinity was to be life-changing for Denis. He arrived in Dublin on Saturday October 5, 1929, for the start of the Michaelmas term[1] in the car of Captain and Mrs Newenham, his aunt and uncle from Coolmore, County Cork. On arrival his college room was not available, so they went to a show at The Gaiety Theatre and stayed in the Standard hotel. The next day he was able to get into his room 16, Botany Bay, which he would share with Barber. It was a large room with an open fire but no other amenities. After settling in he took the bus to Zion Church, in the Dublin Diocese of the Church of Ireland, and after the evening service had supper at the Rectory. It was the church of Uncle Louis, a brother of Gwen, who through his Ministry would have a profound influence on Denis' life. Students usually made their own breakfast in their rooms, but a College buffet was available for meals during the day, and in the evening, students could dine 'on Commons' in the College. Lectures the first week were drawing and maths. Drawing required special instruments, so Denis requested some from James, who was doubtless very happy to see his son following in his footsteps. Lonely, being without Robin for the first time, he wrote to him and also to his school friend Broad. He joined the Harriers, a running club, walked in Phoenix Park and went to the cinema to see 'Lucky in Love' (Fig. 3.1).

In some ways Trinity was just an extension of his school days with lectures instead of lessons, sport, walking, letters home and to friends, church on Sunday and communal mealtimes. But there were important differences. He was now training to be an engineer and studying more advanced maths, physics and chemistry. Latin was compulsory, he wrote an essay on Horace, and he was able to sign up for arts lectures, an interest hinted at during his school days. He had a much more diverse social life with frequent visits to the cinema and theatre and late night

---

[1] Terms at Trinity were Michaelmas, Hilary and Trinity.

© The Author(s), under exclusive license to Springer Nature Switzerland AG 2022   33
J. H. Cummings, *Denis Burkitt*, Springer Biographies,
https://doi.org/10.1007/978-3-030-88563-2_3

Fig. 3.1 Coolmore

discussions in rooms, and James came to see him more often because the journey to Dublin was much less arduous than to Cheltenham.

And there was Violet, whom he had met with her mother Mrs Moore soon after term started. They had invited Denis to a party at the Creightons but he refused because he didn't know them very well. A month later he went with Violet and other girls to see Vagabond King at The Gaiety Theatre and the next day took her to Sunday lunch at family friends outside Dublin. Whilst Denis considered himself to be lacking in self-confidence, he was now easing himself into a larger world than that of Dean Close. The influence of religion on everyday life at home stayed with him. Apart from regular church attendance, he went to meetings of the Student Christian Movement (SCM) and to a debate at the Divinity Hostel on 'Sunday games should be abolished' where he spoke for the motion. On Friday October 25, he was invited to a meeting of Christians called No 40, this being a group of under-graduates who had originally met in room 40 although they were now meeting in room 14 on campus. It was his first encounter with them but was the start of a jour-ney that would change his outlook on and purpose in life completely.

Back at home at the end of the first term, Denis went shooting, which continued to be a favourite pastime in addition to which he now had a .22 rifle. In the early days after New Year, there were fancy dress dances to attend, trips to the cinema, cycling, skating – there was a lot of snow that winter in Fermanagh – playing cards, numerous social calls to friends and family and listening to the wireless. Denis was also learning to play the piano, unsuccessfully, with his mother's tutelage. Returning to Dublin on January 15, 1930, he started to attend the meetings of No 40, most of whose students had fought in the trenches in the World War I and perhaps because of their experiences now met regularly at Trinity to talk about their faith and its implications for their lives. They were predominantly medical students some of whom went on to have distinguished careers in academia and as medical missionaries in Africa. Denis quickly came under their influence, and new names start to appear in his diary of fellow students with whom he dined or went to church. One in particular may have been important, Dick Jones.

In January 31 Denis notes that he had 'Lunch in Buffet with Jones'. This event was repeated in February 7 and again in the 12th. In February 14 he attended a meeting of No 40 and then spent a while in Jones' room. This date is highly significant in the life of Denis because tucked away in the front of his diary, hidden between a couple of handwritten lists of parts required for his wireless and their costs, is a small scrap of paper torn from a notebook on which is written in ink, 'I am going to do my best to follow Christ. Denis Parsons Burkitt. 14/2/30'. It is unlikely that this note, folded almost as if to conceal it, has seen the light of day since the day it was written. It is not mentioned in Denis' autobiographical account of the challenge that Jones made to him, which was probably during the discussions they had in Jones' rooms in February 14, and it is improbable that he told his parents about this commitment when they visited him a few days later. They would have been very pleased and reassured that their elder son was choosing a way of life that they understood and wholly believed in (Fig. 3.2).

On the night of February 14, Jones had challenged Denis with what he describes as an embarrassing question 'whether I had ever responded to Jesus Christ's claim on my life'. Denis' faith had followed conventional lines up to this point starting with acceptance of the beliefs of his family and their traditions. Attendance at

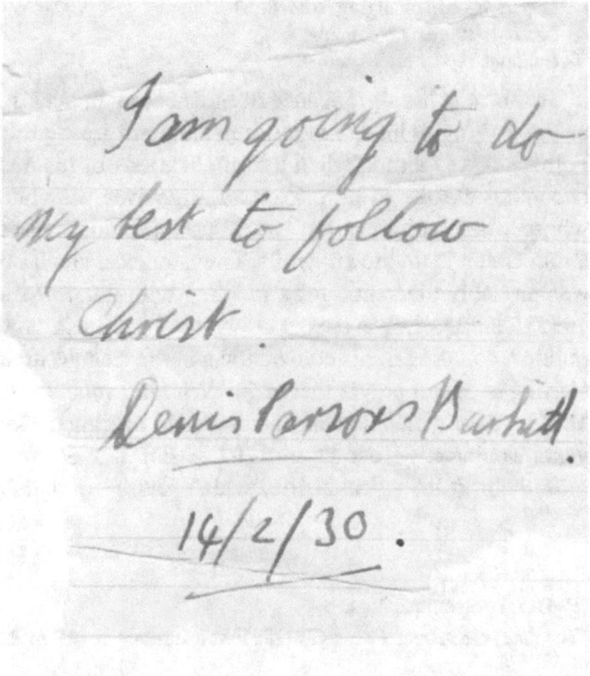

Fig. 3.2 Note to himself, tucked into the front of his 1929 diary, following his conversion experience on February 14, 1930

church on Sundays was expected, and there was a subconscious acknowledgement of a set of standards and behaviour that permeated his daily life and which had been explicit during his schooling. But now he had made up his mind, with little understanding of the implications, to endeavour to follow Christ and to be identified in College as a Christian. His conversion has a defined moment in time attached to it, but the changes in his way of life that were to follow took some years to evolve. In the immediate days after February 14, Denis remarks that for the first time, the Bible became alive to him and studying it became an adventure rather than a tedious duty. But he was to acknowledge in later years that the awakening of a conscious awareness of God is far more often a process than a sudden crisis experience.

Some indication of the evolution of his faith and its practice is given in another note, handwritten in pencil, on a piece of paper torn from a writing pad and folded up so as to fit into as small a space as possible. It must have been written after 1932 and appears to be list of headings for what might have been a talk he had been asked to give on his conversion experience and its consequences for his daily life.

> Dick Jones' Room
> Period of Doubt
> Zion Mission
> Talk with Brian Green
> In doubt sought conversion in Greystones 1930
> Whit weekend 1932
> 1st Testimony Summer 1932 in 40.
> ---P.M's to start attending[2]
> 1931 longing to win souls by _self_
> 1932 "for souls to be won
> Summer 1932. Increase prayer time
> Experiences in changing to Medical
> Xmas 1930 give up dancing
> Summer 1931 " " cinema

In this note the significance of the meeting in Dick Jones' room is acknowledged as the starting point of his journey although it was followed by a period of doubt. Perhaps it was a time when the implications of his decision to follow Christ were becoming clearer to him. Zion Mission was the church headed by Uncle Louis whose whole congregation would now provide real support for the new convert. Brian Green is unknown, whilst Greystones, a small coastal town south of Dublin, was probably the venue for a meeting with his new friends. The Whit weekend in 1932 related to two events, namely, his (first) testimony, which in this context is a public recounting of his conversion and the changes to his life that followed, and his starting to attend prayer meetings.[3] Winning souls for Christ was to turn out to be a challenge for which he was ill suited, and his failed attempts to do so in the ensuing years would only frustrate him. Increasing prayer time suggests he was in need of more help in his spiritual life, whilst giving up dancing and the cinema, both of

---

[2] PM's are prayer meetings.

[3] The date is incorrect. Denis gave his first testimony at '40' in June 1931.

which he enjoyed, may indicate that he was aligning himself more with his peers in their commitment to a way of life in Christ.

Denis' conversion experience was no Damascene revelation such as the Apostle Paul had experienced. Denis already considered himself to be a Christian having been part of a family that practised a way of life based on Christ's teaching as revealed in the Bible, inherited in the case of both James and Gwen from parents both of whom came from families with strong traditions in the church. He was used to prayers and Bible reading at home, church on Sundays and mixing with family and friends who had similar beliefs. This was his comfort zone. What had happened on February 14, 1930, was the realisation that God was not an external being with a rule book for living but instead was in fact part of him and that Christ was the living example of this relationship. He started to understand that a love of God from which follows a love of your fellow men and women was the basis for living and that the Bible contained both a history of Christianity and the collective wisdom of people inspired by God through whom God had worked.

Whatever it was that came into Denis' mind in February 14, it changed his life for ever, went on to permeate everything he did and gave him a clear sense of duty. It came at a time when he needed a new vision of his future. He had arrived at Trinity with a modest academic record, no great sporting achievements and no vocation. Becoming an engineer did not inspire him. Now, with an increasing awareness of the presence of God in everything he did, he began to work through the implications of what he had committed to. Everything he was to achieve in later life must be seen against the background of events in February 14. From that day onwards, he would rise early wherever he was in the world and spend a quiet time (qt) of Bible study, prayer and planning. Meeting fellow Christians on his travels was always a highlight of any day for him and with whom he would share fellowship, pray and read the Bible. He would become someone who, although a family man, was bored by social occasions but intensely focussed on his work. He would go out of his way to avoid conflict or alienate people despite his discoveries arousing considerable controversy and jealousy. He was always at pains to share credit with others for the work he was doing.

After February 14 the Hilary term continued predictably with lectures, practical classes and projects to complete. He was still doing French and Latin as well as more engineering-related subjects such as chemistry, mathematics, metallurgy and physics. He was also attending painting and drawing classes as an extra subject despite a lack of talent as was to become evident in his professional years when a drawing or painting might be required for his work. But his life outside the engineering course started to change. He was developing a new group of friends, who were likely members of 40 and included Jones as well as Lee, Ejerton, Anderson, Harper, Bourne, Scott, Wallace and Dunlop, of the acclaimed family. He now had what he considered to be real friends, an identity, a sense of purpose and direction and subsequently a sense of vocation. The Bible became alive for him and rather than reading it as a duty its study became an adventure. Denis admits to becoming rather intolerant of the views of people who were not committed Christians. His outlook was narrow with unquestioning acceptance of certain prohibitions, such as

dancing, going to the cinema and consuming alcoholic drinks, thought to be manda-
tory to a faithful following of Christ. But he was pleased to belong to the 40 group
and to stand up and be counted. In later years his views mellowed and he became
more tolerant (Fig. 3.3).

Apart from his new group of friends at Trinity, there was one other individual to
whom Denis was indebted. At a 40 meeting in February 28, the address was given
by Uncle Louis. For all the distinguished uncles and aunts that were part of Denis'
inheritance, he always claimed that Uncle Louis was the one who influenced his life
the most. Uncle Louis, one of his mother's brothers, had spent some time as a missionary in Japan and became a leading evangelical preacher in the Church of Ireland
often speaking to the Christian Union at Trinity. Denis and brother Robin spent
many Sundays at Uncle Louis' home, the Rectory at Zion, during their time at the
university. For both of them, this time was to be prolonged by changing from their
initially chosen courses to medicine.

In the weeks after his commitment to follow Christ, Denis sought the guidance
of God through prayer as to what he should do with his life and in particular whether
he should continue with his engineering course. What explicitly made him change
courses is not recorded in any detail, but he must have been influenced by his room-
mate Barber, who was a medical student and with whom Denis would have
exchanged news of their daily activities. The members of 40 were mostly medical
students and committed Christians whose belief and assurance that God had called
them to serve him through service to others. They made Denis realise for the first
time that changing to medicine would give him a sense of vocation. Being a doctor

Fig. 3.3 Glenmalure cycling holiday 1931
From the left, Breaky, Lee, Howe, Jones, Finch, Dunlop, Torrens, Rea

would be more likely to fulfil that vocation than becoming an engineer, although James might have put a different case to him. And the engineering course was not going that well. A few days after he had sat the entrance exam for Trinity in July 1929, his tutor Arthur Ashton Luce had written to James saying:

Dear Burkitt
The enclosed gives all information.
Don't blame me, if he gets stuck and loses his 10gn fee, or if he fails in …1930 and loses the year.
I suppose you have thought it out & I daresay he will pull it off. All the same it is a wish.
I hold this up until I have seen his marks at …
Yours sincerely
Arthur

And then appended later to the letter (Fig. 3.4):

The April fee (10gn) must be paid shortly before he sits for the J.F. The October fee will fall due shortly after.

Maths 62/100
Mech 65/100

Quite good – assuming the first three (…illegible) languages, this removes most of my (…) objection

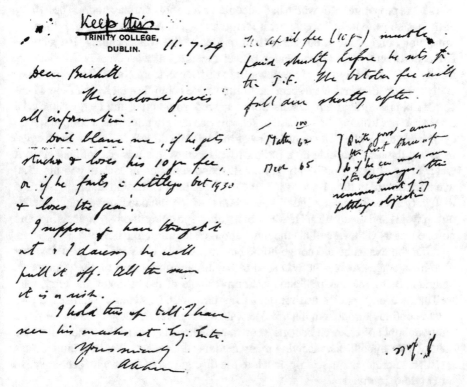

Fig. 3.4 Letter from Dr Arthur Luce, Denis' tutor in 1929, to Denis' father
See text for 'translation' (Trinity College Dublin, Manuscripts and Archives Research Library MS 11268/6/2)

The letter, warning James of the likely failure of Denis to succeed in the engineering course, is very difficult to read, but the exam results, which were appended after Dr Luce had written it, clearly surprised him. Denis had gone up to Trinity to do engineering more out of a sense of loyalty to his father, a school career that had not identified any particular interest nor academic ability and a lack of any other sense of vocation. He was not without ability but had lacked motivation for much of his school days. Many years later he would write that after his February 14 experience, 'This provided me with a sense of identity, value and direction…Finding motivation and direction pushed me near to the top of my class and kept me there without altering my IQ' [1].

The Easter vacation of 1930 followed along its usual lines for Denis despite the life-changing events that had occurred during the preceding 3 months. He had not at this stage decided to change to medicine – so was buying equipment for the next term's engineering practicals – and was at ease with his family and local friends. He was able now to give his father useful assistance with his surveying and bridge-building whilst enjoying making gadgets such as a wireless transmitter with his friend Eric and was enjoying listening to his radio. Photography continued to be a major interest, and once Robin returned from school, they went off together to visit their usual haunts. He returned to Dublin for the theoretical chemistry exam the first of a series, preparation for which had occupied a lot of his holiday hours. The Trinity term was short, barely 7 weeks with lectures on geology, surveying and drawing, mineralogy and physics, but the pattern of Denis' life was changing. He joined a Bible study circle and now attended prayer meetings at least weekly. Meals were mostly taken in the rooms of his new circle of friends from No 40 including especially Ejerton. Sundays were spent attending various churches including Zion where Uncle Louis would preach. Letters were received from home, and Peggy came on a brief visit, but Robin, who was still at school, was clearly missed as a companion on cycle rides and walks. The only record of his daily menu was in June 8 'ate 6 eggs' followed the next morning by a further four. Perhaps this was exceptional since it was worthy of recording in the very small space allocated to each day in his 1930 diary. By Saturday June 14, he was home for the summer vacation although returned to Trinity a week later for end of year exams, which included surveying, mineralogy, physics and mechanics. He was still at this stage committed to engineering, and no-one was as yet aware of his thoughts about changing to medicine.

After the exams he did not go home but travelled to Liverpool on the ferry, this time in steerage whereas he was used to having a proper cabin on these journeys. Steerage, being the cheapest and least comfortable of the births on the ferry, may well have been a choice arising out of his new-found morality. Such self-denial would continue throughout his life. He was to travel very widely once his work on the lymphoma of children became known, and later his role in the synthesis of the dietary fibre hypothesis, receiving many medals and honours. Yet he would always insist on economy class seating, much to the discomfort of those who accompanied him on long journeys.

From Liverpool he made for Bristol and the home of his good friend from school days, Broad, with whom he had kept in touch by regular letter writing. They set off on a 5-day cycling and camping holiday in the south-east of England returning to Bristol for the weekend where together with Dudley Smith they attended an Old Decanians Reunion, including speech day and a school play, Midsummer Night's Dream. Back home in Enniskillen, the summer holidays proceeded with visits to family and friends, church, boating, golf, tennis, ping-pong (table tennis), shooting and trips to the cinema.[4] Denis was now also able to help James in a more competent way using a theodolite, surveying for the building of the new bridge over Upper Lough Erne, Boa bridge, and laying out the new roads for it. Thus, he remained an engineer in training, doubtless with his father's teaching and encouragement throughout July 1930, a real bond between father and son. Peggy was his companion at home until Robin appeared on Saturday August 9 when they played tennis and Denis started to tell Robin about his thoughts on changing from engineering to medicine. Robin would have been a trusted sounding board for Denis. In August 12 Denis records 'Almost decided to do Medical'.

And so it was that he decided to change courses. But there was the problem of telling his parents what he had in mind. Knowing that his father would be very disappointed, he left this encounter for 3 weeks finally deciding to tell them when the family were together having supper on the evening of September 2. After the meal he said, rather bluntly, 'I want to give up engineering and become a doctor'. James did not react immediately, but Gwen with an eye to the future said, 'My son, Doctor Burkitt'. James eventually responded with 'Laddie. It's your decision and your life'. Laddie was the family name for Denis when he was younger, so it must have been reassuring to hear his father use it, confirming the father-son relationship but now handing over control of Denis' destiny to his son. The change in course meant a lot more expense for the family, but that seems never to have been an issue. What reassured his parents more was the knowledge that there had been a number of successful doctors already in the family including Dr W R Burkitt, a surgeon in the Crimean War, Uncle Roland who was in Africa and had come close to putting Denis and Robin off medicine with their tonsillectomy and Gwen's brother, Ernest. A week after this defining moment with the family, he heard from Luce and Major Allen at Trinity that a change in course was possible. Michaelmas term started in October 1 with a lecture on bones, Robin now accompanying him to Dublin (Fig. 3.5).

In the second week of term, Denis was required to sit pre-reg exams, which were in effect entrance exams to the medical course. The papers were chemistry and physics. Perhaps to his surprise, particularly because chemistry had never been one of his strong subjects, he came first and was third in physics. Soon after he was fully engaged with many hours spent in the dissecting room, starting with the arm and a copy of *Gray's Anatomy*. He went to see the Provost and Clerk of Registry of Chambers in October 24 with a view to changing his room. The motivation for this change is unclear, but it was not to be able to share with Robin. Denis was involved in some sort of religious activity most days including attendance at meetings at the

---

[4]At Laragh, where there were 20 acres of fields, a farmyard with large stone barns, workshop, stable, pig sties, gardeners' cottage, orchard and large vegetable garden. There was plenty to do.

Fig. 3.5 Burkitt family circa 1930
Front row, Gwen, Peggy, James
Back row, Robin (Robert), Denis

YMCA, Scripture Union, Crusaders, No 40 group and the College Chapel. Towards
the end of term, in December 14 Denis wrote to his parents that he was giving up
dancing. When home for the Christmas vacation soon after it was clear, they thought
that this was not a necessary part of normal religious observance and Denis watched
with some regrets as he saw Robin and Peggy go off to the hospital dance in
December 30.

The Hilary and Trinity terms saw Denis working at his medical studies much
harder than he had ever done for engineering. His friends remained the same group,
Ejerton, Pratt, Robinson, Lord, Scott and Bourne, but now Robin was again a regular companion on walks and outings. By the middle of February, Denis had been
given new rooms, No 19 on the top floor of the building, which he found to be an
improvement on No 16. He was the sole occupant, which may have been an important motivation for him to move. He was elected to the Committee of No 40 and
began to speak from time to time at church meetings. He was clearly growing in
faith and was at ease with the meaning it gave to his life. In May he spent a residen-
tial weekend at St Valerie, near Dublin, which he described as a glorious time and
the best weekend he had ever spent. At home in the holidays, he helped James build
a greenhouse from scratch, which must have required some considerable carpentry
skills, and on Sundays he was enjoying the preaching in the local Methodist Church,
which he preferred to that of the Rector at Trory – 'Awful sermon' (Fig. 3.6).

Fig. 3.6 Denis' room at Trinity. Probably No 19, which he occupied during 1930–1935. Note picture on wall of mother and water jug with basin for washing

The following 3.5 years (January 1932–June 1935) are undocumented in his autobiography and there are no diaries to be found. This is something of a mystery. The last entry in December 1931 and the next on Friday June 21, 1935, are written with an assurance that the story was continuing. There is no suggestion at any point that he was going on furlough from diary writing. At the time of his death in 1993, all his diaries were neatly arranged on the shelves in front of his quite small desk in the house that he would occupy in later years with Olive his wife. Few people were ever admitted to this holy of holys, and if they had been, why should they even have the knowledge to abscond with these particular years? The only person with the time and inclination to read the diaries was Olive, but she did not read them. The family believe they were lost. As a result, Denis' description of the whole of his medical school training is limited to a couple of sentences in his autobiography. It follows on from a brief note about his need to follow God's will with the change in direction. 'Never have I at any time regretted this decision. My course was set and for the first time in my life I found my work of absorbing interest and managed for the first time in my life to keep near the top of my class, and to win prizes, which I had seldom managed to do at school'.

His class marks were now good, he was awarded the Hospital Prize in Surgery in his fourth year, and there are two 'First Class Certificate of Merit' documents in the Burkitt Archive in Trinity Library,[5] which includes:

---

[5] Trinity College Dublin. Manuscripts and Archives Research Library; TCD MS 11387/7/1-6.

Second Class Certificate of Merit 1931–2 Winter Physiology 69%
First class Certificate of Merit 1932 Histology 74.4%
Second Class Certificate of Merit 1931 32 Winter Experimental Physiology (no marks noted)
Class Certificate of Merit 1932–33 Anatomy Third Year 61%
Class Certificate of Merit Anatomy second year 1931–32 55%
Class Certificate of Merit 1930–31 Anatomy first year $77^{1/2}$% of marks
Further Certificate reads as follows;
Trinity College Dublin
First Class certificate of Merit
Faculty of Medicine
I certify that Denis P Burkitt
Acquitted himself with high distinction as a member of the class of
Histology during the Summer session 1932 and obtained 74.4 per cent of the available marks
Harold (?Pringle) professor

Clearly now better motivated in his studies, he was never a high flyer academically and is remembered by one of his classmates as being 'A nice enough person who worked hard but was not top of the class'.[6] But medicine was a means to an end for Denis who had changed courses after being challenged about his ambitions and commitment to serve his fellow man. There were clearly no highlights worth his recounting from the 5 years of the course, and perhaps he saw even diary keeping as a form of self-indulgence. But he would continue throughout the rest of his life with his diary and with longer accounts of his activities in letters to his family, which would also include his views and sometimes his feelings about events.

Robin, acknowledged by Denis to be brighter than him, had gone up to Trinity a year later than Denis to read modern languages/literature and did well. At the end of his third year, he became a foundation scholar, which meant that he would get his board, rooms and tuition fees paid for his final year, thus relieving his parents of further financial responsibility. He planned to enter the colonial diplomatic service, but despite his success he also had doubts about the future perhaps encouraged by Denis' clear change in motivation and enthusiasm for the medical course, and so he also decided to change to medicine. Remarkably he passed the entrance exam and went on to attend his course in the arts school for his final year graduating with first class honours and also to complete his first year in medicine.

The two brothers would soon step out into the world of medicine, but their careers would follow very different paths.

Reference

1. Burkitt D. Unpromising beginnings. J Ir Coll Physicians Surg. 1993;22:36–8.

---

[6]WA Gillespie, a classmate of Denis, personal communication, 1969.

Chapter 4
Learning to Be a Doctor with a Faith

Denis graduated Bachelor of Medicine, Bachelor of Surgery and Bachelor of Obstetrics (BAO[1]) from Trinity in June 1935 having graduated BA in 1933.[2] He now embarked on the postgraduate training that every doctor follows to turn him into a fully competent and experienced medical practitioner with some specialist training. The first posts to which newly qualified doctors are appointed are the most junior rank in the medical hierarchy and in 1935 would have been called house surgeon or house physician. Junior doctors were expected to gain experience in as wide a spectrum of medical practice as possible before embarking on a career as either a hospital consultant or general practitioner or to follow an academic path with research and teaching. Denis applied for a job at the Adelaide Hospital, the major teaching hospital in Dublin, where he had done a lot of his undergraduate learning. Competition for these teaching hospital jobs is always great, and they would have been given to the brightest students of the year or those who had distinguished themselves in other ways (Fig. 4.1).

For all his self-proclaimed lack of academic leanings, Denis was awarded the Hudson Prize and silver medal at the final prizegiving where he was second to JNP (Norman) Moore, the academic star of the year who would go on to become Ireland's leading psychiatrist. Denis did not get the Adelaide job and neither did Norman Moore. Instead, it went to one of Denis' friends from Portora school days, Robin Pratt[3] who had played rugby for the university and for Ireland earning five caps in 1933–1935. Denis was disappointed but felt that God must have some other plan for

---

[1] *BAO Baccalaureus in Arte Obstetricia* (Bachelor of Obstetrics), a degree unique to Ireland which the Irish universities added in the nineteenth century as the legislation at the time insisted on a final examination in obstetrics. This third degree is an anachronism which is not registerable with either the Irish or British General Medical Council.

[2] Students graduating from Oxford, Cambridge and TCD are all given BA, whatever their subject.

[3] Robin H Pratt, 1912–1997. After graduating in medicine, he joined the RAF Medical Services where he continued to play rugby and remained, rising to the rank of air commodore.

© The Author(s), under exclusive license to Springer Nature Switzerland AG 2022
J. H. Cummings, *Denis Burkitt*, Springer Biographies,
https://doi.org/10.1007/978-3-030-88563-2_4

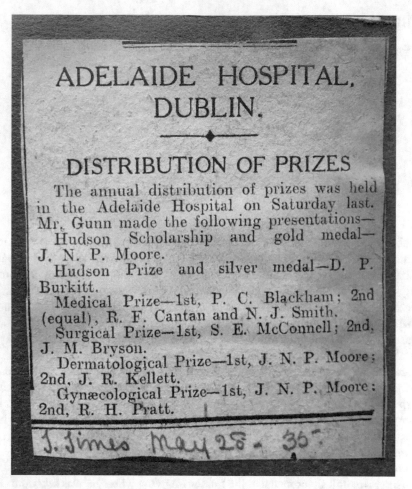

ADELAIDE HOSPITAL, DUBLIN.

DISTRIBUTION OF PRIZES

The annual distribution of prizes was held in the Adelaide Hospital on Saturday last. Mr. Gunn made the following presentations—

Hudson Scholarship and gold medal—J. N. P. Moore.

Hudson Prize and silver medal—D. P. Burkitt.

Medical Prize—1st, P. C. Blackham; 2nd (equal), R. F. Cantan and N. J. Smith.

Surgical Prize—1st, S. E. McConnell; 2nd, J. M. Bryson.

Dermatological Prize—1st, J. N. P. Moore; 2nd, J. R. Kellett.

Gynæcological Prize—1st, J. N. P. Moore; 2nd, R. H. Pratt.

J. Times May 28 - 35

Fig. 4.1 Prizes awarded at final graduation of Trinity College Dublin Medical Course. (Note 'Hudson Prize and silver medal-D.P.Burkitt'. Irish Times May 28, 1935)

him in closing the Adelaide door. Later that same week, a telegram came from Robin, who was still at Trinity whilst Denis was by now at home at Laragh, telling Denis that the dean wanted to see him immediately about a job and that he would wait until 6 pm in his office. Denis made it back to Dublin in time and the dean offered Denis a job in Chester, which he accepted. Denis had to complete an application form, which he did immediately, and just managed to catch the last post on Friday as the postman was clearing the letter box. The next morning a wire came from Chester inviting him to interview the following Tuesday. There were two other hopefuls at the interview both of whom had more experience than Denis, but he was given the job.

Denis saw the hand of God in all this but a more down to earth possibility that might explain Denis' good fortune is that the Dean, having heard the decision concerning the Adelaide job, was looking after one of his more conscientious and able students and operating what would have been a well-established old boy network. In Denis' mind there was no question that God wanted him to go to Chester, and in later years he was to recall that had he been given the Adelaide job, he would never have met Guy Timmis, a fellow medic with whom he would work in Africa and who would become a lifelong friend.

Arriving in Chester on Sunday July 7, Denis was pleased to find copies of Testaments in the bedside lockers of patients and gave out some SCM[4] leaflets. At this time Denis clearly felt he was on mission to save or win souls for Christ and his duties as a doctor on the wards were a God-given opportunity to do this. Even whilst walking the streets of Chester when he was off duty, he would accost people and ask them if they had ever read the Bible. As he settled into his job, he engaged with his fellow doctors and the ward staff who were polite and friendly to him but must have wondered what was in the mind of this young doctor from Ireland. He went into Liverpool on his first free weekend and bought a copy of CT Studd's 'Prove me now'.[5] Denis returned to the hospital and left the book on the table in the doctors' sitting room and then carried it with him into casualty where the nurses asked him about it. On the wards he gave out more leaflets and copies of the Gospels and showed two nurses the book, which surprised them and who remarked 'didn't think you'd read that sort of thing'. On Thursday 18th he gave a dying patient a book of psalms and quoted various texts to him. The patient died the next morning but word spread to the more senior staff at the hospital that he was distributing leaflets. Meeting with Mr Woodroffe, one of the senior surgeons, on Friday morning, he was told that he must stop giving out religious tracts (Fig. 4.2).

For Denis this was a big blow to his ambitions but he was sure that God would work things out for him. In the following days, he felt that the wind was against him, not because of any problems with his clinical work but that he was not able to accomplish the Ministry to which he had been called, 'to testify the Gospel of the Grace of God'. He prayed at night not for the health of his patients but that he might meet someone to whom he could talk about spiritual things. Inevitably word of his beliefs and motivation spread around the hospital, and colleagues or patients who were churchgoers were able to share their experiences with him and receive tracts and copies of the Gospels to read. Mr Woodroffe relented a little and told Denis that if he wanted to go to church on Sundays, he would excuse him. Then, at the end of his first month in Chester, an event occurred that would bring sharply into focus the realities of life and how sometimes those with the strongest beliefs responded.

---

[4] Student Christian Movement.

[5] Studd (1860–1931) was one of the three remarkable brothers, sons of the wealthy Edward Studd, who had been converted during a Moody and Sankey campaign. The boys were educated at Eton and went on to Trinity College Cambridge each in turn becoming captain of cricket for Cambridge. CT (Charles) played for England and was a member of the team that lost to Australia in 1882, the first Ashes test. A visiting preacher to the Studd home led to Charles' conversion, and he went on to become a missionary to China and write a series of books concerning the Christian faith.

Fig. 4.2 Staff at Chester Royal Infirmary September 1935
Back row, Hamilton, Stout, Macaskill
Front row, Mallock, Burkitt

Denis heard from home that 'Peg had a "mental breakdown"'. Peggy at the age of 28 was the oldest of the three siblings. Photographs of her show that she was tall and good-looking and had taken the fullest part in family life with the boys as they had grown up. Although by 1935 it was becoming much less of the norm, she had not married and had stayed at home with her parents. But at the end of July, she developed what is reported as a severe attack of chicken pox complicated by a nervous breakdown [1].[6] Whatever the actual diagnosis, Peggy never recovered. She was cared for at home for the rest of 1935, but then in January 1936, James and Denis drove her to the Bloomfield nursing home in Dublin where she was to remain until the creation of the National Health Service (NHS) in 1948, whereupon she was moved into NHS care in Omagh. It would be 40 years before Denis saw her again.

When back in Chester, Peggy's illness did not seem to be a matter of remark for Denis, but he was having several difficulties in the hospital. He was asked by one of the nurses if he had ever considered entering the Ministry, which he had not, but he did tell a ward sister that he intended becoming a missionary, which he did not. His unease at the lack of progress in his clinical skills was beginning to bother him. He was doing minor surgery such as circumcisions, removing sebaceous cysts, setting fractures and giving anaesthetics but wanted to move on. Being in his first junior surgical post, he was much less experienced than his colleagues, and in October he asked his Chief, Mr Woodroffe, if he could do an appendix operation, to which he

---

[6] Neurological and psychiatric complications of chicken pox are very rare and not usually fatal but include encephalitis, which may have long-term consequences.

agreed. Denis was doubtless a very conscientious doctor and cared for his patients but his zeal to convert the world was never far away. When removing the fluid from a man's hydrocele [2],[7] he talked to him about the Bible and recommended he read Revelations 7. He was still very much in touch with his friends from No 40 at Trinity, who were now all practising doctors in and around Dublin. They told Denis that they were going to start a meeting of believers in the medical community, which stimulated Denis to think about one in Chester. Since he had found the nurses more receptive to his ideas than the doctors, he decided, after talking to Sisters Williams and Quick, that this could work but was advised to approach Matron. She was sympathetic to Denis' suggestion but said she would have to get permission from the Hospital Committee, which duly came, and the first meeting was held in early December with 25 attending, and his good friend Priddy came over from Dublin to talk and give support. Matron also attended.

Despite his faith Denis felt hopeless in the face of his patients' dying. He could not bring himself to offer Bible readings or prayers to them, but their death brought home to him what he felt was the transient nature of life and that he must focus on eternity. He was attending church regularly but preferred the non-conformist, usually Methodist, style. By now he was being asked to speak at meetings in the area, so he decided to try and get another job in Chester at Mildmay Hospital in order to continue his evangelical work and build on the contacts he had made. But he was not eligible because of his lack of seniority and experience so instead applied for a job at the Adelaide and was back in Dublin as house surgeon by February 1936. Here he was much closer to the friends who had so influenced him and supported him during his undergraduate years (Fig. 4.3).

On the wards, again Denis was troubled. He was starting in a new environment to make known his beliefs and to be accepted as much for his evangelical work as for his clinical skills. A patient, McGovern, whom he had catheterised and to whom he had quoted words of encouragement from the Bible wanted to express his gratitude with the gift of a valuable book. Denis was very moved by this encounter but declined the gift in a letter. Although it was now almost 6 years since he committed to following Christ, he did not have the confidence to accept the gift nor an understanding of the sense of fulfilment it would have given the patient. The next day he prayed that a patient who was dangerously ill with signs of obstruction of his gut would recover and took the signs of such recovery later in the day as confirmation of God working through him.

Two months later Denis became worried about money. He believed that money was given to us in trust by God and was persuaded to donate £7 for a fellow student, George.[8] This was a considerable sum since Denis' monthly salary was probably

---

[7] Hydrocele of the testis is an accumulation of fluid beneath the thin membrane surrounding the testis. It is common and readily treated. Its occurrence in Africa was to provide Denis with his first successful foray into the cause of disease.

[8] Cofie George was a Ghanian who struggled financially at Trinity but was helped by his fellow students. He returned to Accra, married well and ran a successful nursing home in the capital. Denis would meet him again there in 1962.

Fig. 4.3 'Adelaide Hospital Resident Staff Jan-June 1936'
Back row, 'S.L.Tomlinson BA, E.R.N.Cooke BA, G.K.Donald BA, G.W.G.Warwick BA'
Front row, 'J.H.Mitchell Sch.(Mod)BA, D.P.Burkitt MB, Kellett MA, E.D. Colier'

only £15. He had apparently been asked to give £14 but declined, his natural thrift overcoming what he believed to be his Christian duty. The next day however, 'God made me give or lend George the whole £14. I fought it but God won'. A few days later, a friend gave him £4 for George, which made Denis feel that giving away the £14 had not turned out to be so painful. Soon after George received a cheque from home for £30 and Denis was returned another £5. Perhaps the relief at getting some of his money back, rather than the thought of having fulfilled his obligation, led Denis to remark 'I am glad from the bottom of my heart that I supplied the whole £14'. But the numbers still did not add up for Denis. A few days later, he bought something for 5/– that he had found elsewhere priced at 15/–,[9] and later he was

---

[9] Prior to the introduction of decimal currency, twenty shillings 20/- were equal to £1.

given a lift back from a wedding in Clonagh, where he had felt like a fish out of water, but it saved him a further 10/– on the bus fare he was expecting to pay. Finally, he received an unexpected £2-2-0 from an insurance policy, thus having recovered in one way or another £12-2-0 of the £14 he was able to conclude 'The Lord will be no man's debtor'. In later life Denis was to be generous with money, to the chagrin at times of his family, but there was also a frugal streak in him, which explained why he derived pleasure from doing things as cheaply as possible (Fig. 4.4).

Rarely do clinical problems or achievements merit a note in Denis' log of his days as a young doctor, but one clinical incident did get through to him. A patient called Brown, who had been ill for some time and was a Christian so merited a long talk with Denis, suddenly went downhill and died. That same evening Denis had to do an autopsy on Brown. He described the finality of this as gruesome.

The Adelaide job ended after 6 months at which point Denis had done the requisite preregistration training and could now become a fully registered medical practitioner with freedom to follow any career path. But without the direct support of the medical school, he had to find a job on his own. He did not have one arranged for August and beyond and was at home at Laragh following the usual social round when a friend, Allan, suggested they both went to Spain where they could work with the Spanish Gospel Mission, following in the footsteps of George Borrow, giving out Christian literature. Denis booked two berths on a cargo boat leaving from

Fig. 4.4 Denis and Cofie George, probably at Laragh 1936

Liverpool and sailing to Bordeaux going out in July 10 and to return on the 31st. They travelled with tents and bicycles eventually making their way from Bordeaux by train to Bilbao in northern Spain. Shortly after their arrival, in July 17, the Spanish Civil War started, and the British Consulate advised them to get aboard a destroyer that was being used to evacuate British citizens. They arrived back in Bordeaux and found passage back to England on the same boat that had brought them out. About one thing Denis was pleased, the whole trip had cost them only £14.

Back at home Denis started applying for jobs. He wrote over two dozen letters to hospitals seeking a post-registration job starting in October 1 but without success. It was not that his former employers held anything against him. A reference from Mr Woodroffe in Chester read:

Chester 27-8-36
Dr Denis Burkitt was my H.S. (House Surgeon) at the Chester Royal Infirmary for the last six months of 1935.
He was a thoroughly painstaking and competent H.S. He has a good theoretical knowledge of Surgery and can use his knowledge to good practical effect.
His work was always thoroughly and conscientiously done, and his notes well kept, and at the end of his time he was beginning to show promise as an operator.
He is a very gentlemanly and likeable man, and I always found him cheerful and ready for more work at the end of a heavy day.
I can thoroughly recommend him for any post of H.S. or H.P.
Signed. H.L.Warren Woodroffe M.D., M.Ch.

The mission to save souls had clearly moderated, and Denis was showing signs of the work ethic, practical skills and character that would carry him through the 30 years of his surgical practice. He would always love operating and would acquire an unrivalled breadth of experience and competence whilst in Africa.

A locum appointment as Resident Medical Officer (RMO) at Guy's Hospital for the month of September was advertised at £6-6-0 a week, but after praying about it, he decided not to apply. Soon after this he read a notice about an ear, nose and throat (ENT) and eyes post in Preston, Lancashire, and immediately felt this job was for him. He applied and a week later and without any requirement for testimonials or interview he was offered the job. The staff in Preston had clearly sought opinion about Denis from his former classmates Bill Gillespie,[10] who was already working there as a house surgeon, and Rouse Mitchell who both put in a good word for him. Meanwhile he was given a job as locum in general practice for a month in Blackwood, a small town in the Welsh valleys north of Cardiff in the county of Monmouthshire. He immediately struck up an acquaintance with the Practice Dispenser Miss Rowlands who told him she thought religion was dying out but was prepared to listen to Denis and read some of his literature. The relationship with Miss Rowlands seemed to flourish. She accompanied him to church, with her sister, and invited him to her house to meet friends. They went on drives together where she told him that he had made her much less worried about her job, and when Denis finally left for

---

[10] William A Gillespie would become Professor of Clinical Microbiology at Bristol University where Tony Epstein would become Head of Department.

Preston, a letter from her arrived the next day saying what a help he had been. But there it seems to have ended.

The Preston job was for 6 months from October 1936 until March 1937. Now, over 7 years since his conversion at Trinity following the challenge of Dick Jones, Denis was completely engaged in spreading the Christian Gospel. On arrival he found that there was 'A lot of selfish, discontented and complaining spirit about; also disparaging gossip'. His response was to pray for grace and for love for those who annoyed him. He fell out with the ward sister, Sister Qualtrough, who had a reputation in the hospital for being difficult, but Denis felt that God would help him to get around this. By the end of the 6 months, she was accompanying him to hear visiting speakers on religious themes, borrowing his Bible so that she could read some texts he had marked, and she even asked him out to tea in Preston where they went to the local art gallery together. Although Denis approached people with one thing in mind, to win their souls, he must have been quite a charismatic and charming young man (Fig. 4.5).

He was also a good speaker. Denis was now accepting invitations to speak at a range of meetings especially Bible classes, Crusader[11] meetings and gatherings in the homes of people where he would usually start with a text or passage from the Bible and expound it. He would lead prayers, in the absence of a member of the clergy, and give his testimony at every opportunity. Severely ill patients, such as a woman in casualty threatening suicide or a boy with a badly bleeding tonsil after surgery, would be prayed over, told that God loved them and given appropriate literature. The highlights of his day or week were a conversation with fellow Christians

Fig. 4.5 Preston Royal Infirmary 1936

---

[11] Crusaders – Christian organisation involving mainly young people.

or a meeting at All Saints Church where he made friends with the vicar Rev Cundy and his curate, Moss. One Sunday when invited to speak at a meeting in Southport, a journey of 20 miles, he wrestled with his conscience because he felt that taking a bus or train would be contrary to the Fourth Commandment 'Remember the Sabbath day to keep it Holy' (Exodus 20:8–11) and in particular that you should not make other people work for you on that day. So he borrowed a bicycle for the journey. On another occasion he declined to go to the cinema to see 'The Life of Louis Pasteur' because, despite it being of medical interest, attendance would compromise his commitment not to go to the cinema because it was of this world and not the next. That same evening he dined with a cinema-going friend and his parents and sitting around the fire with them had 'a wonderfully free talk on prayer and faith'.

Despite Denis' calling to serve God, he still didn't really have any special career path in mind. A few days after the memorable evening with his friend and parents, he was reading the magazine of the Oriental Missionary Society where there was an appeal for money. He decided to give £1 a month. At the Christmas dinner for the doctors and nurses later, he told one of the sisters that he hoped to be a medical missionary. It was an ambition he was never to fulfil.

The Preston job finished at the end of March. Denis had for some time been applying for a further post and was becoming anxious at not getting any replies. A few days later, a job in Poole[12] was advertised, which he liked and applied for. He was called for interview and despite having less experience than other applicants was offered the post. It was in general surgery, which suited him because he had it in mind to sit the exams for Fellowship of the Royal College of Surgeons a necessary step on the road to becoming a fully qualified surgeon. Why had he finally chosen to be a surgeon? He must have enjoyed his surgical house job in Chester and not been put off by the Preston ENT post, despite the local difficulties and perceived enemies. Perhaps a visit to see the operating theatre at the hospital in Enniskillen in February 1931, after he had changed course at Trinity to Medicine, had impressed him. Denis had a natural aptitude for practical things that would have been given more credence by his year's training as an engineer. Maybe Uncle Roland who spent his career as a surgeon in Nairobi was influential, but Denis' encounters with his uncle, especially the tonsillectomy episode when he was 7, were more likely to have put him off. However, the world of the physician may not have appealed to him with its much more cerebral approach to diagnosis and more limited options for treatment in the days before the availability of penicillin. He was really a surgeon by inclination and skill right from the start of his medical training (Fig. 4.6).

Denis arrived in Poole on April 17, 1937, and was immediately pleased with what he saw: not the good facilities for surgery but a Bible on his ward used for regular Sunday services and several tracts lying around. One of the sisters had read CT Studd, so he was in his element immediately. Permeating the whole of his life and work in Poole was his pursuit of God's will for him. His clinical duties are barely recorded and the importance of his training in surgery not worthy of mention.

---

[12] A large town on the south coast with a natural harbour leading into the English Channel.

Fig. 4.6 Denis at a wedding. August 12, 1936. With, left to right, James (Father), Uncle Percy Newenham and William Burkitt, son of Roland and Annie

Faith can of course be challenged, often repeatedly. For Denis the fate of a patient sometimes made him seek reassurance. On Friday July 23, 1937, he operated during the night on a young man, aged 22, with a burst appendix and abdomen full of pus. There was nothing much Denis could do but close up his incision and put a drain in place. Without antibiotics the outlook was very poor, and the next day one of the senior surgeons, Mr Forrest, told Denis that the patient would die. That evening he was reading his Bible and came to James 5:14–18, which includes the lines 'And the prayer of faith shall save the sick, and the Lord shall raise him up; and if he has committed sins, they shall be forgiven him'. He took this to be a promise from God. The patient became worse in the night but believing in God's omnipotence he was pleased when he began to improve. Ten days later when the patient showed Denis a piece of paper from a friend telling him to look up and trust Jesus, he was reassured that someone else was also praying for him. Soon after the young man died. Denis turned to his Bible and found 'Thy will be done'. There was thus no conflict with Denis' faith, and his regret was that the young man had been well enough on only one Sunday during his stay in hospital to listen to Denis' sermons on the ward, but he was nevertheless pleased that he had the opportunity to show him the way of salvation.

What the hospital staff must have made of all this is not recorded but they were very tolerant. Early in his appointment in Poole, one of the senior surgeons, Mr Laker, invited Denis to a dinner where he was expected to wear evening dress, prior to going to the Pavilion Theatre in nearby Bournemouth for a variety show. This invitation was a great honour for Denis, but he knew he could not accept because of his beliefs and declined the invitation telling Mr Laker why. A couple of days later when he met Laker in theatre, he was quite friendly. Denis would also refuse to go to any dances whilst he was in Poole or buy raffle tickets.

Denis gradually gained the confidence of the local clergy, and he was given the job of leading the ward services on Sundays, speaking at open air evangelical meetings, and was allowed to give out booklets and tracts to listeners. He was often thanked, especially by patients, for his reassuring words and attracted the attention of a number of the nursing staff of which one in particular, Sister Newton who worked in the operating theatre, seems to have tried to build up a relationship. This started, inevitably, with attendance together at church services, prayer meetings, walks, a trip to hear the Messiah and invitations to dinner. In November she asked Denis to go to a meeting with her and afterwards, perhaps knowing what might draw Denis' interest in her, told him that she didn't think she was saved. She drove Denis back in her car that evening and on the way they stopped and they talked for an hour. The result of this was for Sister Newton unrequited love. Denis marked several passages for her to read in her Testament. She became known to him as Phyllis Newton, but that was all.

Denis must have kept a firm grip on his emotions. Some time towards the end of his job in Poole, two nurses, Galpin and Ellis, came into his room after midnight. 'We had a long chat sitting on my bed in the dark, which turned to spiritual things'. The girls admired his faith and left. There was safety for all in numbers. Nurse Galpin was to return to Denis' room a couple of weeks later on her own and was given special treatment. She was shown Denis' favourite verses in his Bible.

Also working at the hospital in Poole as a trainee surgeon was Dr Conrad Latto who was to become a great friend of Denis' in later life. Conrad had been brought up in a Christian home but had lapsed in his commitment and told Denis when they met that he had never read the New Testament and had not prayed for 5 years. He was persuaded to attend the ward service, and Denis gave him a pamphlet 'The pull of the unseen'. Conrad became a regular attender at the services, which Denis was now taking weekly, but left Poole in early March. Shortly after he wrote to Denis saying he was now reading the Bible regularly. They were next to meet at the Prince of Wales Hospital in Plymouth in 1940.

The Poole job ended after a year and there was a tearful farewell for Denis on April 1, 1938. He then set off for Edinburgh with Guy Timmis to take part in a postgraduate course leading to the exams for the Fellowship of the Royal College of Surgeons of Edinburgh (FRCSE). Denis had met Guy during their first house jobs in Chester and they became as close as brothers. Guy had not come from a Christian family but had heard the call of God whilst at Trinity and was looking for ways to serve Him. The two of them shared lodgings and took part in the ward services at the Edinburgh Royal Infirmary with Denis preaching and Guy choosing the hymns. The exams were in early July after which they both set off for a few days holiday in Scotland. Denis then returned to Laragh and set about finding another post. He was still not sure what career path to choose but felt guided to get a job on a boat as a ship's surgeon to give himself chance to think about his future. He applied for several without success, but in early October, whilst he was away from home, he received a telegram from his Father saying 'Congratulations, Denis through, Guy fails'. Armed now with his new qualification, he applied to the Blue Funnel line but was told that there would not be a ship until early in 1939 unless an unforeseen vacancy occurred. He nevertheless filled in all the necessary forms, sent testimonials and a few days later was offered a post as ship surgeon on MV Glenshiel sailing in early November, a voyage lasting nearly 4 months. Would 16 weeks aboard ship give him sufficient time for reflection about his future?

References

1. Rudrajit P, Singhania P, Hashmi MA, et al. Post chicken pox neurological sequelae: three distinct presentations. J Neurosci Rural Pract. 2010;1(2):92–6. https://doi.org/10.4103/0976-3147.71718.
2. Burkitt DP. Primary hydrocele and its treatment: review of two hundred cases. Lancet. 1951;1:1341–3.

Chapter 5
War Changes Everything

The voyage, which gave Denis time for reflection, did not help him decide on his future although did provide the first step in opening his eyes to the world around him. The ship, MV Glenshiel,[1] sailed from the Royal Albert Dock London on November 12, 1938, with a crew of many races and around 25 passengers some of whom were missionaries heading for China. The route took them to Casablanca then through the Mediterranean to Haifa, Port Said on the Suez Canal, then Aden, across the Indian Ocean to Colombo in Ceylon, Penang, Singapore, north to Hong Kong, Shanghai and finally the port of Dairen (Dalian) in north-east China. Denis, always a photographer, recorded the journey in detail from which it can be seen that the passengers looked to be comfortably off, middle class, husbands and wives aged 40–60 with one or two children on board. Life for the passengers was agreeable. The food must have been good if the menu for Christmas Day 1938 is in any way a reflection of the chef's skills with a choice of at least a dozen dishes including roast turkey, sirloin of beef, lamb sweetbreads, turbot, salmon in crab sauce and an assortment of hors d'oeuvre, desserts, cheeses and coffee (Figs. 5.1 and 5.2).

Denis' medical duties were not very demanding allowing him to enjoy the journey, his first overseas travel apart from the abortive trip to Spain. He was fascinated seeing countries whose geography and cultures contrasted markedly with what he had experienced. The voyage was clearly an education as he photographed local people, street vendors, camels, dhows on the Suez Canal and women in Islamic dress. The culture of Malaysia was new to him as he had never seen mosques, temples, rickshaws, monkeys in the street and rubber plantations. They arrived in Dairen on January 1, 1939, where it was very cold with snow on deck and icicles on the rigging. They were soon on their way back, stopping again at Shanghai where the poverty and devastation, resulting from the ongoing war with Japan, drew his heart

---

[1] The MV Glenshiel, 9412 tons, a merchant vessel was commissioned in 1924 and sailed between London and the far east until torpedoed and sunk whilst in the Indian Ocean in 1942 by a Japanese submarine.

© The Author(s), under exclusive license to Springer Nature Switzerland AG 2022
J. H. Cummings, *Denis Burkitt*, Springer Biographies,
https://doi.org/10.1007/978-3-030-88563-2_5

Fig. 5.1 MV Glenshiel on which Denis served as ship's doctor from November 12, 1938, to March 2, 1939

out to China. There they took aboard a number of missionaries returning home who introduced Denis to Rev E.L. Kilbourne Vice President of the Oriental Missionary Society. Briefly after this encounter, Denis seems to have made up his mind about his future. He would be a missionary, but this was not to be the case at all (Fig. 5.3).

MV Glenshiel docked in London 6 weeks later with Denis still very worried about his future. Spending hours in prayer and Bible reading whilst on the voyage had not helped him. He was away from his family, friends and colleagues, and there had been no-one on board the Glenshiel in whom he could confide. He was no more settled in his mind but decided that he was really called to the Government's West African Medical Service. Then a letter arrived from his father encouraging him to do non-professional missionary work in part because 'the money is better'.

Stepping off the boat, he visited the Colonial Office in London to enquire about a post with the medical service, but war would intervene. He then set about finding a job. After only a week of looking, he was appointed as a locum in general practice in Chacewater, Cornwall, perhaps securing the job through the good offices of Robin who was working nearby. Denis was pleased with Cornwall where he found that almost every house he visited displayed texts and had a Bible. He then moved to South Wales to Blackwood where he had done a locum in 1936 soon after qualifying. There he met the Practice Dispenser Miss Rowlands once again with whom he had had a platonic relationship at that time. To his chagrin she told him that she thought he was now less keen spiritually. At the Heath Presbyterian Church on Sunday in Cardiff, the preacher talked about personal renewal and the things that

Fig. 5.2 Christmas menu in MV Glenshiel 1938

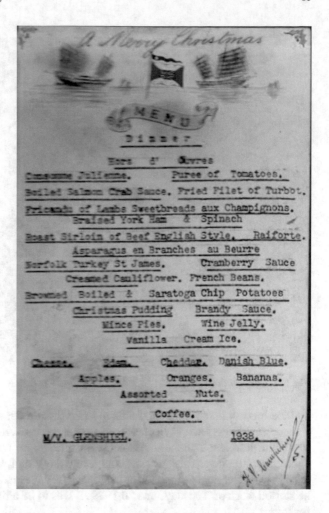

people needed to put behind them, included amongst which were lies. Here Denis had something on his conscience. Whilst in Cornwall the dispenser at the practice had asked him to sign some National Health Insurance Certificates, which before the advent of the National Health Service in the UK in 1948 would have allowed patients to claim the costs of drugs and treatment from an insurance company. He had done so but without him seeing the patients and in some cases with the wrong dates on the prescriptions. He told Miss Rowlands about this who offered a mild reprimand for his actions. This did the trick and he was back on track the next day with his devotions.

Leaving Blackwood on Friday April 14, he drove to Enniskillen where he threw himself for 3 weeks into an evangelical campaign led by Rev Lionel Fletcher. He decided, reluctantly, to sell his stamp collection the proceeds of which he donated to the work of the Overseas Missionary Society. Then turning his mind to getting a job, he applied for a Resident Surgical Officer (RSO) job in Winchester. He also

Fig. 5.3 Denis with passengers in Aden. Mrs Prior, Mr Worskel, Mrs Vincent, Denis

applied for a similar 1-year appointment in Plymouth, but after what he thought was a successful interview in Winchester in June 8, he cancelled his application for the Plymouth job. But Winchester didn't want him, so he wired Plymouth saying he could after all attend for interview, and a few days later, he had been appointed to the Resident Surgical Officer job at the Prince of Wales Hospital in Plymouth where he started work on Tuesday June 20 1939. It was to be a life-changing appointment for him.

He felt rather strange in his new more senior role where he had no direct responsibility for patients on the wards; instead he was available to all the junior surgical house officers as a second opinion. Shortly after he had started work, he received a 'phone call to tell him that Robin had fallen off a horse, cracked his skull and was in the nearby Truro Infirmary. He drove down to see him the next day but was pleased to find him doing well. The outbreak of World War II in early September 1939 signalled by the invasion of Poland by Nazi Germany merits two diary words 'War Declared'. He does note being busy in October 14 with casualties from two French ships that had been torpedoed by a submarine, but aside from that the world outside did not seem to impinge on his consciousness.

Sister Wright caught his attention early on during his time at Plymouth and soon they were going on walks together. She invited him home where they had 'a very pleasant sing-song'. She told him that she was converted, perhaps a strategic move on her part, and they were invited together for supper with friends. On the way back to her house, they stopped in a park where they said some prayers together. The relationship progressed no further, but she was to prove to be a valuable ally in Denis' main ambition in the hospital to spread the Gospel. Outside of this quite narrow focus, two events occurred that were worth Denis' attention. George, the TCD medical student from Ghana, whom Denis had been supporting, mainly with money, had applied for a junior house job at the nearby Truro Hospital and was given it without an interview. He had not told them he was African. A week later Denis heard that his appointment at Truro had been cancelled on account of his 'nationality'. In October Dr Kearney, a medical missionary from Sierra Leone, arrived at the hospital in order to watch and learn from the medical work being done in Plymouth. He told Denis that there was much more need for government doctors in West Africa than for missionaries.

The first 6 months of 1940 were busy. He was doing a lot of emergency surgery and spending time with various nurses, especially Sister Isobel Brown who was captivated by the young surgeon from Ireland. They became close but when the time came to reapply for his post he did not. Instead, he went for interview for a surgical job in Barnsley and started work there on July 15, 1940. There was some sadness at leaving Plymouth particularly at saying goodbye to Isobel, whose relationship with Denis had survived the Bible reading, prayers and church meetings. A letter from her arrived on his first day in Barnsley but events were to overtake this embryonic courtship.

The outbreak of the war and an unstable family life persuaded young Olive Rogers to apply to train as a nurse. Wishing to start in London, she was persuaded by her sister to apply to the Prince of Wales Hospital in Plymouth where her sister's fiancée David was stationed at the nearby naval barracks. Arriving in April 1940, Olive was a shy unworldly 20-year-old but with a Christian faith. The third of six children, her mother had died following the birth of the sixth child when Olive was nine resulting in the family being split up. Olive with her older sister Lucy was sent to live with her aunt and uncle who never missed a chance to tell the girls what a hopeless man their father was. It was an unhappy time, which ended when their father remarried, to the women who had nursed her mother through her first five pregnancies. She was a very good nurse and ran a very successful nursing home in Seaford, Sussex, to which the family moved. Olive did not settle at her new school but was taken under the wing of the local Baptist minister and his wife where Olive felt secure and was able to grow in their faith[2] (Fig. 5.4).

A short while after arriving at Plymouth, Olive met a lady doctor on the corridor and asked her if there was a local church. She gave Olive directions and said that she would meet her there and mentioned that she would also be meeting Mr Burkitt who was the resident RSO. They all went into the church together, and afterwards Denis asked Olive if she would like a lift back to the hospital in his car. Olive was pleased with the offer but on arriving back at the hospital was called to the matron's office

---

[2] Rev Laurence and Mrs Wilson-Haffenden. It was his second marriage. His three daughters were all away from home, so the couple looked upon Olive as their own.

Fig. 5.4 Nurse Olive
Rogers, Plymouth,
circa 1941

where she was told in no uncertain terms that she had behaved extremely badly.
Olive, as the most junior nurse in the hospital, had been seen getting out of the
RSO's car, and if that was how she intended to conduct herself, she should leave
immediately. No explanation was allowed and Olive left the office in tears. However,
she did continue her nurse training at Plymouth and thus had taken the first small
step in a relationship, between Olive and Denis, that would last 53 years and be
compacted of love and tears.[3]

The naval base at Plymouth was an important target for the German Luftwaffe
which, from early on, was intent on destroying Britain's naval capabilities. The
hospital in Plymouth was a major receiving station for casualties. Olive had been
nursing for less than 3 months when the RMS Lancastria, fully laden with fuel and
soldiers off the coast of France, was bombed by the Luftwaffe which then dropped
lighted flares into the oil floating on the surface of the sea so that the men were

---

[3]With apologies to Studdert Kennedy. 'Compacted of laughter and tears' from 'Woodbine
Willie' [1]

swimming in burning oil.[4] Some of the casualties were brought to Plymouth. Olive records:[5]

> I was appalled by the condition of those who were brought in. Their burns were horrific. I had never before heard men screaming for their mother…or lying motionless on a stretcher. Most of those poor men died.
> This whole episode had a profound effect on me.

Plymouth was on the front line in the war, and Olive, young, sweet-natured, sensitive and in many ways naïve, repeatedly experienced horrors that were to affect her for the rest of her life.

The next memorable occasion for her was the night of January 13, 1941, which she recounts in a letter to Denis, who by this time had moved to Barnsley Hospital:

> Mr Burkitt – or may it be Denis?
> Now something about our "blitz" night. The raid didn't last long – not more than about three hours.
> Lots of incendiaries dropped – about 3,000 I am told.
> We had two direct hits from H.E. bombs.[6] One went through two rooms in P.P.(Private patients) and the other through the P.P. kitchen – that is where Sister Brown (Isobel) and Nurse Wilson were. Luckily all the patients had been taken down to the basement - and they (Brown and Wilson) had returned to fetch some things from the kitchen, when the bomb came – and so they were trapped beneath all the debris.
> Poor dears! They are in a terrible state. Mr Latto has worked day and night for them.
> I have heard people say that Sister Brown is so badly injured, that they hope she will not recover – but I simply can't think like that – Oh Mr Burkitt, I don't want her to die – she is young and so full of life.

Thanks to the efforts over many weeks of Conrad Latto, now on the staff in Plymouth as an RSO,[7] both Nurse Wilson and Isobel survived but would remain invalids all their lives. Isobel never lost her affection for Denis and in later years would be a regular visitor to the Burkitt household in Gloucestershire. From now on it would be Olive that would occupy his attention, although, at first, she was doubtful about the relationship and writes in her biographical notes:

---

[4] RMS Lancastria was a requisitioned ocean liner who on June 17, 1940, was sent to the French coast south off St Nazaire to help evacuate members of the British Expeditionary Force cut off in France, as part of Operation Ariel. Although her normal complement of troops was 2180, she was asked to load as many as possible and by the middle of the afternoon was ready to depart. At 13:50 the air raid began. She was hit by three bombs and sank in 15–20 minutes. Estimates of the death toll range from 3000 to 5800 making it the worst single-ship maritime disaster in British history. The loss of life was such that Churchill, Prime Minister, suppressed news of the catastrophe [2].

[5] Autobiographical notes of Olive in the family archives

[6] H.E. A large high explosive bomb, which could be carried by aircraft, designed for general demolition

[7] Conrad Latto was very able, qualifying from St Andrews University with first class honours and a gold medal. He was a vegan and conscientious objector during the war, but his work organising surgery for the casualties at Plymouth won him high praise from his colleagues, and when the tribunal met to consider his application to be excused military service, he was excused with the condition that he continued to serve as a doctor [3].

I admired Denis hugely but I certainly was not ready for a lifelong commitment and tried
very hard to keep it at a platonic level. I wanted to enjoy my freedom! I was also a little
afraid of his "rigid" faith: it seemed to be as negative as that in which I had been brought
up. I felt I could never live up to his expectations of the wife he wanted.

Olive continued to be upset by the air raids and the resultant casualties she had to
manage, recording her feelings in a tiny notebook:

> January 21st. Our blitz has made an awful coward of me and when I hear the siren I long to
> fly away.
>   March 20th & 21st. 'These two nights have been hectic nightmares. Plymouth was
> blitzed badly both nights running – Our city is destroyed.
>   We had casualties pouring in – never have I seen such awful wounds. I cannot put down
> here just everything that happened – or describe the utter horribleness or grim realities
> that we saw.
>   At the City the children's ward was hit…killing about 30 children and six nurses…that
> I think was one of the most tragic sights – the little twisted and broken cots, sticking out
> from the piles of debris' (Fig. 5.5).

Those who work in hospitals, doctors, nurses and all the support staff, and those
who have served in the forces during armed conflicts are not supposed to be affected
by seeing mortally wounded people often crying out with fear and pain with disfiguring injuries, blood or body parts and people dying or dead. Such experience is just

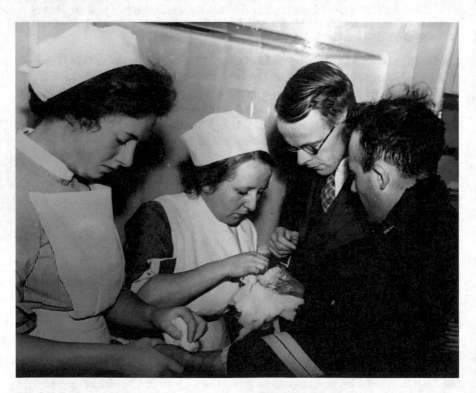

Fig. 5.5 Denis and nurses attend to the burnt hands of a sailor in Plymouth

one aspect of the job. Olive, young and impressionable, wrote about the horrors of war she experienced, and within a year, by 1942, she was beginning to get episodes in which, briefly, she felt 'down'. These would become regular and more disabling as she grew older and clinically were regarded as bouts of depression. Although seldom reported Florence Nightingale, witness to the carnage of the Crimean War, suffered similar bouts of physical illness and depression in later life. Looked at with present-day understanding of mental health, Olive, and Florence, would almost certainly have been diagnosed as having post-traumatic stress disorder (PTSD) and managed appropriately and sympathetically. For Olive the burden of her wartime nursing experiences would continue throughout her life.

Meanwhile Denis wrote to his parents to say he was definitely called to serve abroad, to which they responded positively. He completed an application for the Colonial Medical Service but very quickly received a reply to say that they could not consider his application 'on account of my having lost an eye'. Denis was stunned by this but had sent with the application 'a clear uncompromising statement of my motives in going abroad'. His statement of motives, including as it must have done, an expression of his desire to win souls for Christ must have made the hierarchy at the CMS wary of taking him on. The lack of an eye was a good excuse for them. He then applied to the Sudan Medical Service but was told that they did not take people over 30 years old, and an application to the Southern Rhodesian Medical Service produced the reply that all vacancies were filled. He therefore decided to apply for a commission in the Royal Army Medical Corps (RAMC) (Fig. 5.6).

With Isobel still recovering from her injuries, Denis' acquaintance with Olive grew into affection, and a weekend in Plymouth in June 1941 followed during which he and Olive went for a walk. On Sunday, instead of the usual devotions that had characterised his relationships with women up to now, they went for a picnic with friends. Olive was quite pleased with this but wrote:

> June 1941. Denis has grown too fond of me I'm afraid – he is a dear fellow – but I would never feel anything like a strong affection for him.

She may have been in love with Conrad Latto at this point describing him as a fellow with the highest ideal and that she very was fond of him. He became a special friend, and they would often go out together, and in the hospital Conrad would defend her against the vicissitudes imposed by the nursing hierarchy.

Back in Barnsley Denis received a letter from the Central Medical War Committee saying that they would send his name forwards to the War Office provided he resigned from his job. This was a turning point for Denis who in deciding to apply for a commission told himself that:

> 1. Men are badly needed.
> 2. I, having no ties ought to go before those with ties.
> 3. I don't feel I should do more hospital jobs, in which case I would be recruited in any case
> 4. The need for Christian testimony in the Army.
> 5. It will teach me much, and be good training, no matter what kind of work I have to do.
> I can see no obvious alterative considering these facts.

Fig. 5.6 Letter from the
Colonial Office in London
declining Denis'
application to join the
service ostensibly because
of his lost eye

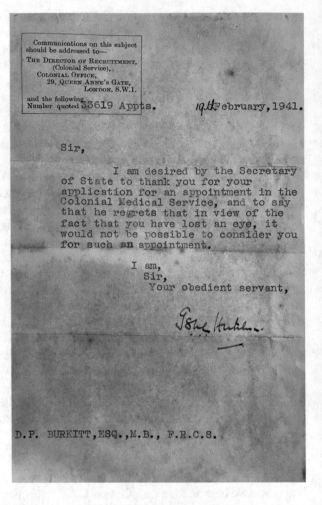

Communications on this subject
should be addressed to—
THE DIRECTOR OF RECRUITMENT,
(Colonial Service),
COLONIAL OFFICE,
29, QUEEN ANNE'S GATE,
LONDON, S.W.1.
and the following
Number quoted 33619 Appts.   19th February, 1941.

Sir,

           I am desired by the Secretary
of State to thank you for your
application for an appointment in the
Colonial Medical Service, and to say
that he regrets that in view of the
fact that you have lost an eye, it
would not be possible to consider you
for such an appointment.

           I am,
           Sir,
           Your obedient servant,

D.P. BURKITT, ESQ., M.B., F.R.C.S.,

He was soon called to York for an army medical and in May 15 left Barnsley for Plymouth. He obtained a locum job there and was able to spend more time with Olive on walks, picnics, attendance at church and meetings with friends. He described these occasions as happy, and of particular note is that there are no references to mutual reading of the Bible, prayers or any such activity that had characterised his liaisons with the many nurses who had tried to attract his attention. In May 19 he received notice that he had been accepted for service in the RAMC, and in July 27 he started as medical officer to the 342 S/L Battery, Royal Artillery at Herstmonceux. He was disappointed not to have been given a surgical grading.

What the army thought of this Bible-reading teetotaller who knelt by his bed at night to say his prayers is not recorded. But rank was all-important in protocol, and Denis was a commissioned officer, lieutenant at this point, which protected him from the frank opinions of the foot soldiers. His unusual habits, particularly that he did not take alcoholic drinks, was a matter for discussion in the officers' mess and allowed Denis to take the first very tentative steps in this new environment to explain his beliefs and spread the Gospel. He managed to attend the local church and responded positively to an appeal from the army for volunteers to serve overseas but hoped it would not be too soon. Brief, 48 h, leave was spent in Plymouth with Olive and friends, but in September 25 he was posted to Norwich as graded surgeon. His medical orderly, Bishop, who was a Christian accompanied him, and they were stationed at 219 Field Ambulance at Witton, north of Norwich and close to the Norfolk coast. Here he met Douglas Black who was attached to the unit briefly and who would go on to become President of the Royal College of Physicians and one of the leaders of English medicine. After 2 weeks' leave, which he spent in Ireland, he returned to Witton and then early in November was sent to the Radcliffe Infirmary, Oxford, to take a short course in accident and emergency surgery, which he was pleased about (Fig. 5.7).

At the end of January 1942, his unit, 219 Field Ambulance, was posted to Somerleyton Hall, near Lowestoft in Suffolk. He quickly became acquainted with Lady Somerleyton, who with her husband Lord Somerleyton owned the magnificent stately home that had been requisitioned for war service as a hospital. Denis now resumed his photographic activities from which it can be seen that their training for what seemed likely to be an overseas posting included cooking, learning to cross rivers, improvising stretchers, laying cables and recovering casualties in the field. After a 30-mile route march in April, he developed migraine, his first attack of what would trouble him intermittently throughout his life. His army life kept him moving about the country with Denis regretting he could not have a proper surgical job. Back on home leave in Ireland in October, Robin came with his new wife, Violet

Fig. 5.7 Army days. Denis leads on exercise, Somerleyton. Early 1942

who had been the sister on the children's ward in Truro where Robin had done his first house job. Denis regretted not being in such a relationship. Life for him was rather mundane and, as far as his army medical career was concerned, frustrating.

Then came an extended 48 h leave, which he decided to spend by visiting Olive in Plymouth. He had seen quite a lot of her in 1941 with frequent visits for walks, picnics and times with their friends, but over a year had passed since they had last been in communication, so what had made Denis make this journey is unclear. He travelled by train and was met by her at the station early on the morning of Saturday November 7 from where she took him to Rivermead, which was probably a residence where nurses could entertain family and guests. They had breakfast, went for a walk and spent the afternoon in the drawing room talking. Denis was somewhat worried when Olive told him that she had changed a lot since coming to Plymouth, but he was reassured when walking after dinner they were able to talk about spiritual things. The next morning Olive told Denis that her brother had rung up and was coming down to see them both. To Denis' sorrow he arrived with his mother, because he had hoped to have Olive to himself for the weekend. But it was soon clear that mother and brother Alan had already heard about Denis, and in turn Denis was glad to find them both keen Christians. After lunch they walked and Alan took the first photograph of Olive and Denis together. The family left on the 8:20 pm train after which Denis and Olive walked hand in hand back to the hospital. Olive changed into her uniform and went to the station with Denis to see him off on the night train. Denis notes, 'Saying good-bye at the station I ventured to kiss her good-bye and she responded with a kiss. I have never kissed a girl before, except my sister, but it seemed quite natural'. For Denis it was a memorable weekend and the start of a lifelong love and partnership. For Olive it was 'one of the happiest weekends I have ever had. Now I know that all this time that I have been trying hard not to let myself love Denis – nor let him love me – has only made my love for him deeper and stronger' (Fig. 5.8).

Denis' request to the assistant district medical officer (ADMO) for a surgical posting was turned down, but the colonel offered him a posting with 'B' Company, also based in Norfolk, which Denis accepted. Olive wanted him to move nearer to Plymouth, but they had to be content with regular correspondence in which Olive explained her family background and Denis warned that they might have to build a home together overseas. Olive made it clear to Denis that the only true friendship is that based on mutual love for Jesus, which reassured him, and she promised to send him a photo of herself for Christmas. A visit to Plymouth was shortened by Denis having to attend a gas course in Leeds, whilst their respective duties frequently kept them apart. A week's leave for Denis early in January found them together in Plymouth walking, talking, having meals together and attending church on Sunday although Olive was on duty for part of the time allowing Denis to meet former colleagues in the hospital. On returning to Norwich, he was asked to do a 2-week surgical locum at Addenbrooke's Hospital, Cambridge, which he was pleased to do if for no other reason than it gave him an opportunity to keep his surgical skills up to date. Whilst at Addenbrooke's he received a letter from Olive's parents suggesting that they ought to be officially engaged. On returning to Norwich, he was posted to the

Fig. 5.8 First photograph
of Denis and Olive
together taken by Olive's
brother Allen on Sunday
November 8, 1942, near
Plymouth

military hospital at Hatfield but managed to get a weekend leave for February 20/21.
On the Saturday he caught a train at 5:45 am for Paddington, arriving at 7:30 am
where he was met by Olive who had travelled up from Plymouth.

> DPB was on the platform waiting for me so it was with all the cacophony of the hooting and
> hissing of steam trains around us and the steam like mist enveloping us that he asked me to
> marry him. In all the noise I missed what he said and had to say" What did you say?" and
> the poor, shy man had to repeat the question.

Olive accepted and he bought her a now fashionable platinum engagement ring at
Mappin and Webb, gold being unobtainable during the war, whilst she bought him
a pair of hairbrushes. They set off together for Reading to visit Olive's parents
whom Denis found to be real saints. They spent to rest of the day talking and on
Sunday going to church. Denis was back at Witton on Monday 'after a wonderful
week-end. To God be the praise and the glory for ever'. Olive back at Plymouth was
studying for her final nursing exams.

Although the war had been going for 3.5 years by this time, there seemed to be
no hurry to use Denis' surgical skills on the battlefield. He attended courses on
abdominal and on chest surgery whilst keeping in touch with Olive by letter and

visits to Plymouth. Late in May he received a letter from his parents, both apparently quite independently suggesting that Olive and he should marry soon. He saw the hand of God in this and 2 weeks later took Olive to stay with Aunt May and Miss Knight at their school in Poole, which had closed down for the war, to discuss wedding arrangements. They opted for July 28, so Denis asked for 3 days' leave for the occasion, which he was given. Before the great day, Denis was posted to Netley on Southampton Water, a lot nearer to Olive in Plymouth who was now about to take her final exams. Getting married posed a problem for Olive in that she would be required to give up her nursing duties. Matron was not pleased to hear about this because Olive was contracted to do a year as staff nurse after qualifying. Olive was told that she would have to reimburse the hospital for breaking the contract. She resigned in July 11 with mixed feelings, but knowing she was to become Denis Burkitt's wife outweighed the sadness of having to say goodbye to her friends and her career in nursing (Fig. 5.9).

They married at 2:30 pm on July 28, 1943, in the coastal town of Seaford, east of Brighton. It was a wartime wedding, so celebrations were limited, as were the resources for travel at that time. No-one from Denis' family was able to attend but a few friends from Plymouth were, including Captain Walter Thomson RAMC who was best man, Conrad Latto, the ever-faithful Isobel, Norman Gray and Alan, Olive's brother. The bride was given away by her father Rupert Vere Rogers. The reception was at a nearby hotel at which all the speakers wished for God's blessing on the couple, who left at 5 pm by bus for Brighton, then Midhurst and a final walk of 2 miles to a hotel in the small village of Iping. They had 3 nights together before

Fig. 5.9 Denis marries Olive Rogers, July 28, 1943. Best man, Captain Walter Thomson RAMC; bridesmaid, Olive's sister Maureen

Denis was due back at Netley. During August they saw each other most weeks but the time spent together was limited by Denis' duties. Early in September came the news that he had been expecting, but not wholeheartedly wanting, that he was to be posted abroad. He was given 2 weeks' embarkation leave, which gave him the opportunity to take Olive to Ireland to meet the family where she was warmly welcomed and made to feel a daughter of the house. Denis took Olive to Mullaghmore, scene of many happy family holidays, for a couple of nights and then down to Cork to meet his mother's family. Back at Laragh Denis wondered how long it would be before he returned (Fig. 5.10).

Olive, having now resigned her post as a nurse, was living at Reading with her parents where Denis visited as often as possible in the final days before his departure. She gave him a book marker on which was written 'Mispah – the Lord watch between me and thee – while we are absent the one from the other' and wrote a reference on it from Genesis 'My presence shall go with thee and I will give thee rest'. There is no doubt that they both shared a belief in God and sought his guidance through prayer and Bible reading on how they should live their lives. It is unlikely that Denis would have married Olive without this commitment on her part, but their shared faith was to see them through the difficult times ahead marked by long periods of separation, which took their toll on Olive, the raising of a family and Denis' pioneering work that was to turn him into a household name. Their final time together was in Reading with Olive's family. The next day he travelled to Glasgow where on Saturday October 23 he boarded a ship bound for Africa, although he was

Fig. 5.10 Major Burkitt. Portrait photograph sent to Olive circa 1944

unable to tell Olive of his destination. The route included 10 days at a transit camp in South Africa before sailing to Mombasa in Kenya, arriving in December 12. The following day Denis was taken by train to Nairobi where he would be stationed. It was the start of a 22-year attachment to Africa, which was to fulfil Denis' wish to serve God and allow him to serve his fellow man through his medical skills and where he was to develop a lifelong interest in the cause and, ultimately, the prevention of disease. He would not see Olive again for 2½ years.

References

1. Studdert Kennedy GA. The unutterable beauty. London: Hodder and Stoughton; 1927.
2. https://en.wikipedia.org/wiki/RMS_Lancastria
3. https://livesonline.rcseng.ac.uk/

Chapter 6
Doubts and Frustrations. 1944–1946

Denis was already almost 32 years old but his role as medical officer in the army was not allowing him to develop surgically. Nor had he seen anything of the war, despite being in the army for over 2 years. On arrival in Kenya, familiar to him from Uncle Roland's stories about his life there, he was immediately posted to Mombasa, on the coast. Attached to No 6 General, there was no vacancy for a specialist surgeon so he had to be content with everyday medical practice. He was cheered by letters from Olive, who had taken a job at the Princess Alice Memorial Hospital in Eastbourne. Starting as staff nurse in the children's ward, Olive moved to casualty and outpatients. Then, after a year, she was offered the post of sister. The long separation from Denis now allowed her to pursue her nursing career ambitions. Denis started to learn Swahili.

On March 26, 1944, he was posted to Mogadishu the principal town in British-controlled Somaliland,[1] which he did not like at all. The heat and glare of the sun was relentless, his kit failed to arrive for weeks, and there was very little work for him. What surgery there was he had finished by 10 in the morning. He felt demoted and in fact he discovered that he had lost his specialist surgical rank when he looked at his monthly pay slip. Depression was added to his usual anxiety; he was lonely and in his darkest moments hoped for consolation in his reading of the Bible. He reminded himself that he had given his life to the service of God and should not complain about earthly matters such as his kit. After a Sunday service late in May, which Denis had conducted, an officer with the rank of major began to talk to him about the church and how any religious revival in East Africa was considered dangerous and that men had been imprisoned for testifying about their beliefs. Probably a 'quiet word' of warning to Denis in the inimical style of the British Civil Service. Fortunately, word came through in May 24 that Denis was to leave Mogadishu, that

---

[1] Somaliland, recaptured from the Italians in 1941, was a British protectorate bordering the Gulf of Aden with a population of less than 2 million.

© The Author(s), under exclusive license to Springer Nature Switzerland AG 2022
J. H. Cummings, *Denis Burkitt*, Springer Biographies,
https://doi.org/10.1007/978-3-030-88563-2_6

his kit was at last on its way and there were prospects for promotion. He left in June 14 arriving in Nairobi 6 days later.

Why had he been posted to Mogadishu? He was a highly competent, qualified and experienced army surgeon, and there was a war going on not least in North Africa. When in 1941 he had been turned down by the Colonial Office for service in Africa because he had lost an eye, Denis felt that the real reason might be his Christian convictions, which he had made plain in his application for a post. Later that year when he had received his commission in the army, it must have been very apparent that Denis was no ordinary soldier and that his religious beliefs determined how he lived and were expressed openly in his everyday work. This and his lack of enjoyment of and therefore participation in the usual social life of an army officers' mess may have made it difficult for the more senior ranks to place or promote him. Yet many doctors, nurses, chaplains, rabbis and other religious leaders with strongly held religious convictions gave dedicated service to the armed forces during the wars. Denis was clearly different and, perhaps in the minds of the army hierarchy, unsuitable for war.

His stay in Nairobi was brief, only 3 days. Then army life became more agreeable. He was promoted to major, having been made captain in 1942, and on Sunday June 25, 1944, was driven by Colonel Conolly to his next posting at Gilgil. He liked it immediately remarking that it was good job in a good climate with a good room and friendly mess. Gilgil, about 80 miles north of Nairobi on the main road from Nairobi to Nakuru, had a population of less than 20,000, but there was a large military presence in an otherwise agriculture-based town (Fig. 6.1).

Gilgil was also host to a British Internment Camp to which Irgun and Lehi men, members of a movement for the establishment of a Jewish State in Palestine, were moved in 1947. The existence of this camp does not get any mention by Denis, but his meeting on his first day at Gilgil with a fellow member of the Officers' Christian Union, Philip Blakely, gave him an immediate introduction to like-minded people. Life in the Rift Valley was pleasant; he was busy with his job, but was still not certain about his long-term future, except that it would be in East Africa. He had a talk with Colonel Copley one evening who praised Denis' surgical skills and suggested he get a post on the staff of a large teaching hospital such as the Manchester Royal Infirmary. Then a letter from Olive arrived in which she asked 'Darling, do you really want to make your home abroad?' They had not seen each other for just over a year at this time. This caused him considerable pain, but he felt Olive did not understand that it was God's will not his own personal preferences that was leading him in this way. For Denis if the Christian life was not causing pain, even to those whom he loved, then it was probably not being truly lived. Perhaps he should have mentioned this to Olive before he married her. The frequent and sometimes long periods of separation in their marriage certainly hurt her and compounded her depression. As the months went by, he became anxious that Olive's letters did not mention spiritual things, but then one arrived in November 2 from her saying how much the Bible had meant to her of late – how she had grown to love and understand it, referring to her spiritual fights. Later, in his autobiography, Denis would remark that despite separation becoming increasingly traumatic for them both as his career

Fig. 6.1 Denis with Captain Thomas (anaesthetist) and Captain Copeman (dentist). Outside the mess of the Station Hospital in Gigil 1944

developed, and that goodbyes were a price to be paid for his success, 'It was always my wife who paid the greatest price'.

On November 16, 1944, Denis started 3 weeks of leave, his first in Africa. By train, boat, bus, lorry and car, he journeyed west from Gilgil to the shores of Lake Victoria, then to Kampala in Uganda and on to the foothills of the Rwenzori Mountains, south of Lake Albert, that are the border between Uganda and the Democratic Republic of the Congo. The return route followed a more northerly path. There is a brief reflection on the events of the 3 weeks in Denis diary/note-book, which reads:

> On Nov 16th I joined Norman Miller for leave.
>
> I have never known such Christian fellowship, with Colour and Class forgotten in a wonderful Christian brotherhood. Prayer, reading and sharing Victories and defeat with Africans was a new experience. I got a new vision of what God can do for a man. I have never been so challenged in my life as when Nsimbambi spoke to me and asked how I was spiritually – to date and was God using me. William Nagenda was also a mighty challenge and many more – Bertha and Ralph Leech – Bill and Nancy Butler. As African houseboys expounded scripture I saw that it wasn't a matter of scholarship and intellect, but the work of the Holy Ghost. I never met men who so evidently saw all men as bound for heaven or hell and who saw all as measured against eternity. In every way it was a blessed time, and a mightily challenging time (Fig. 6.2).

There is much more to the record of this journey than Denis' first encounters with the African interpretation of the Gospel. The certainty with which the African Christians he met and lived out their lives contrasted very much with Denis' chronic anxiety about his faith and the hours he spent in prayer and Bible reading looking

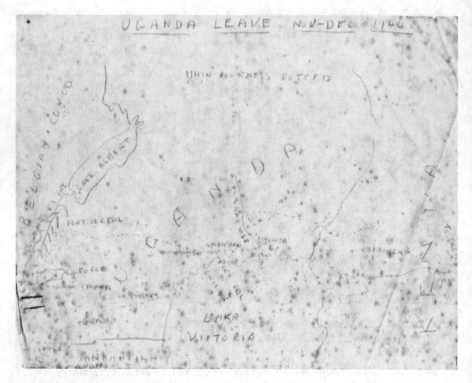

Fig. 6.2 Map of route taken by Denis on his first army leave November 16–December 21, 1944, during which he visits Uganda for the first time

for guidance. Their passion and unquestioning faith were a long way from the formalities of Sunday services at Trory Church. It was challenging, an education and an awakening for him that would be invaluable in his dealings with patients and staff at the hospital in Kampala where he was to spend many years. Later, when travelling in the so-called Bible Belt of the USA, he would meet Christians with a similar certainty of the Resurrection and life hereafter.

Denis' brief reflection in his diary of his travels whilst on leave makes no mention of the real account of his journey. During his travels he typed 18 pages in which he described every aspect of his experiences. Around 13,000 words in total. How did he manage this whilst constantly on the move? He must have been a fast and competent typist although did not take a typewriter with him for the last sentence of the account that reads 'Forgive me for making this so long, and excuse mistakes. Different type-writers have been used and very old carbons in the first pages'. It was the first of many such accounts of his travels in Africa. Who were they written for and why did Denis make such detailed records of his journeys? Did he already have a sense of destiny that in the future people would want to read about his life? Unlikely, as he constantly reminds himself of the dangers of pride and the need for humility in everything he does. Perhaps he was just fascinated by Africa, its geography and its people. This early account provides a cameo of Denis and the White well-educated professional men who joined him, journeying through Kenya and Uganda in the last years of the British Empire. Along the way he became convinced that his future lay in Uganda.

The record of the journey, or 'safari' to use the Swahili word, is very much about the people he met. Norman Miller, who joined him on the early part of the safari, was an Officer in the Highland Light Infantry and described by Denis as 'a man jealous for the Word of God'. People he met en route were admired particularly if they were Christians but also if they had been to Trinity College Dublin. They were mostly from the army and were doing missionary work or running the government hospital services often in remote areas. Aside from aspects of people's faith, Denis took an interest in farming and the crops grown, which included cassava, soya, earth nuts (peanuts), oranges, bananas, tomatoes, guavas and pawpaws. There were tea and coffee plantations and fields of sugar cane as well as cotton growing. He was disappointed at the bribery and corruption he saw notably in the tea industry but also to his surprise in the hospitals.

Two days into the journey, at Kisumu, their host Peter Akehurst, a government agricultural officer who had trained at Cambridge and rowed for Trinity Hall, arranged for Denis to see around the local hospital and meet the government doctor. Denis notes 'When cases come up they are first seen by orderlies, the more severe cases only referred to the doctor. This eases the doctor's work but is open to one big snag – graft. Unless the patient gives some money to the dresser he won't be referred to the doctor, or not for days. The business of graft is rampant out here'. Clearly this was Denis' first encounter with the way business is done in some parts of Africa as was his first observation of racial prejudice. 'Indian doctors are on a different status than European even though they might have better degrees from a British University – such is the colour bar here. They are called sub-assistant surgeons'.

Crossing Lake Victoria by boat from Kisumu to Port Bell in Uganda, Denis made for Kampala and was immediately at home. He was regaled with tales of Uncle Roland and found a great Christian fellowship there. He felt Uganda was more like England and was struck by the girls in Kampala whom he thought were much smarter than those in Gilgil. On Friday November 24, he operated at Mengo Hospital on the outskirts of Kampala, which had been founded by Sir Albert Cook and where he would return, to the neighbouring hospital of Mulago, just over 3 years later as surgeon to the Uganda Ministry of Health. He stayed for a few days in Kampala and saw Makerere University for the first time, and when Norman Miller left, he decided to get a driving license, hired a car and set off west for Fort Portal where he spent a week. After climbing Mount Karangara, one of the lesser peaks in the Rwenzori range, with Philip Blakely and a local guide, he turned and headed for home. The journey gave him further insights into how Africa worked. The bus he caught initially broke down, so he had to overnight in a hotel. He then caught a lift in a lorry but could not get onward transport so spent the night in Kabale. He left on Sunday morning bound for Kampala reaching there at 9:30 pm on Monday. The lorry had no lights, so a boy had to sit on the front holding a hurricane light all night. Out of Kampala he was given a lift on a milk van at 4 am eventually catching a train to arrive at Gilgil at 11 am on Thursday December 7. Uganda had captured his heart.

After leaving the army, Denis would go on many safaris usually in his role as district medical officer, sometimes with his family or with other colleagues. There are detailed, mostly typed, records of 15–20 of these journeys, running to many tens of thousands of words, the most notable of which is known as 'The Long Safari' undertaken in 1961 that formed the subject of a book by Bernard Glemser published in 1970. Some were undertaken in pursuit of his ideas about the childhood cancer that would take his name but do not expect to find in these accounts any insights into his research or even much mention of it and only scant references to his clinical work. They record where he went, what he ate, whom he met especially if they were Christian folk, the countryside he passed through and the problems he encountered on his journeys. Some are addressed as letters to Olive, whilst others are more in the form of a diary of events and places. The time spent writing or typing these must have been considerable, usually early in the morning during Denis' 'quiet time' or last thing at night. In addition to often busy and taxing days of travel, coping with surgical problems at the hospitals he visited and meetings with key people running medical services in far distant outposts of the territory, Denis must have been very determined to record what was happening in his life. Why he should do this, especially at an early stage in his career before he had even started the work for which he became known and whilst he remained continually anxious and uncertain about his future, nothing of this is recorded in any of his writings.

Back in Gilgil he received notice that he was to be posted overseas, which didn't please him. He left on December 27, 1944, for Nairobi where he met up with Sister Osborne and various friends. Two weeks later he was told that he was off the draft for the moment and could return to Gilgil. This he did on January 16, 1945, but on arrival he found that the officer commanding the station was away and that he was in charge. He was not happy about this, and it may have been one of a number of matters concerning his army life, his recent experience on his travels and the thoughts about his future that were emerging from his quiet times that made him decide to leave the army and fill in the application forms for the Uganda Medical Service. These he posted to the Colonial Office, and 3 weeks later Olive wrote to him saying that she was willing to come out to Africa if he was able to get his release from the army. He felt that his future was starting to take shape at last. Soon after this he was marching at the head of a column of men to the station in Nairobi at the start of his posting to active service in Ceylon (Sri Lanka). After a voyage of 8 days, they landed in Colombo, transited to the Base Reinforcements Camp where a few days later Denis was ordered to go to '48 I.G.H SEAC' where he was to do a surgical locum. To his amazement he met Major Lee-Spratt, who had been a member of Grandpa's church in Athenry, Galway. Nothing commended someone more to Denis than a Christian who also came from a part of Ireland he knew well and who was acquainted with his family. All boxes ticked.

In Ceylon, very much as in Kenya, his army duties, whilst they must have always been fulfilled adequately, always took second place to his church-related activities. A rather longer diary entry than usual for Sunday April 15, 1945, reads:

> Got lift back in R.A.F. lorry that brought chaps to church. Overheard a discussion in which an officer said 'Is Christianity a philosophy of life or is there more to it than that?' He gave the impression that he was determined to find out. I butted in and said it was a personal relationship to a person who <u>enabled</u> us to lead straight lives. Unfortunately I had to get off just as we were getting going. I wondered afterwards whether I should have stayed on

Maybe, when he had more time to think, Denis might have refined his definition of Christianity. It was somewhat unconventional but reflects his ongoing anxieties about the direction his life had taken up to that time and the uncertainty, very much brought into focus by the African Christians he had met, of the exact nature of his relationship with God. And there was conflict, between the austere religious practice of the Church of Ireland at the start of the twentieth century, in which he had been brought up and the much more evangelical brand to which he had been introduced by the '40' club at Trinity. In addition, Denis was not by nature a clubbable person, so the guidance he sought from God concerning his everyday life, by prayer and recourse to find appropriate texts from his reading of the Bible, was somewhat moulded to his natural instincts. Thus, on attending an ENSA[2] concert around this time, he tells himself 'My last, unless advertised as a good musical programme. It was one of the most vulgar and sexy shows I have ever seen. I walked out before the end'. But supping with the devil was not always to prove a bad thing for him. Later in the year, he went for the first time into the Officers Club in Colombo, not by inclination but to wait for someone to pick him up and take him back to base. He started a letter home in which he wrote that he was hoping to meet Robin, who was on his way back to England from Singapore and was expected to call in at Colombo. He looked up from his writing and there was Robin standing in front of him. They had a good time together for a few brief hours, which made Denis feel homesick. It was not long afterwards that news came that his unit would be moving out and starting the journey back to England. They left Colombo on December 12, 1945, and travelled via Singapore, Rangoon, Calcutta and Tel Aviv to arrive on February 7, 1946. The great events of 1945, which brought the World War II to a close, get no mention at all in any of Denis' writings at the time. He clearly never felt he was part of it (Fig. 6.3).

Two days after arriving in the UK, he was able to meet Olive. They had been apart from October 21, 1943, until February 9, 1946, a period of almost 2 ½ years. A few days later, they left on the Stranraer to Belfast ferry and onwards to Laragh for reunion with Denis' parents. Olive was not well on the boat and Denis asked their local GP to see her. She was admitted to hospital the next day, underwent minor surgery and was back with the family a few days later. Shortly after this Denis heard from the Colonial Office that he was to be interviewed in March 27 for a post in East Africa. Olive and Denis had 2 weeks of freedom in Ireland before Denis returned to Belfast where he was demobilised from the army in March 12. He had served for 4 years and 8 months, attained the rank of major and had provided a range of medical and surgical services to the army, which would serve him well in the future. The interview at the Colonial Office went well but was followed by a period

---

[2] Entertainments National Service Association.

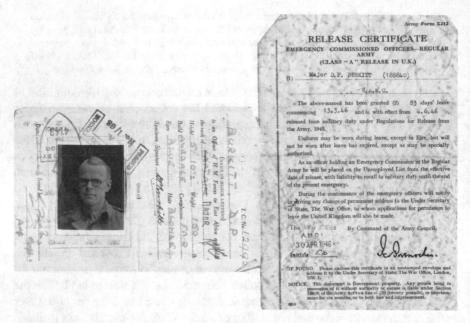

Fig. 6.3 Major Burkitt Army ID card and demobilisation certificate, 1946

of great anxiety for Denis. He was worried that Olive might be losing interest in spiritual things, but having dismissed this as his own weakness, he started to worry once more about his future and whether he had made the right decision to go to Uganda. Olive was unwell at this time, early May, and surely Denis must have realised that her morning sickness indicated that she was expecting their first child. How would he and Olive manage the inevitable periods of separation, what about their children's education, would he be effective in his new role? In May 11 he confesses to feeling depressed, but Olive brought him a letter from the Colonial Office to say that he had been accepted for Uganda. This brought him considerable relief as did the news a few days later that the MD he had submitted to Trinity had been accepted.[3] He had written this during times when in Ceylon he did not have much to do (Fig. 6.4).

In July 1 Denis was in Dublin receiving his MD. A few days later, the Colonial Office wrote saying they were hoping to find a passage for him early in September. Olive was to follow at a later date, travel now complicated by her pregnancy. They decided that she should remain in Ireland with the family until after the baby was born. Together she and Denis talked about names for the child, which was to be Neil Dermot Parsons if a boy and Judith Marion if a girl. Marion was a family name, whilst Parsons was the name handed down from Denis' grandmother, daughter of the notable non-conformist preacher James Parsons of York. For the rest of July and August, Denis oscillated between Dublin and Laragh doing locums in general practice. In

---

[3] The subject of his MD was 'spontaneous rupture of abdominal viscera'.

Fig. 6.4 Family group, probably Laragh, circa 1946
Back row, Olive and Denis, Robert (Robin) and Vi
Front row, Uncle Roland, Gwen and James, Aunt Norah
Child at front of Robin, son of Robin and Vi

September 3 Olive and he had a final tearful parting. They had enjoyed over 6 months together in which they had laid the foundations for the future of the Burkitt family and Olive had done her best to assure Denis that the choices they had made regarding Africa were the right ones and what God wanted them to do. Denis remarked that 'Our lives were to be punctuated by tearful good-byes, but the pain was a source of gratitude being the inevitable outcome of a closely-knit partnership'. He was very much aware of the pain he was causing Olive both at the time and in the future.

Complicating all this there was the continuing doubt in his mind that he had chosen the right course. He was giving up the opportunity of a successful surgical career in the UK and also the opportunity to be the community general practitioner in the nearby Fivemiletown, which he had been offered. Colonial life had a reputation for parties and drink, club membership and sport, all of which were contrary to his own inclinations. But he felt completely assured of his sense of calling having spent months if not years praying about it. For Olive these partings never gave her much sense of gratitude, more a sense of loss and loneliness. But their relationship was to survive despite the challenges of lives spent apart for long periods, the cares of family life that Olive had to bear alone at times and Denis' complete immersion in the work that was soon to start in Africa. He sailed from Southampton at 5 pm on the Alcantara, a ship of the Royal Mail Lines bound for Mombasa where he is listed as 'Gov. Official'. He described the occasion as 'the greatest adventure of my life, but not alone'. He was borne up by reading Matthew 28:20 '… and lo, I am with you always, even unto the end of the world'. Olive was left at Laragh to make preparations for the birth of their first child, Judith.

Chapter 7
Lira. The Start of a Great Journey

Denis was now a District Medical Officer for His Majesty's Colonial Medical Service heading to a part of the world with which he was already familiar through his army service. He was posted to Lira, the administrative and commercial centre of the Lango District in the Northern region of Uganda, where he had sole responsibility for the health of about 250,00 people living in an area of 7000 square miles. There was a 100-bed hospital with one locally qualified African doctor for assistance, Dr Kununka, who later went on to lead one of Uganda's main political parties.[1] Denis' stay in Lira lasted only 15 months, from October 3, 1946, to January 4, 1948, but it was a major turning point in his professional life. During this time, he would make his first observations of the geographical distribution of a disease, that of hydrocele of the testis, which were to set him on a path of discovery that would occupy him for the next 40 years. In Lira he also began to question the rigid application of his Christian faith in the way that it affected those with whom he came into contact. He was now able to take more control of his life (Fig. 7.1).

On the voyage out to Uganda, aboard the Alcantara, were a broad mix of people including missionaries and a number of doctors and nurses, several of whom had been appointed to the Colonial Medical Service at the same time as Denis. There were Christians from all strands of the church who had a common bond in their commitment to serve in Africa, which brought home to Denis an important lesson – that 'spirituality and devotion are not the sole prerogative of evangelicals' and those he considered to be high church, who had offered the hand of friendship to him, were in fact deeply humble and spiritual people. Sixteen years after his encounter with Dick Jones and the 40 Club at Trinity, he began to realise that religion could take many forms all of which were acceptable in the sight of God.

Arriving in Mombasa in September 23 'In the land of my calling', he went by train to Kampala via Nairobi, accompanied by Austen and Jean Best who had been

---

[1] Dr Baranabas N. Kununka who became a leading figure in the Uganda National Congress party and was awarded the Obote Independence medal by President Museveni.

© The Author(s), under exclusive license to Springer Nature Switzerland AG 2022
J. H. Cummings, *Denis Burkitt*, Springer Biographies,
https://doi.org/10.1007/978-3-030-88563-2_7

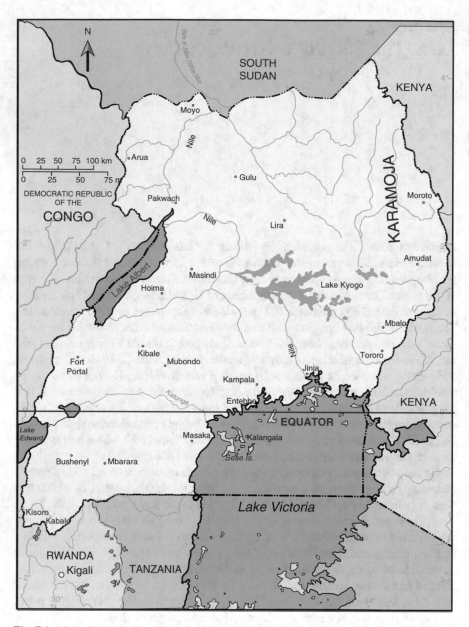

Fig. 7.1 Map of Uganda showing major places visited by Denis for his work

on the boat from England. Austen had been appointed to the Colonial Medical Service at the same time as Denis but was less well qualified. At the station in Kampala, Denis was met by Dick Drown, whom he had known since the days when he was a Resident Surgical Officer at Barnsley Hospital in 1940. Dick's father had been the vicar at the local parish church. Dick had graduated from Oxford and after a period teaching in Wales had been appointed to King's College Budo about 9 miles south-west of the centre of Kampala, regarded as the Eton of Uganda. Dick and his wife Gwyneth would become lifelong friends of the family and provide support, especially for Olive, in the turbulent days that lay ahead. After an informal interview at Entebbe[2] about his proposed posting, he met Roy and Dora Billington[3] who also were to become family friends and had supper with them. They introduced Denis to the European church in Kampala, where in future years he would take an active part. The journey to Lira took 2 days by train where on arrival he went to the Church Missionary Society station at Boroboro, about 4 miles from where he was to be based, to meet Alf and Gladys Peaston,[4] who were working with the Africa Inland Mission (AIM).

Denis' role as medical officer was to take responsibility for all aspects of healthcare in the region and at the hospital, which was badly understaffed. Similarly, equipment was in short supply, and a single room served as an operating theatre, anaesthetic area and instrument cleaning, washing up and sterilising room. Anaesthesia was by ether dripped onto a facemask of gauze, and the anaesthetist was a locally recruited man with only primary education. For surgery that was required for problems below the waist, such as hernia repairs, caesarean sections and hydroceles, Denis took to doing spinal injections giving him 2 hours of sensation-free operating without the risks of full anaesthesia, albeit with the patient fully conscious. At night when operating on emergencies, Denis would take his own oil lamp and hang it over the operating table for illumination (Fig. 7.2).

To manage healthcare beyond Lira, Denis would visit, on safari, major towns such as Gulu in the north, Amolatar in the south, Tororo in the east and as far afield as Ibuye in Rwanda to the west where there was a 36-bed hospital built by Kenneth Buxton, a medical missionary [2]. Initially he had to travel in the ambulance that was visiting these areas, but after 4 months, with medical priority, he was able to buy a Ford pick-up truck for £140. He would set out with the truck loaded with people discharged from the hospital, only to return with it equally full of patients that he had selected for surgery from the clinics. Despite the limitations of the job, Denis remarked that 'it was a veritable paradise for a young man passionately interested in the science of surgery'. But once more Denis was in a job for which he was overqualified, lacked appropriate support and facilities and was completely isolated from colleagues with whom he would normally have discussed patients and their management. Despite all this or maybe because of it, he was to start his pioneering research, the ideas for which and the practice of it were entirely his own (Fig. 7.3).

---

[2] Entebbe. Administrative capital of the protectorate until independence in 1962.

[3] Dr Roy Billington (1870–1951) went to Uganda as a medical missionary for the Church Missionary Society (CMS) in 1937 and in 1950 became Medical Superintendent of Mengo Hospital in Kampala. [1].

[4] Wilf and Gladys Peaston were missionaries with CSM and AIM and were based at the church in Boroboro.

Fig. 7.2 Hospital at Lira showing theatre block and a ward with cross on roof

Fig. 7.3 Lira. Bringing patients back to hospital in Denis' pick-up. 1947

The arrival of newly born daughter Judy was a real boost for him. Judy was born at Laragh on December 24, 1946, requiring an immediate telegram from Olive to Denis to announce the event. A few days later at the turn of the year, he listed the events of 1946 that he called 'big things'. They included his release from the RAMC effective from June 4, 1946, his being accepted by the Colonial Office for duty in

Africa, the awarding of his MD by Trinity and the 'awful parting' once again from Olive. He went to the New Year's party for medical staff in Lira but did not enjoy it at all, leaving just after midnight. He took refuge in his Bible, but when he was told a few days later that a man much younger than him, who had received his fellowship only the previous year,[5] had been appointed to a post at Mulago[6] in Kampala, effectively the main teaching hospital in Uganda, he was openly jealous with his colleagues about this. He hoped God would forgive him. It was not the last occasion for this reaction. Denis now had ambition. A cable soon arrived from Olive saying that they would be sailing in February 21 on the RMS Winchester Castle. The ship had been used as a troop carrier during the war and for the voyage to Africa had been loaded with people. Olive and Judy shared a cabin with 11 others, 5 mothers with their 6 children. Space was limited and there was little room even to move around. It was a grim voyage for them and disease was rife. Thus, there was mutual delight when she and Denis met on the quayside in Mombasa on March 10, 1947.

The journey from Lira to Mombasa was about 725 miles (1164 k). Denis had set off in his car for Tororo on the border with Kenya and boarded a train there for Nairobi and then Mombasa. On the train he met for the first time Dr Hugh Trowell, a physician from Mulago, the main hospital in Kampala, who was to become a close colleague and who would have a major impact on Denis' life at Mulago and, later on, his thinking about dietary fibre. The return journey with Olive and Judy seemed quite daunting for them after the privations of the sea voyage, so they all spent 3 nights in a hotel in Mombasa sightseeing and shopping before getting the train, arriving in Lira on Friday March 14, 1947.

Olive's commitment to join Denis in Lira brought her some clear benefits. They would now be together as a family for the first time and have their own home. This was a square colonial-style house built of brick with corrugated iron roof and comprised four rooms with one opening into the next and a wide veranda around three sides of the house, the whole elevated slightly on pillars because of the danger of termites. The toilet was a bucket placed under a wooden seat that was emptied daily, usually at night, by men with an oxcart collecting what was known as night soil. Hot water was available in the evenings by lighting a wood fire under a corrugated iron tank and then piped into a metal bath. Cold water was carried in old petrol cans from an outside tap. Oil lamps provided light but attracted lots of flying insects. The area was known as being high risk for malaria, Denis had already had one attack, so the family used mosquito nets at night and had suitable netting over the windows. Cooking was done in a separate building blackened by years of smoke from the wood-fuelled stove. But there were some definite benefits to counteract these

---

[5] Denis has passed the FRCS exams 8 years earlier.

[6] There were two hospitals in Kampala, Uganda, that would be prominent in Denis' life. Mulago, on a hill to the north of the city, opened in 1921 next to which would be built the Medical School of Makerere University. Mengo Hospital founded by Sir Albert Cook for the Church Missionary Society in 1897 lies on Mengo Hill south of the city. There was also a European Hospital serving the ex-patriot community and similarly a hospital for Indians. Staff at Mulago tended to live close by in hospital accommodation.

privations. Denis employed six servants. Yusufu was the 'head man', and when they moved to Mulago in Kampala, he went with them and was to remain with them until they left Africa in 1966. Yusufu had a boy to help him in the house, and there was a cook, who also had a helper, and a gardener, and with the arrival of Olive and Judy, Denis took on an ayah to look after the baby. Olive was somewhat taken aback by this extravagance and teased Denis reminding him that this was quite out of character given his own background of frugality. But Denis justified the expense by saying that with his experience and qualifications, he was now receiving a salary of £600 a year whilst he paid Yusufu only £1.10s a month and the garden boy 11s. Moreover, his conscience was kept clear as he told himself that given the high numbers of unemployed people in the area, he had a duty to provide jobs for as many as he reasonably could (Fig. 7.4).

Olive was quickly introduced to the Peastons at Boroboro where Judy was taken to be christened in March 23 just over a week after their arrival. It was Olive's first experience of an African service, which Denis was very keen to show her. It was immediately clear to her that he was very involved in the mission at Borboro attending Bible study there and taking Sunday services and that his commitment to evangelism was still the driving force in his life. Having Olive with him enhanced this for Denis who noted 'How happy Sundays now are with Olive. The house seems a home. Formerly it was my loneliest day'. A few days later, a letter arrived from Guy Timmis, his long-time friend from Trinity who had preceded Denis by only a few weeks into the Colonial Medical Service and was district medical officer for the adjacent area. Guy had met a dentist who had recently visited Denis and who had remarked to Guy that Denis was 'trying to be a missionary and a DMO[7]' Denis rejected this, writing 'If Christ can't keep a man in a secular calling the Christian Gospel has no message for the ordinary man.' That wasn't quite the point that Guy was making.

Fig. 7.4 Yusufu

---

[7] District medical officer

After Olive's arrival two aspects of Denis' religious observance began to trouble him – his disinclination to take part in any social occasion that involved dancing and his complete avoidance of alcoholic drinks. Whilst these self-imposed restrictions on his life could be seen as a natural component of his inherent lack of sociability, they were also a matter of conscience adopted by him at the time of his conversion whilst at Trinity and under the influence of the 40 group. Prior to this Denis had enjoyed going to dances at home in Laragh, but now when the first Club[8] Dance after Olive's arrival was held, he was 'Much exercised in thought on this matter. Felt definitely I couldn't participate. It is hard that my decision has to involve Olive. Very hard on her. It is going to cause much pain'. He rationalised it all by thinking that his role as a leading Christian in the community of African brethren in Lira would be compromised if they learnt he had been to a dance. As time went by, Olive succeeded in getting him to attend events where there was dancing, but he was always pleased to leave as early as possible. He may have been tolerably competent on the dance floor, but his dislike of social occasions and especially the small talk that went on at them was an abiding characteristic.

Abstinence from all things alcoholic had a sounder basis. Excessive consumption and drunkenness were a significant problem amongst both the ex-patriot and local African people. This allowed Denis to turn his teetotal lifestyle into a one-man campaign for sobriety. But Olive had other ideas. As senior members of the community, they had a responsibility to entertain, apart from Olive's wish to make their home a welcoming place for all. It was customary on such occasions to offer guests sherry at the start of a meal and beer after a game of tennis. This was a problem for Denis. Was he being a fanatic and would allowing alcoholic drinks in his house mean he had compromised all his principles? Whilst in the army, he had seemingly no problem with fellow officers accepting his position on alcohol; in fact he felt that they respected him for it. Since arriving in Uganda, he had seen some bad examples of drunkenness. He recounts one such encounter in his autobiography:

> On the train up the coast to Kampala I had shared a compartment with a senior official in the medical department. The boat on which we had travelled having been dry he had relieved his enforced abstinence by overindulgence, with the result that the African servants on the train had to carry him to bed. Unable to walk to the toilet he relieved himself on our cabin floor. This despicable behaviour so disgusted me that I felt more convinced than ever of the dangers of over-indulgence in alcohol.

Discussion on the topic of alcohol with Olive continued, but once they moved to Mulago, the pressure to entertain would increase, at which point they agreed to offer sherry to guests, wine with meals and beer after tennis. Rather than have drinks parties, they would entertain people to dinner, which was a lot more work, but started each meal by offering thanks to God. They never kept spirits in the house. Whilst their missionary colleagues were able to take a clear position on alcohol, as determined by the church, it was not so easy for Christians in government service. Denis

---

[8] 'The Club' was the social centre for British expat officers and their wives. The Club Dance would be an important event. As DMO, Denis would have been a highly respected member as would Olive as his wife.

felt that in spite of the condemnation of situation ethics, the Bible made it clear that the application of these principles can differ in varying situations.

Their stay in Lira was quite brief, only 15 months, but before they moved to Mulago, Denis took, unknowingly, his first steps into history. In the archives of the library of Trinity College Dublin lies a map rarely seen and never publicised even by Denis. It is in colour, hand-drawn on a page torn out of a 'Return of Revenue and Expenditure' leger, undated but almost certainly drafted in 1947, which shows the prevalence around Lira of a disorder of the male testis, known as hydrocele[9] (Fig. 7.5).

Hydrocele of the testis presents as a swelling, which occurs due to the blockage of the fine lymphatic channels draining the testis, usually by a small worm, *Wuchereria bancrofti*, leading to the accumulation of fluid within the membranes surrounding the testis. It is the commonest manifestation of a disease that occurs

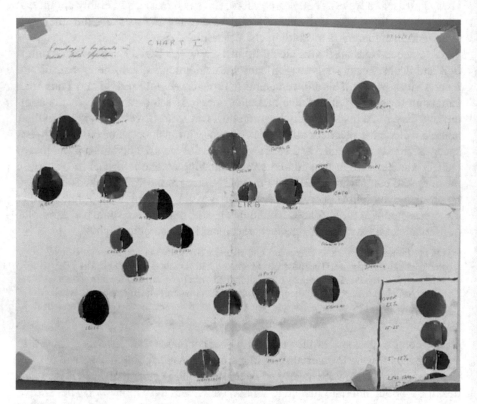

Fig. 7.5 First map of the distribution of a disease drawn by Denis in 1947 when in Lira. Shows 'Prevalence of hydrocele in adult male population' 'CHART 1'. The red colour indicates high prevalence to the east, the black low prevalence in the west
(Trinity College Manuscripts and Research Library TCD MS 11268/2/1)

[9] Prevalence in this context is the number of cases of a condition in a defined population at a point in time usually expressed as per 1000 or 10,000, etc. of the population being studied.

with great frequency in tropical and subtropical regions of the world, known as fila-
riasis – spread by mosquitos, which when biting their victim, ingest tiny microfilariae that are an early stage in the life cycle of adult worms living in the human
lymphatics. In Lira hydrocele was the outstanding surgical problem Denis had to
deal with and was unusual because it was much more severe than he had seen during
his surgical training in England. He thought it remarkable enough to write a case
report for the *British Medical Journal* in 1947 in which he noted that he and his colleague Dr B.N. Kununka were operating on six to eight cases a week and the patient
he reports presented with a large abdominal swelling, which was the hydrocele
extending into his abdomen and which contained six pints (3.4 litres) of fluid [3] – a
truly enormous swelling.

Denis' interest in hydrocele was further stimulated because of the places from
which came the cases he saw each week in his clinics. He noticed that the majority
of the men with the condition lived in villages to the east of the town. This was a
crucial observation. Having had his curiosity aroused, Denis decided to get more
information about the distribution of the condition, so ahead of his visits to the village clinics, he would warn the chief and ask him to line up 100 men so that he could
examine them for the presence of hydrocele. On the map there are 24 villages
marked. To the east of Lira, the prevalence is shown as mostly over 25%, whilst in
the west it falls to less than 5%. It is entitled, in Denis' own hand, 'Chart 1 Percentage
of hydrocele in adult local population'. Thus, one of the earliest maps of the prevalence of a disease in Africa was drawn, and Denis had taken his first step on a career
as a geographical pathologist.

This interest in hydrocele gets no mention at all in his diaries except that he notes
on Friday September 6 'Read a paper on hydrocele and demonstrated a case'. He
had driven to Kampala for a 3-day meeting sponsored by the British Medical
Association (BMA) to mark the 50th anniversary of the arrival in Kampala of Sir
Albert Cook, a legendry physician of the early days of Ugandan medicine who
founded Mengo Hospital.[10] Denis would have done case presentations whilst working towards his FRCS in England, but this was his first in Africa. Whether he showed
his map to the assembled doctors in Kampala is not recorded. His main motivation
at this time was to suggest a new surgical approach to hydrocele that he had been
able to develop. He tried four different techniques and went on to write a paper
entitled 'Primary hydrocele and its treatment. Review of two hundred cases' [4] that
was later published in *The Lancet*,[11] reporting his results and recommending the
procedure that allowed patients to return home at the earliest date and with the least
complications. Denis is the sole author of the paper, which was the first of many that
he was to write about the cause, treatment and prevention of conditions he encountered in Africa and later, when back in England, about diet and Western diseases.
The Lancet article comprises a description of hydrocele with illustrations of the

---

[10] Mengo Hospital, known today as Namirembe Hospital, is about 3 miles (5k) from Mulago and is
the earliest surviving hospital in Uganda.

[11] *The Lancet*. Leading journal publishing articles from all areas of medicine since 1823.

surgical technique he was recommending and of the post-operative care, which required the scrotum of the patient to be encased in plaster of Paris slung from a bandage around the waist. The size of the hydroceles he encountered is remarkable containing on average 'a little over a pint' with the largest ten pints. There is a table that shows percentage of cases ready to be discharged at various days post-operatively, but there are no other statistics, a feature characteristic of his papers in the ensuing years. There are no references to other papers on the subject although that was probably because Denis would not have had access to a good library at the time. Today surgery remains the main treatment for hydrocele although there are drugs that can be valuable in eradicating the filariae. In his paper he mentions that microfilariae were found in 17 of the 130 cases where they were looked for, but he does not mention transmission by mosquito.

Why did Denis decide to investigate the prevalence of hydrocele, and why did he draw the map and then write a paper for a prestigious medical journal? What first brought the problem to Denis' attention was the huge size of some of the hydroceles and their frequency of occurrence. You could not miss this condition. What Denis noticed was the big difference in cases occurring in the east where it was present in more than 25% of the adult men when compared with the rate in the west of less than 5%. He began to make arrangements to survey the men in the villages, which suggests a curiosity about the condition that was not directly related to his main concern, that of finding the best operation to cure it. By acquiring this information and then drawing the map, he was setting down for himself in a simple but explicit format the information he had and the question that it posed. What precisely was the geographical distribution and why should it occur? What was the cause? Largely isolated from the medical world in rural Africa, Denis was a true pioneer.

Denis was not the first person to draw maps of disease prevalence, so from where did he get his inspiration? The earliest examples of such maps are credited to Valentine Seaman who plotted the distribution of yellow fever in New York in 1798 followed in 1830–1840 by Shapter and then John Snow with their maps of cholera cases in Exeter and London [5]. These landmark studies of disease may well have been part of the curriculum at Trinity. Perhaps more pertinently how much was he influenced in this mapping exercise by his father's maps of bird territory, which led him to make original observations about bird behaviour? It would seem obvious that this was something that Denis would have seen as a child, but in his autobiography written 40 years after all this, he gives little credit to James saying only 'I can now recognise the similarities between these sketches (that James had drawn) and the maps of tumour distribution which I was drawing in Africa nearly half a century later'. From this one might conclude that the motivation to draw the map of hydrocele was not a consequence of his childhood experiences, but it is difficult to ignore an environment where a major figure in Denis' life, his father, was mapping events that occurred in the world around their house with a view to explaining the behaviour of a bird that Denis would see frequently. That Denis should not give credit to his father for the mapping concept is surprising because in later life he would go out of his way to give credit to others who had worked with him or were active in the same field.

Whatever it was that led Denis to draw the map in Lira, the context was of a man working in a remote area of Africa with few colleagues to whom he could talk. We must credit him with having an original and enquiring mind. With regard to Denis writing a paper on the subject and sending it to *The Lancet* rather than a surgical journal suggests a new self-belief and ambition. Did he get the confidence to write for *The Lancet* from his father who is credited with authorship over 50 papers and other communications about his studies of birds? Perhaps not, because most of James' work was done whilst Denis was quite young. Maybe the confidence to write about his hydrocele work developed in the more academic environment of the medical school at Mulago to which he would move shortly.

About a week before Denis journeyed to Mulago for the BMA meeting at which he presented his hydrocele paper, he was told by the Principal Medical Officer Dr Rainsford that there might be a job for him at Mulago. Surprisingly he had mixed feelings about Mulago and felt if he had been given the choice, he would have chosen to go to Masaka, a large town in the central region west of Lake Victoria that he had visited and liked during his army safari of 1944. However, when the chief pathologist from Mulago visited Lira a few weeks later, Denis was told that the job in Kampala might be given to a young surgeon Dr Tucker if he passed his FRCS exams. This made Denis quite jealous, anxious and very determined to go for the Mulago job. He drafted letters to the Director of Medical Services (DMS) and to Professor Ian McAdam, head of surgery at Mulago, but Olive advised him not to send them. Reading his *Daily Light*[12] enabled him to process this recurrent anxiety.

Almost 2 months passed before he had any further news, but then a telegram arrived from the DMS to say that his transfer to Mulago was imminent. Four days later in December 16, there was another telegram to say that his transfer was to be immediate. He asked for Christmas leave and went to Kampala without Olive to sort out accommodation for them. He spent Christmas in Budo with the Drowns and Robinsons and on Boxing Day met Professor McAdam, professor of surgery at the university, who told Denis that he was to be lecturer in surgery and government surgeon. At last he was to have a job that recognised his surgical training and skills. On returning to Lira, Denis found that there was some problem with housing for him and family in Kampala, eventually sorted out by the district commissioner, so he began to hand over to his successor. They left Lira on Sunday January 4, 1948, and were settled in their new house at Mulago by Wednesday 7th. Denis was almost 38.

The time in Lira, although brief, allowed Denis to move forwards with his life. He now had a career to which he felt called, a wife and family with whom he was living and had taken the first steps in what was to be a remarkable journey of discovery. His diary entries, although still almost entirely a record of his evangelical and devotional activities, now no longer detailed his chronic anxiety about his future and need for guidance from God. He continued to be vitally involved with the church, to

---

[12] *Daily Light*, published since 1875, gives passages of scripture from the Bible for morning and evening devotions.

spend time each day in devotions, his 'quiet time', and was pleased that Olive was content to attend church with him and to accept and enlarge the circle of friends that shared Denis' beliefs. But there were challenges ahead that would engage him in a completely new way. Reflecting on his time in Lira, Denis considered his main achievements to be a 20-fold increase in the number of operations performed although in retrospect he felt that doing more to prevent disease than treat it would have been of more lasting benefit to the people. He did try to reduce the incidence of one of the commonest diseases, yaws,[13] but in trying to do so, he was to learn a pertinent lesson. The campaign to eradicate the condition, which started after Denis left Lira and was organised by the World Health Organization (WHO), consisted of inoculating the whole population with penicillin. Yaws virtually disappeared from the community but also disappeared from neighbouring districts where no penicillin had been given. As with the prevalence of many diseases, both infectious and non-infectious, environmental factors are important, such as the provision of clean water, sanitation, better housing, nutrition and the relief of poverty. But from his earliest days in Africa, Denis was looking beyond the surgical management of the diseases from which his patients suffered to understanding their cause and developing a strategy for their prevention.

References

1. http://www.mundus.ac.uk/cats/17/271.htm
2. https://en.wikipedia.org/wiki/Kenneth_Buxton
3. Burkitt DP, Kununka BN. Enormous intra-abdominal hydrocele. Br Med J. 1947;4517:175.
4. Burkitt DP. Primary hydrocele and its treatment. Review of two hundred cases. Lancet. 1951;257(6669):1341–3.
5. Howe GM. Some recent developments in disease mapping. Perspect Public Health. 1970;90:16–20.

---

[13]Yaws is an infectious disease caused by *Treponema pallidum* ssp. *pertenue* characterised by small ulcerating lesions on the skin of the lower limbs and buttocks. It can be treated with penicillin or recently with azithromycin but may become chronic.

Chapter 8
Mulago 1948–1956; A Busy Surgeon

Denis' early years at Mulago were spent dealing with a huge clinical workload, leaving little time for considering the cause of any of these problems. He did little research although continued to observe the differing patterns of disease he saw in Africa compared with those he had managed as a surgeon in England. He was also able to turn his engineering skills to help some very vulnerable patients.

The only fully qualified surgeons in Kampala, in fact in the whole of Uganda, were Denis and Professor Ian McAdam,[1] who had been so overwhelmed with responsibilities at Mulago that he had developed a duodenal ulcer and hence the sudden call to Denis to move from Lira (Fig. 8.1).

Denis had been well trained, by now had first-hand experience of Africa's problems and was able to cope with the job. He had always risen early in the morning, and apart from family there were few distractions to take up his time. He wrote his paper on the hydrocele and some case reports, but for the first 2 or 3 years, his mind was very much on his clinical work. He was also more certain about his future and what God had called him to do. This change in outlook was noticed. Whilst on his army safari in 1944, he had been met in Fort Portal, in the south-west of Uganda near the border with the Congo, by Bill and Nancy Butler who were teachers, spoke Swahili and had been through missionary training. He had spent a week with them and climbed Mount Karangara (10,000') with Bill. The Butlers had moved to Mulago and Denis met them at the European church, where Bill preached. The next day, Sunday, they came to dinner with Olive and family, and as they were getting into their car to leave, Bill said, 'On Saturday Denis I felt you weren't quite the same Denis as I knew at Fort Portal'.

---

[1] Sir Ian McAdam (1917–1999) was a South African who graduated from Cambridge University, trained in surgery at Edinburgh and was sent out by the Colonial Office to Mulago succeeding Sir John Croot as professor of surgery. He played a major part in establishing the reputation of the medical school and attracting young scientists from all over the world. He invited Denis to Kampala in 1948. The McAdams and Burkitts lived in adjacent houses at Mulago.

© The Author(s), under exclusive license to Springer Nature Switzerland AG 2022   97
J. H. Cummings, *Denis Burkitt*, Springer Biographies,
https://doi.org/10.1007/978-3-030-88563-2_8

Fig. 8.1 Professor, later Sir Ian, McAdam at his desk in Mulago Hospital. In the background left to right, Roslyn, his daughter; unknown; Mr Tommy Millar; Dr Kywalianga; Dr Alec Alderdyce. (By kind permission of Professor Keith McAdam)

A group of Christians who brought home to Denis how much his faith had progressed were the Balokole. They were African fundamentalist Christians, a movement that had started in the 1930s, who regarded themselves as 'The saved ones'. Denis felt that they had only one message, which was along the lines of 'Believe on the Lord Jesus Christ and thou shalt be saved', with which he now found hard to identify. When Bill brought two Balokole to see him, he noticed Denis' reticence and hence the remark that Denis was not as keen as he used to be about matters of faith. Denis nevertheless respected their open and straightforward approach to the practice of their beliefs. However, he had moved on. But it was his unease about what the Balokole would think of him that made Denis so reluctant to agree with Olive about keeping alcoholic drinks in the house for guests.

1948, their first year at Mulago, passed uneventfully. There were family holidays, usually touring in the car, and Saturday trips for picnics to Hippo Bay, Entebbe, on the north shore of Lake Victoria with the Drowns, Taylors and Guy Timmis. He was still happy to go shooting and often went out with Guy to shoot kanga.[2] His Christian duty provided a legitimate escape from social functions. After dinner with Dick Savage, his fiancée and Colin Campbell, Denis remarked 'How superficial is most dinner talk'. He had not really changed from the days when as an undergraduate he had done his best to avoid parties, drinking and especially dancing. So absorbed was he in his daily clinical work that Olive having a baby early in 1949 gets only a passing reference: 'Sunday January 30 Spoke at evening service at Mulago to enable Roy to deliver Olive. Carolyn born at Namirembe at 10.30. Both well'. And that was it.

Although he wrote little about it, Denis' medical work was very important to him and his work as a surgeon a matter of pride. So much so that he kept a record of the operations he performed from his early days as a trainee surgeon at the Beckett Hospital in Barnsley and at the Prince of Wales Hospital in Plymouth. These may have been required to show the examiners for the FRCS that he had sufficient experience to be awarded the fellowship. But the records of the surgery he performed during his first year at Lira and then at Mulago were for other reasons. He was developing his interest in the pattern of diseases (Table 8.1).

Table 8.1 indicates what an astonishing range of surgical skills and competence surgeons in the countries of Central Africa, mainly below the equator, developed of necessity. For Lira he records 335 operations and for Mulago 316 each over the course of a year. In Lira he was on his own and, having general medical responsibility for a whole region, would have travelled regularly quite long distances to do clinics, whilst at Mulago the local population was much greater, but he could share the workload with Ian McAdam and junior medical staff who did simple operations such as strangulated hernias. It is the range of surgery he was having to perform that is so striking. Apart from what would today be considered general surgery such as appendicectomy, hernia repair, laparotomy[3] or excision of perianal fistulae, Denis was also doing paediatric, orthopaedic, genito-urinary, plastic surgery and some gynaecology although there were specialist services for this at Mulago. Some of it is for conditions seen less commonly outside Africa such as the recto and vesico-vaginal fistulae, possibly following trauma at the time of childbirth, whilst the frequency of urethral strictures and penile cancer surgery may reflect a greater incidence of sexually transmitted diseases in the city [1]. The amount of difficult orthopaedic surgery was probably the motivation for him to learn more about this.

After nearly 2½ years in Uganda, Denis was due for home leave. To gain more training in orthopaedic surgery, he booked himself into a 5-month postgraduate

---

[2] Swahili name for Guinea fowl.
[3] Exploratory operation of the abdomen.

Table 8.1 Surgery performed by Denis Burkitt during his first year at Lira (L), 1946/1947, and Mulago (M), 1948/1949

Operation	L	M	Operation	L	M
Hydrocele	16	12	Thyroid adenoma	1	4
Inguinal hernia	65	44	Femoral hernia	1	3
Elephantiasis of scrotum	33	1	Excision of elbow	1	3
Amputation of leg or foot	16	7	Amputation hand or arm	3	3
Enucleation of eye	3	4	Perineorrhaphy	2	–a
Urethral stricture	2	37	Ventral hernia	1	2
Repair wounds to bowel	2	2	Excision of knee	1	1
Laparotomy	8	10	Haematocolpos	1	0
Strangulated inguinal hernia	3	21	Prostatectomy	1	1
Bowel resection	0	8	Intussusception	1	1
Volvulus	1	11	Hysterectomy	6	–a
Suprapubic cystotomy	1	43	Branchial cyst	1	0
Amputation of penis	3	5	Kidney cyst	1	0
Transplantation of ureters	1	9	Myomectomy	2	–a
Pedicle graft	1	14	Acute appendicitis	2	3
Trephine	0	2	Closure of enterostomy	2	5
Splenectomy	0	2	Vesico-vaginal fistula	3	–a
Subtrochanteric osteotomy	0	2	Open reduction shoulder	1	0
Intestinal anastomosis	0	3	Orchidectomy	3	1
Mastectomy	0	2	Parotid tumour	2	0
Colostomy	0	4	Rib resection	1	3
Recto-vaginal fistula	0	3	Ramstedt	0	1
Undescended testis	0	2	Osteotomy femoral shaft	1	1
Perforated gut (typhoid)	0	2	Plating fracture	0	2
Tumour of mandible	1	2	Phrenic avulsion	0	5
Hydrocele of cord	1	1	Condyles of mandible	0	2
Axillary dissection	0	1	Gastroenterostomy	0	2
Groin dissection	0	2	Ligation common carotid	0	2
Repair hair lip/cleft palate	0	1	Ileostomy	1	2
Glass button for ascites	0	2	Ruptured ectopic	1	–a

aDone by gynaecologists in Kampala
Others, one or two cases done only in Mulago, included PDU, excision of femoral head, tracheotomy, closure of faecal fistula, removal of deep fascia of leg for elephantiasis and submaxillary tumour. Data are incomplete in that Denis reported a series of almost 200 cases of hydrocele in his Lancet paper of 1951
(From original data collected by DPB now at Trinity College Dublin, Manuscript and Archives library)

course at the Royal National Orthopaedic Hospital in London commencing October 1, 1949. Olive and he took some time to pack up many of their belongings for what was to be a 10-month absence from Kampala. They left Mulago in July 30 with many warm goodbyes from the staff some of whom came down to the station to see them off. The day before they left, Denis and Olive had lunch with the DMS

(Director of Medical Services) who told them that it was probable that they would be returning to Mulago. It was well known that working in the Colonial Medical Service meant that they could be sent wherever the need was greatest at the time. They set off for Nairobi and then on to Mombasa with their two children and gifts for the family in Ireland where there were still post-war shortages of many foods. They sailed on the TSS Modasa, with Guy and Dawn Timmis also on board. All was well until 10 days later when a few hours out of Port Said Olive developed the clinical signs of acute appendicitis. Although very rare in Uganda, Denis had seen plenty of cases during his training in England and in servicemen in the army, so he knew that surgery was indicated, but facilities for this were effectively nil. So, Denis had to manage Olive conservatively. There may possibly have been antibiotics available and something for pain relief. She survived and they arrived at Plymouth in August 24 and were driven to Portsmouth where Olive's father and her brother Alan were waiting. Three days later Denis, Olive and the girls were back at Laragh where they found mother and father to be well and Robin also there with Vi (Violet). The journey had taken them 28 days.

Robin, [2] who had married Violet Laker, a nursing sister he had met in Cornwall, was at this time working as a surgical registrar at Ashford Hospital in Middlesex. He was near enough to London to be able to meet Denis over the next few months whilst he was attending his orthopaedics course, which finally ended in March 17. The family then stayed on for a few more weeks in London visiting friends including Conrad Latto, a surgeon in Reading whom Denis had first met when they were both working in Plymouth and with whom he would develop a close friendship over the coming years. The return journey to Uganda started in April 27 and, for the first time for Olive, was by air. Unlike today when such a flight would take only 12–14 hours, the journey, which was clearly memorable, is described by Denis. 'We were driven by the Air Company[4] to a magnificent hotel in the New Forest where we were accommodated prior to our early morning departure from Southampton. The terminal was a wooden hut and other passengers merely the few dozen on our flight. The plane was a double decker and exceedingly comfortable. Only the elect few flew in those days. With an early morning start we made Sicily by nightfall to be conveyed to our hotel for the night. An early start enabled us to land at Alexandria for lunch. By our second night we had reached Luxor on the Nile, again to be accommodated at Air-line expense. Next day we landed on Lake Victoria, a pleasant but doubtless uneconomical means of transport gone forever'. They were met by the McAdams, Drowns and Goodchilds and stayed overnight with the Drowns. The journey had taken nearly 4 days. It was Sunday April 31, 1950.

Now back at work, clinical duties continued to be the overwhelming activity with Denis remarking that he could never remember being so busy. But, having had his curiosity aroused by the geography of the occurrence of hydrocele, he noted that the emergency cases on which he operated every week were very different from those he had experienced whilst working in England. He wrote to Robin, working in

---

[4] BOAC.

London, asking him to keep a note of the nature of the surgical emergencies, exclud-
ing gynaecological emergencies, on which he operated saying he would do the same
in Mulago. They each collected a series of three hundred cases over a period of
2 years, the ones at Mulago being exclusively of the local Buganda people. On
reviewing the data, Denis' early observations were confirmed. Major differences in
the reasons for acute abdominal surgery were evident. In London there were 223
cases of acute appendicitis compared with only 4 in the Buganda, whilst Denis had
operated on 219 cases of strangulated hernia with Robin seeing only 27.[5] Other
major differences were 31 cases of perforated peptic ulcer in London vs. none in the
Buganda, whilst for volvulus[6] there were 29 operations done by Denis vs. none in
London. The contrast is very striking. Ninety-three per cent of African cases were
obstructive or mechanical in origin, whereas 86% of the English cases were inflam-
matory. In the discussion of the findings, Denis makes some unexpected suggestions. For appendicitis he writes 'This is believed to depend on the nature of the
intestinal flora which reflects dietary habits'. Who, in the early 1950s, had given
Denis the idea that the gut flora were influenced by dietary habits? Dietary associa-
tions with disease were not to come into Denis' consciousness until 1967. That diet
was a key factor in controlling the metabolism of the intestinal flora would not be
demonstrated for another 25 years [3]. The lack of peptic ulcer in the Africans he
ascribes to their slower tempo of life. Robin is not a co-author of the paper, which
was published in 1952 in the *East African Medical Journal*, Denis' second major
paper [4] (Fig. 8.2).

Despite the focus on his busy hospital work, Denis made overtures to the Director
of Medical Services, Dr Hennesy, about being allowed to do private practice. Why
he should want to do this on top of his university and hospital workload is surprising. Clearly, he would be able to earn more money, but this was not something that
motivated him. Perhaps Olive encouraged him with the needs of a growing family
and their recent agreement to tithe[7] their income. Maybe he was being asked by the
ex-patriot White community in Kampala to provide a service that would allow them
to be seen and treated separately from the indigenous people. The DMS declined
Denis' request and he was very disappointed. A few weeks later, when in conversation with Dr Hennesy again about other more mundane matters, the DMS suggested
that there might be some hope for Denis' request. Clearly the matter must have been
passed up the office for a view. By the end of February, Denis had the go ahead to
start his private practice.

Olive, now 31, and Denis 40 did a lot of entertaining at home, at least twice
weekly. Friends such as the Drowns, Timmises and Butlers came to stay and they
would play tennis or go for picnics. There was a brief excursion to climb Mt Elgon,
10,660', on the Uganda/Kenya border in May, which Denis found tiring, but no

---

[5] Probably inguinal hernia although the exact type is not given. Significantly 29% of the African
cases were gangrenous.

[6] Twisting of a loop of gut, in these cases the sigmoid colon. Seen here only in men.

[7] Tithe – to give a fixed proportion of income usually for charitable causes.

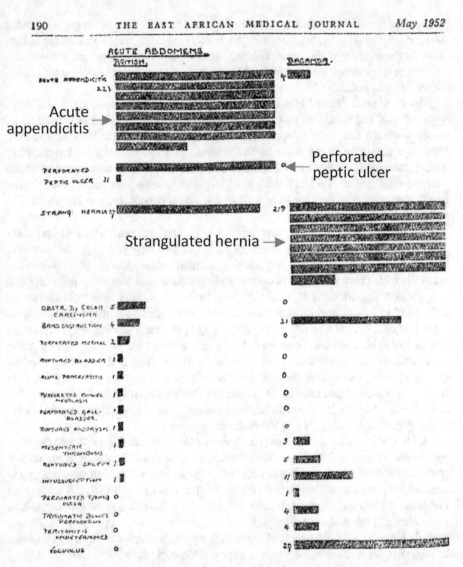

190   THE EAST AFRICAN MEDICAL JOURNAL   *May 1952*

Fig. 8.2 Acute abdominal surgery performed in London by Robin Burkitt and in Kampala at Mulago Hospital by Denis Burkitt; circa 1950–1952. (Modified from Burkitt DP. Acute abdomens – British and Buganda compared. E Afr Med J. 1952;29:189–194 (With the permission of the *East African Medical Journal*))

other family holiday on safari during the year. They went most weeks to the cinema or to a play and found a Passion Play at Easter put on at Mulago quite moving. But 'All Quiet on the Western Front', which would have been the 1930 version of the movie about World War I, was 'realistic and harrowing'. Olive could not persuade him to take her to a dance and alcohol remained a point of contention. By the middle of August, Olive began to get morning sickness and was now starting her third pregnancy. She would remain unwell throughout leaving Denis feeling guilty about the relatively small amount of time he gave to looking after Judy and Carolyn, especially in the mornings, and when reading them stories. But he was not at ease with children and remarked after taking a children's service in June that he found speaking to children difficult.

Was he a good father? His daughters say that he loved them. He made them wooden toys and took them on picnics and on adventures climbing mountains, cooking over a fire, bike rides, swimming and camping. Once they had started boarding school, he would write them frequent, long and affectionate letters. Judy would share in his research by illustrating his lectures with cartoons, some of which have become part of the Burkitt legacy. 'We also thrashed out faith ideas, shared books, and regularly went for long walks with maps. He encouraged our choices of career and believed in us'.

On the clinical front, Denis was having to manage patients who had had a leg amputated. This operation was done following severe injury, which was common especially on the roads, or for congenital deformity, severe ulceration or elephantiasis of the feet. There were also people with limbs paralysed following poliomyelitis who needed support in the form of callipers and crutches. Such appliances were not available in Uganda meaning that the patients were rendered immobile, unable to work and condemned to a life of poverty and dependency. During his orthopaedic course in London, he had been introduced to a type of plastic developed for the construction of aircraft that could be moulded into any shape and was being used by the orthopaedic surgeons in the construction of artificial limbs. Denis' compassion for these people led him to develop simple but serviceable artificial legs constructed from malleable plastics, and crutches and callipers for legs paralysed by poliomyelitis, made from iron rods used to reinforce concrete.

Of interest is the malleable material that was key to the making of these prostheses called 'Durestos'. It consisted of layers of asbestos felt impregnated with a phenol-formaldehyde resin. What became, in later life, of the full-time and part-time African technicians who were making eight limbs a month with this dangerous material is unknown. His work on the artificial limbs gave him great satisfaction, to say nothing of the patients he helped. It allowed him to use his practical skills and brought his work to the attention of his colleagues in the medical school and beyond. To describe the technique, he wrote a paper, which was now becoming his normal practice for anything he felt worthwhile, quite long at 15 pages, well-illustrated with his own photographs. It included an introductory review of the previously known literature, very detailed descriptions of the methods for making various type of artificial leg to fit above or below knee amputations, variations on this, and an appendix listing the equipment and tools required [5] (Fig. 8.3).

The world was not yet aware of Dr Burkitt working in Africa, but Denis was now laying the foundations and acquiring the skills that would enable him to become a medical scientist of international repute and to communicate his ideas. He chose to solve important problems, initially in the context of where he was working in Africa, but eventually, he would take on major issues in world health. He came at these without any preconceived notions but with a wide-angle vision that ruled nothing out. He

Fig. 8.3 'First artificial leg. This man had not walked in 6 years'. The writing is Olive's and the photograph by Denis

was very practical, a talent he had demonstrated since childhood but probably enhanced by his year at Trinity doing engineering and subsequently his surgical training. He listened to the ideas of his colleagues and was always a willing collaborator. He was never what he would describe as a laboratory doctor in a white coat, so when technical skills or tools were required, he relied on the knowledge of the experts. And he was good at writing, perhaps a skill learned from his father James. Also photography, his daughter Judy remarking that 'Daddy went into hospital every day with a stethoscope and camera round his neck', and mapping, which were an integral part of the way he looked at the world. All this was in addition to his clinical and administrative duties, private practice, the practice of his faith, family duties, diary and letter writing and the extensive records of his travels. Denis worked hard and alone. Denis was to remember this time for another reason. He developed toothache and went to see the hospital dentist. 'Molar out with hammer and chisel. Painful' (Fig. 8.4).

Family matters were quite pressing during the first half of 1952. There were repeated illnesses for Carolyn and Judy, whilst Olive struggled with her pregnancy. Returning from a surgical conference in Nairobi in February, Carolyn was sick in the car, and Denis had to ask Hugh Trowell, who was the main paediatrician at Mulago, to see her. She recovered but Olive became unwell with a fever and bile in

Fig. 8.4 'Trick' photograph by Denis of him fitting an artificial limb to himself

her urine and clearly had jaundice of pregnancy. She remained in bed more or less continuously for the next 4 weeks going into labour in March 20, which was protracted but resulted in the successful birth on the 21st of Rachel. The next evening Carolyn started vomiting again, and Judy also became unwell with a fever, so Denis asked Philip Hutton to come out at 10 pm to see them, who recommended that Judy be admitted to the European Hospital, the local hospital for private patients. Philip told Denis that he thought Judy might have polio. With Olive in the maternity unit at Mengo Mission Hospital, Judy in the European Hospital and Carolyn unwell at home, Denis was stressed and anxious. Judy had a lumbar puncture, which whilst not ruling our polio looked hopeful. At no stage did she develop any paralysis and went on to recover, as did Carolyn. Olive was transferred to the European Hospital with baby Rachel making it easier for Denis to visit, but he kept the problems with Carolyn and Judy from Olive. Although she and Judy were in the same hospital, Denis did not take Judy to see Olive until almost 3 weeks after Judy had been admitted. Denis eventually took the two older girls to Budo to stay with the Drowns, but by the middle of April, it was clear that Olive had developed post-natal depression.

Meanwhile Denis had had a letter from the Chief Secretary at the Ministry of Health offering him the post of surgical specialist. Another ambition achieved. His private practice at the European Hospital was successful so much so that he thought he might have to give more than 10% of his earnings to good causes. The family moved back to their former house in Mulago, which they had upgraded and to which were very happy to return (Fig. 8.5).

Fig. 8.5   No 18, Mulago Hill. The Burkitt family home in Kampala

Visitors were frequent especially Dick and Gwyneth Drown from Budo, the Trowells who lived about 7 miles from Mulago and also Robin and Vi who had moved from England to take up a post in Uncle Roland's former practice in Nairobi. Robin had felt that after nearly 5 years as surgical registrar at Ashford Hospital, obtaining his MD, FRCS (London) and FRCS (Ireland), the time had come to find a more senior role. Good though it was to be able to visit Denis and family, the move did not work out well. There proved to be not enough work in the practice for Robin and Vi did not like Nairobi. With the Mau Mau Rebellion developing, they decided to return to England where Robin was appointed to a senior registrar post at the hospital in Slough but had to wait for a consultant post at Ashford until 1963 when he was 51. A big contrast to the rate of progression of doctors through the training years that would follow, but the war of 1939–1945 intervened in the careers of many young people.

Early in 1953 Olive became ill with what Denis thought was 'flu but it rapidly became apparent that she was acutely depressed. It was 10 months since her episode of post-natal depression following the birth of Rachel, which interval of time made Denis realise that Olive now had a serious problem. She took to her bed, ate little, slept badly and wept frequently. She was prescribed quite large doses of a sedative but remained ill causing Denis to cancel much of his work to look after her. He sent the children away again to stay with friends, but Olive grew worse, and on the advice of the psychiatrist Dr George Campbell-Young, she was admitted to the European Hospital. Failing to improve he suggested convulsive therapy (ECT). Denis was to remember this day in future years. He was not keen on ECT so moved into the hospital to be near her where they prayed together and sought reassurance from the Bible, but her illness had come to dominate both of their lives. Olive liked to have someone sitting with her all day, so Denis was able to undertake only limited clinical duties and was unofficially relieved of his work by the middle of March. They were due to go on leave at this time, but Denis had to cancel all the bookings. Shortly after this Olive began to improve and came home to recover, and 3 months after she had taken ill, they were able to set off for England with the girls. No-one at that time related Olive's illness to her experiences in Plymouth during the war.

Leave, which was to last almost 7 months, started in London where Denis collected a new Vauxhall Velox car, attended the Association of Surgeons meeting and visited friends finally arriving at Laragh, to the delight of James and Gwen, on May 20. James was suffering from prostate trouble for which he had surgery later in the year, but this did not stop them all having outings to visit the favourite places of their childhood and spend a week by the sea at Mullaghmore on the north-west coast and a few days visiting Gwen's family in Cork (Fig. 8.6).

Denis decided to sit the exam for MCh, a higher degree in surgery, but failed. After all he had been through recently, he managed this failure well, remarking 'Quite satisfied and quite at peace. I wouldn't have done better with more work'. With Olive much better and off all her tablets, they set off for a tour of Europe in their new car with Denis taking colour photographs (transparencies) for the first time. They were accompanied by the Drowns. Their itinerary took them through France, then Lake Geneva in Switzerland, through the alpine passes to St Gallen, then Austria, Innsbruck, Kitzbuhel, Oberammergau, Freiberg, Titisee, the Rhone

Fig. 8.6 Mullaghmore, July 1953, L to R, James, Gwen, Denis, Olive, Judy, Cas, Rachel

Fig. 8.7 Laragh 1953

Valley, Cologne, Bruges, Liege and finally back to Boulogne arriving at Laragh 3 weeks after they had set off. The trip was a success with normal life more or less resumed for the family and Denis back to writing an extensive account of the holiday, which he sent to James and Gwen. The only incident worthy of mention occurred when they were leaving Bruges. The boot of the car had not been properly shut and a case fell out containing their passports. Remarkably the case was soon returned to them by someone who was presumably travelling the same road. Denis notes 'How wonderfully God's hand has been with us.' A now much less frequent reference to the hand of God in their lives. Despite everything that had happened in the past 18 months, Denis was more at peace with himself and the world around him (Fig. 8.7).

They finally left Laragh a couple of weeks later and headed south to join the SS Uganda sailing for Mombasa on October 23, 1953. Olive had 'some trepidation' about returning to Uganda, associating it with her depressive illness, the return of which she dreaded. The reality was that it would be with her for the rest of her life. On board their cabins were luxurious and comfortable, and sailing down the Red Sea, they passed a boat heading in the opposite direction on board which were Robin and Vi, finally having left Kenya. They were able to exchange marconigrams[8] as the vessels passed each other. Back in Kampala normal life resumed, but on November 30 an event took place that was clearly more important for Denis than the ending of World War II or the recent coronation of Queen Elizabeth II. A state

---

[8] Wireless message sent using the telegraphy system that Marconi had invented.

of emergency had been declared; the Kabaka was arrested; hospital patients were discharged. The Kabaka was King Mutesa II who was ruler of the Kingdom of Buganda, centred on Kampala, the largest territory in a bigger area corresponding to today's Uganda. There were rumours coming out of London that the British Government would like to see a much larger federation of East African states, which the Kabaka totally opposed on the grounds that it would destroy their culture and identity. The Bugandan Parliament sought complete independence, which despite several weeks of discussions with the Governor of Uganda Sir Andrew Cohen, they refused to give up. An ultimatum was issued to the Kabaka who was then arrested and exiled to England where he chose to live in The Savoy Hotel, London. Although strong views were held on both sides, there was no bloodshed, and eventually an agreement was reached in 1955. King Mutesa returned to Kampala having achieved his goal of protecting Bugandan identity within the greater Uganda. The disturbance at Mulago proved to be minimal for Denis and the staff, but the Bugandan problem was not fully solved and became acute again when Uganda, a protectorate,[9] gained independence from the UK in 1962 and became a republic in 1963. Changes were to follow in the running of the health services, which would lead to Denis moving out of Mulago and eventually back to the UK in 1966 (Fig. 8.8).

Meanwhile Denis' duties included paying regular visits to hospitals in areas that were far removed from the concentration of people and skills that were present at Mulago and the European Hospital in Kampala. They were always interesting trips.

Fig. 8.8 Main street Mulago, early 1950s. Post office steps on R and in the distance Namirembe Cathedral

---

[9] Protectorate, in distinction from a colony, is a region or country controlled and protected by a larger country but wherein, for example, the ex-patriot community was not permitted to own property.

Denis set out for a Northern Province safari on April 11, 1954, on his own. He drove north to Nakasongola on the shores of Lake Kyoga and then west to Masindi where he stopped for the night sharing a room in the local hotel with a Polish barrister who had been deported to Siberia, escaped through Central Asia to East Africa and was now a dealer in crocodile skins. The next day he headed towards Lake Albert, but at the Pakwach ferry across the Nile as it emerges from the Lake, he was held up for 2 1/2 hours because there were a lot of elephant around. The following day he spent at his first main destination, the hospital at Arua in the far north-west of Uganda on the border with the Congo. He stayed there for 3 days assessing the surgical services and probably operating on some difficult cases. He met District Commissioner Gibson and that evening went to Bible study at the mission station in Kuleva (Kuluva) where he met the Williams family. In Kuluva, Ted and his brother Peter, both of whom were doctors, and their father, an engineer, had built a hospital under the auspices of the Africa Inland Mission (AIM) (Fig. 8.9).

With work done Denis went walking in the bush looking for rhino with Ted, but they found only tracks so decided to drive to the Rhino Camp, a protected sanctuary on the border with South Sudan where in the evening Denis, ever the keen photographer, showed the guests some of his new colour transparencies. Ted and Denis enjoyed an early morning game drive on the Saturday, and on Easter Sunday Denis spoke both at the morning service, which was attended by lepers, and at the evening service for Europeans ending the day with a prayer meeting at Ted's house. His relationship with Ted Williams would be an abiding one. From Arua he drove to

Fig. 8.9 Hospital at Kuluva built by Ted and Peter Williams

Gulu but was very disappointed with the hospital. They were doing very little surgery, so he headed south to Lira where he had a more interesting time. He met the Tse Tse fly officer in the bush and had buffalo for supper with him. Friday was spent operating all day, and on Saturday he returned to Mulago where he received a wonderful welcome from Olive and was soon back on his own wards (Fig. 8.10).

Olive's welcome may have been something of a reassurance for him. She had been having the occasional week when she was unwell again, and although nowhere near as depressed as she had been a year earlier, she was seeing her psychiatrist Dr George Campbell-Young regularly. In fact, he had become a frequent visitor dining with them and attending their church and had developed appendicitis, which Denis had to treat. Denis began to feel that George was in danger of getting overfond of Olive, so he spoke to her about this and felt much better after they had discussed it. Some weeks later, after George had been to supper with them, Denis' anxiety about the relationship returned. George, being a psychiatrist, became aware of Denis' anxiety and reassured him by saying that Olive had helped him spiritually. Shortly after this Olive took George to Entebbe airport to see him off for a trip, which left Denis very worried. 'This was one of the unhappiest days of my life. I had been labouring under a growing anxiety over George's very strong attraction to Olive. I wrote dear Olive a note explaining how I felt, finding it hard to say. She seemed to resent the note and there seemed to be an estrangement between us, which was most painful. I am weary with worry'. Relationships soon improved although George kept coming to tea.

The visit of the Queen and Duke of Edinburgh to open the Owen Falls Dam and hydroelectric scheme (Nulabaale) on April 29, 1954, provided the excuse for a family outing. Denis and Olive were invited to the Royal Garden Party, and they all

Fig. 8.10 Denis doing ward round at Mulago with sister and junior doctor. Early 1950s

stayed to see Her Majesty and the Duke board a plane to take them to the Kazinga National Park,[10] which they would also visit later in the year. There were frequent visitors to the Burkitt home, mainly of the ex-patriot community, for dinner now preceded by a glass of sherry. Denis' skills as a surgeon were in demand with these people, and a dialogue developed between him and Dr Hennessy the Director of Medic Services. He asked Denis to spend more time at the European Hospital (EH) and less at Mulago, which was a challenge to what Denis felt was his true calling to serve in Africa. Eventually the DMS wrote a very sympathetic latter to Denis saying that he would relieve him of his administrative duties at the EH and thus presumably give him more time for clinical work there.

As the months went by, Denis' life became more settled. He remained very busy with his clinical duties in Kampala and went on regular safaris to hospitals in the major regions of the country where he would quite often do surgery that had been waiting for his wide experience and skills. He was no longer in pursuit of the cause of the diseases he had to treat and published no research papers in 1955, 1956 and 1957. Now in his mid-40s, he was treading water. It is hard to believe, and certainly Denis had no inkling of this, that over the next 10 years, he would become known as one of the great names in cancer research and be feted wherever he went in the world. Meanwhile the family occupied more of his time. The girls were growing up, and Olive required regular support although her episodes of depression were usually short-lived, cyclical in nature and managed by a day or 2 in bed. They entertained frequently, trips to the cinema were now in order with dinner at a restaurant or the Uganda Club where Denis had been made a member, and they went on short recreational safaris often with the Drowns or Timmises. Mubende, west of Kampala, was a favourite destination where Guy Timmis was in charge of the hospital and would join them on safari to Lake Albert. In Mulago Olive made her own life, became a member of the Uganda Council of Women, helped at the YWCA,[11] took a weekly Sunday school class at the English-speaking church, which became affiliated with the Crusaders, ran a senior Girl Guide company for nurses at Mulago Hospital and was a member of the Young Wives Group belonging to All Saints Church. A 3-week family holiday to the east coast in early August took in the Amboseli Game Reserve, south of Nairobi where they met William,[12] and in Tanganyika (Tanzania) where they visited Arusha, Moshi and the coastal towns of the Indian Ocean including Tanga, then back through Mombasa. It was a success as holidays go with remarkable varieties of game roaming wild across the countryside, although nothing out of the ordinary for a family working in Uganda to encounter. It must have been memorable because the trip prompted Denis to type a 16-page account 'Coastal Leave', for his parents, James and Gwen, and moreover Olive, quite exceptionally for her, wrote a 66-page record, 'East Coast Holiday', with map.

---

[10] Kasinga National Park, renamed after visit of Queen Elizabeth in 1954.
[11] YWCA, Young Women's Christian Association.
[12] Dr William Burkitt. Son of Uncle Roland Burkitt.

Home leave was taken from March 22 to June 14, 1955, and again from November 22, 1956, to February 11, 1957. These journeys always took them to Laragh where James and Gwen would provide the base from which they could all revisit places that were a comfortable reminder of earlier years. Parting from them at the end of the holiday was now accompanied by tears because by 1956 James was 86 and Gwen 80. But they would live until 1959 and 1960, respectively. Denis and Olive had already decided to change the shape of their time in the UK. In 1953 Professor Arthur Rendle Short who had been Professor of Surgery at Bristol University, and owned a cottage in Padstow, Cornwall, had died, and word was passed to the Burkitts in Uganda that the house was now free. They rented it for a year and made their first visit to it during home leave in 1955. It provided an ideal base for holidays in the south of England from which Denis was able to reach London more easily than from Enniskillen, and to visit Robin, who was still waiting for a consultant post. Denis made use of his time whilst on leave to visit the main surgical centres and was now happy to give talks to students and staff using his new colour transparency slides. He was a very good speaker having honed his skills in the pulpit, at Bible study classes and at evangelical meetings. It was not long before he was receiving invitations from the major London medical schools who were interested in his African experiences. On his 1956 leave, he gave talks in London at Great Ormond Street Children's Hospital, St Marys, the Middlesex and Hammersmith Hospitals.

Back in Kampala Denis continued with activities related to his faith one shared also by Hugh Trowell, who was a good colleague. After lunch one day, Trowell told Denis that he was making plans to be ordained once he retired from government service. Denis felt this was a challenge to him also and discussed it with Olive who was remarkably receptive to the idea. Hugh was to become ordained in 1959 but Denis never pursued this path (Fig. 8.11).

Hugh Trowell (1904–1989) was an important person at Mulago. Denis had already met him on the train heading for Nairobi in March 1946 when he was en route to collect Olive arriving in Mombasa with baby Judy. Trowell was 7 years older than Denis and already senior physician and paediatrician having moved there from Kenya in 1935. They were colleagues, and after leaving Africa, Hugh in 1958 and Denis in 1966, they would work closely together on the importance of dietary fibre for health with Hugh providing a crucial physician's viewpoint, in addition to his writing and editorial skills, to complement Denis' surgical outlook. Hugh had studied medicine at St Thomas's Hospital Medical School. During his time there, he became President of the Student Christian Movement (SCM) and through this met his future wife Kathleen Margaret Sifton, known as Peggy. She was a very talented artist and founded the Margaret Trowell School of Fine Art at Makerere College in Kampala [6]. Hugh graduated in 1928 and was awarded the Mead and Bristow Medals. Sixty years later St Thomas would inaugurate the Hugh Trowell Lecture in his honour. Arriving in Africa in 1929, the Trowells were posted up country to a small town, Machakos, in the Kamba reserve in Kenya. Conditions in the local hospital were very primitive, and Hugh was responsible not only for the patients but also for training the African nursing orderlies. He was an energetic and literate man who set about writing a training manual *Handbook for Dressers and Nurses in the*

Fig. 8.11 Dr Hugh Carey Trowell (1904–1989) circa 1950. (By kind permission of Thorold Masefield)

Tropics [7], which became widely used in many African countries. He did pioneering work on the treatment of kwashiorkor [8] with powdered milk, writing the definitive book on it for which work he was awarded the OBE in 1954. Hugh was also very observant of the nature of other diseases he saw in the hospital and clinics and like Denis sought to understand their cause. His book *Diagnosis and treatment of diseases in the tropics* [9] is valuable not only for its practical approach to managing these conditions but also for the notable absence of any reference to the emerging diseases of 'Western' countries such as coronary heart disease, diabetes, obesity and various cancers. Moreover, there is virtually no mention of diet anywhere in the book. Forty years later the names of Burkitt and Trowell would be synonymous with the hypothesis that these conditions were due to a type of diet that contrasted with the diet and disease patterns they had seen in Africa [10].[13]

For Denis his life was about to become less ordinary. An event early in 1957 would be a turning point, a moment for which Hugh Trowell was responsible.

References

1. Lyons M. Sexually transmitted diseases in the history of Uganda. Genitourin Med. 1994;70:138–45.

---

[13] Wellcome Library GC/198/B/2/6. Bray E. Hugh Trowell:Pioneer Nutritionist 1904–1989. Wellcome Library, Papers of Hugh Trowell, PP/HCT/A.5.

 2. Royal College of Surgeons. Plarr's lives of fellows online https://livesonline.rcseng.ac.uk/
    client/en_GB/lives/search/results?qu=Burkitt&te=ASSET
 3. Stephen AM, Cummings JH. Mechanism of action of dietary fibre in the human colon. Nature.
    1980;284:283–4.
 4. Burkitt DP. Acute abdomens – British and Baganda compared. E Afr Med J. 1952;29:189–94.
 5. Burkitt DP. A simple serviceable artificial leg. E Afr Med J. 1953;30:177–91.
 6. Trowell M. African tapestry. London: Faber and Faber; 1957.
 7. Trowell HC. Handbook for dressers and nurses in the tropics. London: Sheldon Press; 1937.
 8. Trowell HC, Davies JNP, Dean RFA. Kwashiorkor. London: Edward Arnold; 1954.
 9. Trowell HC. Diagnosis and treatment of diseases of the tropics. London: Balliere Tindall
    Cox; 1939.
10. Burkitt DP. The founders of modern nutrition Dr Hugh Carey Trowell OBE MD FRCP
    (1904–1989). London: The McCarrison Society; about 1990.

Chapter 9
Jaw Tumours, Kilimanjaro and Looking Down at the World

In Denis' own words [1]:

> One memorable morning in 1957, Dr Hugh Trowell, the physician in charge of the pediatric ward at Mulago Hospital, who was by that time well-known for his pioneering work on Kwashiorkor, called me, the surgeon on duty, for consultation on a child. His face was massively swollen, with bizarre lesions involving both sides of his upper and lower jaws. I had never seen anything like it. The teeth were loose and the features grossly distorted. If a single jaw quadrant had been involved, I might have considered it to be an infective process such as osteomyelitis, but not with all four quadrants affected. This unusual distribution also seemed to rule out any form of neoplasia (cancer). Results of the biopsy had suggested some form of granuloma. I was totally baffled but photographed the child and considered this to be another of the curiosities one had become accustomed to seeing from time to time in Africa.

The child, a 5-year-old boy called Africa, died soon afterwards. The bedside meeting with Hugh Trowell was probably within a week or Two of Denis' return from leave in February 10 (Fig. 9.1).

> A few weeks later, I was doing a ward-round in another hospital which I visited regularly. Looking out of the window, my attention was attracted by a child with a swollen and distorted face sitting on the grass with its mother. I immediately went outside to look at him[1] and, to my surprise, recognized precisely the same features as I had observed in the Mulago Hospital ward a few weeks previously.

'A few weeks later' was March 14 when he was doing a teaching round in one of the surgical wards of the Jinja Hospital, located where the Nile leaves Lake Victoria near the Owen Falls Dam. It was where he and Olive had seen the Queen inaugurate the dam in 1954. The boy he saw out of the window proved also to have tumours in all four parts of his jaw, and Denis realised that this was a similar case to the one that Hugh Trowell had asked him about. He drove the child and his mother back to Mulago where further examination revealed that he had tumours in his abdomen. A

---

[1] Autobiography Chapter 3 says 'her'.

© The Author(s), under exclusive license to Springer Nature Switzerland AG 2022
J. H. Cummings, *Denis Burkitt*, Springer Biographies,
https://doi.org/10.1007/978-3-030-88563-2_9

Fig. 9.1 Child with
'sarcoma' involving all
four quadrants of the jaws.
Taken 8 weeks after the
appearance of the tumour.
Original photo by DPB in
family archives and in
Burkitt D. A sarcoma
involving the jaws in
African children. Brit J
Surg. 1958; 58:218-223.
(With the permission of
John Wiley and Sons)

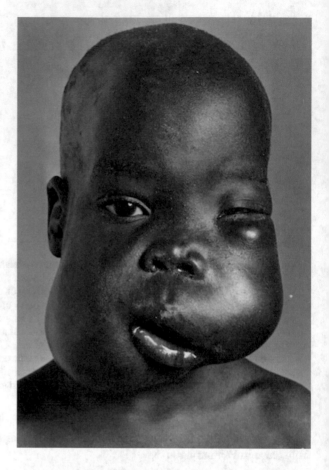

critically important novel observation. Seeing these two boys changed the course of
Denis' life, the outlook for children with this tumour and brought a new dimension
to cancer research.

The occurrence of jaw tumours in young people had been recognised for over
50 years in Uganda. An early description of the condition can be seen in the case
notes of the pioneer Dr Albert Cook[2] who had arrived at the port of Mombasa on
October 1, 1896, with a party of 12 Church Missionary Society people bound for
Uganda. They journeyed overland arriving in Kampala 4.5 months later and had set
up a clinic in a small shed at Mengo within a few days. Soon they were overwhelmed with patients, and Cook obtained the promise of the Katikkiro, or prime
minister, that a hospital would be built. By the end of 1897, they had 3 wards with
28 beds, an operating theatre, a store and dispensary. The first year they saw 17,000
patients and Cook, the only qualified doctor, undertook 314 minor and 140 major
operations. The standards of medicine were exceptionally high, and Cook kept

---

[2] Sir Albert Ruskin Cook (1870–1951).

detailed clinical notes on all patients admitted, which were preserved and eventually bound and indexed [2].

These notes, from 1897 to 1956, were read and analysed by Professor JNP Davies, Professor of Pathology at Makerere, and his colleagues as part of their study of the patterns of cancer incidence in Uganda, published in the *British Medical Journal* in 1964 [3]. They cite Cook as stating that in 1901 'cancer in Uganda was common, jaw tumours being particularly frequent'. Davies and colleagues' careful analysis of these records shows that between 1897 and 1956, there were 83 cases of jaw tumours reported being 8.6% of all cancers but the most common cancer in children (Fig. 9.2).

The case notes of Jejefu Omukalazi, admitted on May 18, 1910, are particularly informative.

> Disease. Sarcoma[3] of jaw.
>   Swelling began underneath the jaw one month ago. The eye began to be affected two weeks ago.
>   Present Condition. The growth appears to be nearly half the size of the childs head. Its characters are:
>   It is firm and elastic. Fixed to the left side of the jaw. Extends upwards to zygoma.[4] Downwards nearly to clavicle. To the right it extends beyond midline. The gum is involved on the left side and the teeth are displaced and dropping out.
>   The child was carefully fed and cleaned up for one week.
>   May 25th. Operation – child died on the table soon after the operation had commenced.

It is hard to believe that Denis had not come across similar cases. He had been in Uganda for over 10 years, had travelled extensively and had become one of the principle people to whom difficult surgical problems were referred.

The list of operations that he had performed (Table 1, Chapter 8) records one case of a tumour of the mandible whilst he was in Lira and two similar cases during his first year in Mulago, 1948/1949. A blank page in his 1956 diary, which has entries only up to April 2, Easter Monday that year, has a simple drawing in red biro of a patient with a probable jaw tumour, skin lesion on the forehead, enlarged liver and spleen and other abdominal lumps (Fig. 9.3).

What must have impressed Denis in the two cases seen early in 1957 was the presence of tumour in all four parts of the jaw in both children, a distressing sight even for a surgeon inured to the effects of cancer, and, most importantly, the occurrence also of tumours in the abdomen. These two children made him stop and think. He could not just dismiss two almost identical but graphic clinical presentations with such catastrophically awful outcomes. They made him realise that this was a problem he must try to solve. He started, as he had done in the past with the problem of hydrocele, by finding out as much as he could about the tumours. His first step was to go back through the hospital records looking for similar cases, from which he found that the tumours were occurring in children mostly between the ages of 2

---

[3] Tumours of the connective tissue, which occurs between cells, binds tissues together and includes bone.

[4] The boney ridge that can be felt running forwards from in front of the ear above the cheek.

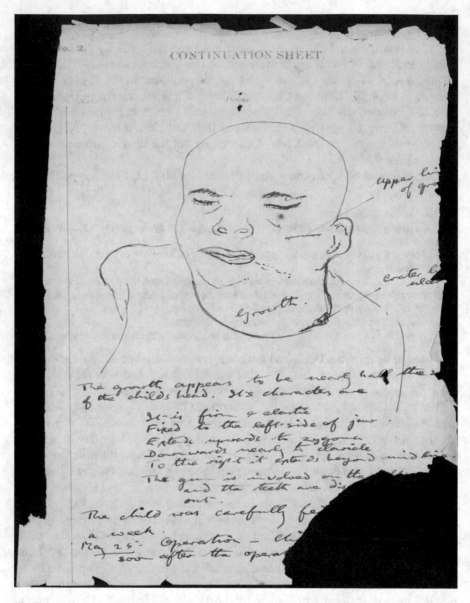

Fig. 9.2 Drawing and case notes, likely by Albert Cook, of patient Jelefu Omukalazi with a jaw tumour. May 18–25, 1910. Mengo Hospital, Kampala, Uganda. (Courtesy of the British Library, EAP 617/1/60 and the Albert Cook Medical Library, Makerere University)

and 10 years old. Within 3 months he had enough information to present his 'Jaw tumours' to his colleagues at the Tuesday afternoon clinical meeting after which Dr Hutton, at that time chair of the Saturday staff meeting, asked Denis to present the

Fig. 9.3 Drawing by Denis of patient with jaw, liver, spleen and other tumours in diary for April 12, 1956

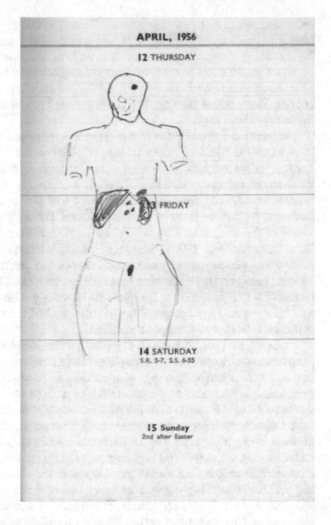

cases to them, which he did in June 1. Denis would go on to present his work on the tumour countless times over the ensuing 35 years.

To proceed any further with his research, he needed the help of the Pathology Department to give him information from autopsies carried out on patients that had died from the tumour and histology[5] from biopsies done in clinics and on the wards. He needed Professor Davies (Jack) with his longstanding interest in cancer in Africa and who, with Dr H J Croot, the first professor of surgery at Makerere, had in 1951 set up a committee to start surveying and recording cancer incidence in Kampala [4]. Davies believed that cancer was different in Africa and that when considering its cause, the incidence and behaviour of it in Africans, as well as people in other

---

[5] Microscopic examination of tissue using various staining techniques to identify the nature of cells involved and their disposition.

countries, should be taken into account. He had a good record of research and col-laboration with colleagues in Kampala and by 1957 had already published 47 papers, letters and case reports including work on kwashiorkor with Hugh Trowell. He was a specialist in the study of endomyocardial fibrosis and of cancer, especially liver cancer although not specifically on the jaw tumours. It is possible that he did not hear Denis talk at the Saturday staff meeting because he was either on leave in England or about to go.

The question that needed to be answered was the nature of the cancer at a cellular level so that it could be given its proper place in the classification of tumours. Because of its frequent involvement of the adrenal glands,[6] widespread organ involvement and occurrence in children, it was initially thought that it might be a neuroblastoma,[7] and histology slides from 14 of Denis' cases were sent to Prof Davies, at the time on leave in Worthing, Sussex. Davies was able to show the slides to one of the world's leading authorities on neuroblastoma, Dr Martin Bodean at Great Ormond Street (GOS) Children's Hospital in London. He had seen more neu-roblastomas than anyone else and was of the opinion that these tumours did not fall into this category. He had never seen anything like it at GOS but thought they were more like a lymphosarcoma. He made various suggestions to follow up including that the epidemiology (geographical distribution) of the tumour be studied, with particular reference to sources of radiation.

Radiation was not an issue in Uganda, but Denis did not at this stage rule out neuroblastoma, despite the expert opinion from London. But he now had informa-tion on 20 cases that had come to post-mortem all of whom had involvement of the liver, kidneys, adrenals and pancreas with or without jaw tumours. He considered this to be a single clinical condition with the implication that all the tumours would have a similar pathology and similar cause. This important concept, that all these tumours in the body were part of the same condition, was the first step towards understanding the cancer and is clearly and rightly credited to Denis. Moreover, in characteristic style he had started plotting where they had occurred geographically and observed that 'They come from every part of the country, even from over the border in Western Kenya and also from Rwanda'. There was no simple geography at this stage. He enquired of major children's hospitals around the world and found from their experience of this condition that it was very uncommon. Great Ormond Street, on reviewing their records, had seen 6 cases in 6 years, Paddington Hospital for Sick Children 4 cases in 20 years and the Mayo Clinic in the USA had 20 cases with histology over a similar period. Childhood leukaemia was much more of a problem. He received encouragement for his investigations from John Croot and Ian McAdam whom Denis recorded as saying 'that when I get my teeth into something that aroused my interest I was like a terrier with a rope and nothing would induce me to let go till answers were found for the questions that had gripped me' (Fig. 9.4).

---

[6] Small glands above each kidney, which produce hormones including adrenaline and steroids.
[7] Arising from specialised cells of the nervous system.

Fig. 9.4 Sir John Croot (1907–1981), first professor of surgery in Uganda and Minister of Health from 1958 to 1961

Relations with the Department of Pathology, and especially with Professor Davies, then deteriorated.[8] After a difficult session with him in October, Denis wrote to Professor Croot saying he had discussed the 'jaw/abdomen syndromes'[9] with Professor Davis putting forwards the facts and arguments on which he, Prof Croot and Ian McAdam had agreed earlier but that these were not received sympathetically. A key issue was that Professor Davis was unwilling to accept the arguments in favour abdominal, jaw and tumours at other sites being part of a single syndrome. Davies thought that if they were published together, no sensible person would accept them and was apparently 'considering the publication of the jaws as a separate entity under signatures chosen by him'. This did not please Denis at all. The idea that the jaw and other tumours were all part of the same condition was essential to Denis' new concept. The notion that Davies might write a paper without Denis' name on it he saw as an attempt to poach what he knew was work that he had originated, presented at the staff meeting earlier in the year and was actively pursuing. Denis had since then collected and scrutinised all available records and had sought assistance from district hospitals. 95% of biopsies had been examined by the government pathologist not Professor Davies. Denis wrote to Davis saying 'Arguments have been developed which would appear to gather into a single syndrome several categories of cases, previously of uncertain etiology (cause). May I suggest that we,

---

[8] Correspondence in family archives.

[9] Syndrome, a collection of symptoms, signs and pathologies that amount to a single condition.

representing the Surgical Division, are prepared to invite and accept co-operation in this work, but are not prepared to let any department publish or claim credit for what is apparently an original observation'. Gone were the days of Denis tentatively seeking a role in the world. This was his territory and he was going to defend it against even the most senior members of the hospital hierarchy.

November saw Denis off on another 12-day duty safari to the Northern Province. He enjoyed these trips. Away from the busy day-to-day activities in Kampala, he could relax. He liked driving especially on the almost deserted roads in the far north, which were often just cut out of the bush with the surface of natural soil and sand. The peace of the countryside, the varied landscapes, which included Lake Albert, the mountains of the Belgian Congo (DRC), the rivers of the Nile and the wild animals all helped to relieve the stress and anxieties of his daily life at Mulago. At one point hydraulic fluid began to leak from the car's breaks meaning that they were no longer effective. This didn't worry him at all. He saw only three other cars that day and was able to use the handbrake when required. The journey was over a thousand miles in 11 days during which he was accompanied by Mr Savdhan, an Indian evangelist who intended to bear witness to the Christian faith amongst the Indian community. The journey took them west to Hoima, north, using the ferry on Lake Albert from Butiaba to Pakwach and on to Arua to call on the Williams, then into the far north at Moyo, south to Gulu, which he had not visited since his army days, and then to Kitgum and finally Lira. He visited eight hospitals where he did ward rounds, and at Hoima, Arua and Lira, he was busy operating arriving home in November 16. These journeys were something of an ego trip for Denis. He was now one of the most renowned and experienced surgeons in Uganda and was able to tackle the more difficult surgical cases for the local staff, advise on patient care and meet officials such as the district medical officer. The trip was funded by the government, so he stayed in hotels or rest houses, which were basic but where he had the luxury of his own bathroom and toilet. Having breakfast in the rest house at Gulu, he sat next to a senior police officer whose wife recognised Denis. She had been a patient of his in earlier times. In conversation Denis asked her where she came from and she replied 'Enniskillen'. Her father had run the local jeweller's shop. That was clearly an encounter worthy of a note in Denis' usual typed account of the journey. Now absent are labels attached to people like 'good Christian' or 'with a deep faith'. Denis' own faith had matured, so whilst he would attend such church activities that were happening in the places he visited, the giving out of copies of the Gospels and pamphlets with a religious message in the hope of winning souls was no longer important for him.

And there was a recreational component to these safaris. In the Northern Province lived the Timmises, Guy and Dawn, who ran the mission hospital at Masindi, 30 miles north-east of Hoima, and also the Williams, Ted and Peter, at their mission hospital in Kuluva, just south of Arua. Denis played a round of golf, 12 holes, with Guy at Hoima where there was no fee to pay, no need to book or ask permission and no-one else playing at the time. But they had to caddy for themselves. Three days later, after a busy Friday operating at the hospital in Arua, Denis set off with Ted, Peter and another friend from the Department of Geology for a day in the Belgian

Congo (DRC). They left early and were able to cross into the Congo about 50 miles south of Arua at Mahagi and then further south drove up the mountains to about 3000' from which they had magnificent views of Lake Albert to the east and to the north the Victoria Nile entering at the most north-easterly part of the lake and the Albert Nile leaving from the northernmost end. With a telescope they were able to see the Murchison Falls they thought, which must have been over 50 miles away. A picnic lunch comprised bully beef and baked beans on bread after which they climbed one of the peaks because Denis was anxious to get some exercise to compensate for his sedentary life in Kampala and in preparation for some leave at the end of the year when he would climb Mt Kilimanjaro.

Meanwhile, back in Kampala, to try and manage the ongoing problem of the dispute between Denis and the Pathology Department, Professor Croot wrote to Professor Davies, with a confidential copy to Denis, saying:

> Dear Jack,
>
> I am anxious that Denis Burkitt should get his observations on jaw tumours in children published soon. The clinical aspect is already complete enough for full scale publication and I should be sorry to see him forestalled by someone elsewhere who had collected a large series, especially if an association with abdominal malignant tumours had been noted.
>
> I know you have views on this latter aspect…Nevertheless, from the clinical point of view, something significant has already emerged, entirely due to Denis…for which he ought to get the credit.
>
> I cannot think that publication of a clinical series would in any way prejudice the subsequent publication of detailed pathological studies…
>
> Denis would, of course make full acknowledgement of the help of the Cancer Registry
>
> yours sincerely
>
> December 24th 1957

There can be no doubt, reading this letter, that the lymphoma syndrome was first recognised by Denis and that he rightly deserved credit for bringing it to the attention of the world, and more especially for the subsequent work he did on its geographical distribution, cause and treatment. Professor Davies had legitimate claims to be an expert, even a pioneer, in describing and recording cancer in East Africa but had not recognised Denis' tumour syndrome [5] and was probably somewhat miffed at having it brought to the attention of the medical school by someone he viewed as just a hardworking surgeon with no reputation at all as a scientist (Fig. 9.5).

For 3 weeks after Christmas 1957, Denis and family were on leave, visiting Kenya and Tanganyika. It was a time for Denis to recharge his batteries for the battles ahead. The ultimate aim of the safari was that Denis and companions would climb Mt Kilimanjaro, the highest mountain in Africa at 19,341' (5,895 m). The party,[10] apart from the Drowns, met in Nairobi at the end of the first day having driven 410 miles from Kampala. The road between Tororo on the Uganda/Kenya border and Nakuru rises to 9000' at the equator before descending into the Rift

---

[10] William Burkitt (son of Roland Burkitt), Dick and Gwen Drown and their daughter Jenny, Ivor Chance a surgeon at Mengo, his wife Lorna and their three children Carolyn and two boys about 5 years old and the Tewfiks, Gerry the psychiatrist, Tessa his wife and their daughter Cherry Ann and the family of five Burkitts.

Fig. 9.5 Photograph of Mulago Hospital staff taken for Mr A J Boase on his retirement. June 1956
Middle row left, Mr Denis Burkitt
Front row from left, Miss Murray, Prof John Croot, Dr Hugh Trowell, then centre Mr Boase and,
at the far right, Professor Jack Davies

Valley and on to Nairobi. Denis described the drive as the easiest they had ever
done. On the Sunday they all went to the Cathedral then drove to Namanga on the
Kenya/Tanganyika border where they met the Drown family, bringing the party up
to 17. Deciding that the women and children should rest, the men set off for the
Amboseli Game Park at 6 am the next day armed with cameras. The range and numbers of game they encountered and their proximity to William's Land Rover were
exceptional leading Denis to remark that it was one of the most interesting and
exciting days he had ever spent.

The next day they drove in convoy to Arusha, a large town in the north of
Tanganyika where they bought enough provisions, for £15, to sustain 17 of them for
the 3 nights they were to spend in the Ngorongoro crater, part of the Ngorongoro
Conservation Area, a world heritage site. Denis left his car with the DMO and joined
William Burkitt in his Land Rover for the 70-mile journey to the crater where there
was accommodation in log cabins at 7,500' with views across the crater, 2000'
below, to the far rim 20 miles away. Supper, prepared by the men, comprised soup,
cold meat and salad, bread and jam and tea (Fig. 9.6).

The next morning the Kilimanjaro party, which comprised Denis, Dick Drown,
Ivor Chance, Gerry Tewfik and William, climbed down into the crater where Denis
describes a wonderful experience walking amongst great herds of game. He climbed
back up to the log cabins apparently without difficulty, despite the altitude, and
arranged for the whole party to descend into the crater the next day by Land Rover,
a journey of 27 miles that took them 2 1/2 hours. In the crater 'we found ourselves
in the thickest and vastest concentration of game I have ever known'. They were
able to approach lion quite closely on foot for photographs. 'These animals take no

Fig. 9.6 Five men en route to Kilimanjaro. January 1958 just before Kilimanjaro expedition. In Amboseli Park, Kenya; L to R, DPB, Dick Drown, Gerry Tewfik, William Burkitt, Ivor Chance

notice of one at all. They treat you with complete contempt. We had a most interesting ten minutes in the midst of them and then went off for our lunch'. They returned via Arusha to the Kibo Hotel in Moshi where everyone relaxed for 3 days prior to the four men, Gerry Tewfik declined the expedition, setting off to climb the mountain on Tuesday January 7, 1958.

The Kilimanjaro party included a guide (Bwana Thomas), cook (Herman) and nine porters to support them. Food, blankets and eating utensils for the expedition were provided by the hotel who had booked huts on the mountain for them all. They left at 9:45 am, and after what Denis described as an easy climb, a steady ascent to 9,000' over 10 miles, they reached the Bismarck hut 5 hours later. The hut was dirty and there was litter around outside which they cleared up using the points of their walking sticks. The bunks were bare boards with a ½" thick rubber mattress covered in hay. They had four blankets each. Denis, typically, had brought with him a thermometer and altimeter. Supper was soup, liver, potatoes, carrots, coffee, bananas, bread, jam and cheese. By 7 pm it was dark, the temperature outside was approaching freezing, and they decided to turn in, all of them taking a sedative to help them sleep. William read to them from the *Daily Light* and Dick led them in prayer. Of the first day, Denis wrote 'Below us on the plains little wisps of smoke curl up here and there. We are above civilisation, alone and away from all the problems of our work. There is something very curative about being right away from things like this'. Perhaps a defining moment for Denis.

On Wednesday they walked up to 11,000' where they rested and picked flowers familiar to them from home such as anemones, harebells and violets, before climbing to Peter's hut at 12,300'. They once again cleaned up and replaced the hay with heather and changed into warm clothing as clouds gathered and the temperature fell to 38⁰F. Supper was soup, potatoes, cauliflower, gravy, chicken, rice, apples and bananas. Denis was very pleased that he had been less breathless than the others during the day's walk and remarked 'It is wonderfully enjoyable - absolutely 100%. We are enjoying every bit of it'. During the night they were joined by a veterinary student who left to continue his descent the next morning. They stayed at the hut for a full day on Thursday to try and acclimatise walking up to about 14,000' in the morning and on the return met several members of the British Kilimanjaro expedition, including the leader John Tunsell, who were doing research on the mountain. That evening two South African students arrived at the hut and would join them on the ascent. In the hut next to them were three girls who were from teacher training college. Friday was important because they needed to ascend to the Kibo hut ready for their attempt on the summit the next day. They were already above the clouds, and with the sun shining on the mountain, Denis felt the whole trip had been worth it just for these views. By 11:00 am they were on the saddle between Kibo and Mawenzi, two of the three volcanic cones that make up Kilimanjaro with Kibo the highest. The terrain by now was loose stones covered with volcanic ash, but the peak of Kibo was covered in snow, a challenge for the next day Denis thought. They reached the hut, 15,500', at 2:45 pm where there were six bunks in a room 12x10'. Denis felt good and was pleased at his fitness. 'I am thankful that I have felt perhaps the fittest of the party today, and carried the pack belonging to the South African students. I still reached the hut first'. Although the guides warned them that they might not feel hungry at this altitude, they had a big tea/supper at 4:30 pm and then put on all the layers of clothes they would be wearing for the ascent before going to bed at 5:30 pm ready for a 1:30 am departure the next day.

Saturday, before leaving the hut, they had some tea and put Vaseline on their faces, and the party, four men and two girl students, with the two guides were soon starting to climb.

> The going was terribly hard. First steep very loose shale, volcanic ash. It was like walking up a sandhill, half a step back for one forward. With lack of oxygen we took slow short steps and had to stop every 50 paces. Later I was stopping after 2 or 3. It was too much of a grind to talk, we just plodded on. Before very long our guide suggested that I might go ahead of the others. I struggled doggedly on and after 6 hrs plodding and panting I reached the summit with a European from another party at 7.30am feeling I really had had enough. The sun had risen about 6. The views across the crater with ice and snow walls were magnificent. I took several photos, and then we started down. (Fig. 9.7).[11]

> Not only going up but also descending was a very slow process, for one can very easily slip on the frozen snow.

---

[11] Personal diary (dated 1956).

Fig. 9.7 Denis and climber from another party at the summit of Mount Kilimanjaro, Saturday January 11, 1958

> Ivor got more than halfway up this final snow face and I met him (as Denis descended) utterly exhausted. He might even then (have) struggled further but I felt he should turn round and he agreed. This proved a right decision, for he found that he had virtually lost the use of his legs, and although he realised the danger of sliding down on his back he felt there was no alternative as he could'nt (stet) walk. He was thoroughly frightened and wore out the seat of one pair of trousers but got down safely, a good way ahead of me. I passed him later while he rested for a time in the cave. William turned back exhausted, I think about half way up the snow, and found Dick, as he says, not quite knowing what he was doing. William wisely told Dick to return and helped him down. Dick cant quite remember what happened.[12]

They were given tomato juice and orange juice on their return to the Kibo hut and lunch but then had to walk 8 miles back to Peter's hut. Denis felt tired but made the effort and they were all back at the hut by 4 pm. They had not had any water except for drinking whilst on the final ascent, but Denis was now able to wash his hands. They were also running out of supplies but the members of the British Kilimanjaro expedition left them some biscuits. In bed by 6:30 pm, they rose early on Sunday morning to enjoy bacon and eggs, porridge, bread and tea before leaving for the trek to the Bismarck hut. Looking back at the mountain, which had been a trial to them the day before, Ivor, William and Dick said that it had been one of the worst experiences of their lives, but Denis had not found it so bad and now felt fit. They reached the hut by 11:10 am, picked some flowers for their wives and had lunch after which

---

[12] Account written on January 13, 1958, at the Kibo Hotel, Marangu.

the three who had made it to the top of the mountain, Denis and the two students, were garlanded with wreaths made out of the wild flowers by the guides. A local tradition. They were back to meet the waiting wives by mid afternoon, after 6 days on the mountain, and spent the evening telling the story of the expedition and enjoying their first bath and shave for a week. Also staying in the hotel were the members of a large expedition from a Japanese university. They had a cine camera, the first Denis had seen, and Japanese 'motor vehicles superbly equipped…brought all the way from Japan…with refrigerators and every conceivable modern device.' Another new experience for him.

Denis' final remarks on the trip, written on the day of their descent were:

> The 22 mile walk down-hill today was more tiring than any of the individual stages up to Kibo. We are so thankful for the weather, and for the happy fellowship and companionship we had together. Surely one of the things that count most in life is relationships.

Back at Mulago on January 22, 1958, Denis was refreshed and with a new confidence. His fitness for the expedition to Kilimanjaro had exceeded his expectations; he had done better than his colleagues on the mountain and had seen the world down below him from a new perspective. Being 'above civilisation, alone and away from all the problems of our work', he had found very curative. It was a different Denis who came back down the mountain. He wrote immediately to Prof Davies saying that he was not happy with the situation regarding the jaw tumours and suggested that they meet to talk things over. He writes 'I must frankly admit that I have interpreted your attitude as claiming to "adopt" this work, and make it equally, if not primarily a pathological rather than a surgical and clinical entity' resenting the suggestion that they, together with Prof Davies' brother Glyn who was a radiologist at Mulago, write a joint paper. Denis saw the jaw tumours as a clinical syndrome that he had described and put a lot more time and effort than anyone else. He was not prepared to hand it over to anyone to claim the credit.

An immediate reply came not from Prof Davies but from Glyn, whom Denis knew well. Having likely discussed matters with his brother, Glyn suggested that Denis and his surgical colleagues were amateurs and unlikely to produce a paper worthy of the subject. Glyn said that he was concerned that Denis had not given enough time and thought to what cases he intended to include, what readers he hoped to interest in the paper and that hardly any patient had had a full investigation. He went on to say that clinical work requires sound histological background and that Denis should have his paper scrutinised by someone accustomed to writing papers. He and Jack did not think this a new disease but a very rare type of cancer.

Denis' response to this is unknown, nor is it to a letter that followed 3 days later from Prof Davies himself. In this he took Denis to task over some of the statements in his letter and said that he had 'no desire to "adopt" this syndrome…only ..that any publication should reflect credit on the institutions with which I have the honour to be associated'. He restates the need for a multidisciplinary approach and that clinical conditions without the backing of good histopathology were 'a hodge-podge of syndromes of doubtful validity'. Remarking that he had collaborated with many workers in a variety of fields where all information was shared and departmental considerations played no part, for him it was now clear 'that mutual confidence between us has not been established and this being the case it is obvious that our association over this matter must come to an end'.

He goes on to say:

> This is a matter of regret to me as it is a subject of very great scientific interest. I hope you will proceed with your publication as soon as possible to get the priority of publication to which you are most certainly entitled and which I have mentioned in my report to the British Empire Cancer Campaign.
>
> Meanwhile my brother and I are proceeding on that general study of jaw tumours which we were working on before your recognition of this entity, and which we were invited to consider presenting at the International Cancer Conference in July in London. You can be assured that there will be no publication by us of any jaw tumour material before July next, which should give you ample time to get your paper published.
>
> Under these circumstances it does not seem to me to be worth our pursuing these matters in discussion as our attitude and approach to the subject is so different.

It took only 2 or 3 weeks for normal relations to resume – they needed each other. There would be only two papers with the names of both Davies and Burkitt on them. That they were both former pupils at Dean Close School didn't count for very much [6]. Professor Davies would eventually move to Albany Medical College, New York [7], but there were further battles ahead.

References

1. Burkitt DP. The discovery of Burkitt's lymphoma. Cancer. 1983;51:1777–86.
2. Foster WD. The early history of scientific medicine in Uganda. Nairobi/Kampala/Dar Es Salaam: East African Literature Bureau: Printed by The English Press Ltd Nairobi; 1970.
3. Davies JNP, Elmes S, Hutt MSR, Mtimavalye LAR, Owor R, Shaper L. Cancer in an African Community, 1897-1956. An analysis of the Records of Mengo Hospital, Kampala, Uganda: Part 1. BMJ. 1964;1:259–64 and Part 2 BMJ. 1964;1:364–41.
4. Croot HJ, Davies JNP. Cancer in Africa. Lancet. 1952;1:158.
5. Davies AGM, Davies JNP. Tumours of the jaw in Ugandan Africans. Acta Unio Int Contra Cancrum (Int J Cancer). 1960;16:1320–4.
6. Whitney CE (2009) At close quarters. Dean Close School 1884–2009. Herefordshire: Logaston Press. p. 329.
7. Davies JNP. Lymphoma in Africa. New Eng J Med. 1964;270:374–5.

Chapter 10
From Africa a Cancer Syndrome

Back into routine Denis had immediately to start preparation for a very important meeting. From February 5 to 8, 1958, the Association of Surgeons of East Africa was to meet in Kampala where, apart from organising the clinical day, Denis was due to present a paper on his jaw tumours, almost a year since he had seen the first case on the ward with Hugh Trowell. He was by now an accomplished speaker having not only given presentations of lesser moment on many occasions to clinical meetings over the 23 years since he graduated in medicine but also was a well-known leader of and preacher at church meetings. But he prepared very carefully for this knowing that if he was to convince the world of the unique nature of the tumour syndrome he had discovered he would need the endorsement of his colleagues. Unusually for him he typed out the text and structure of the talk in detail[1] paying particular attention to his introduction and, a sign of an experienced speaker, the exact words he would use to conclude the talk. Drawing on his photographic skills, he illustrated the talk with 24 slides (transparencies) and covered the history of the tumour including records of isolated cases reported previously, its clinical presentation, the radiology and pathology, early results from treatment with nitrogen mustard, a chemotherapy used in other cancers, and finally a discussion of whether this really was a lymphosarcoma. He had by now collected 41 cases, which he notes had come mainly from the Baganda, Lango and Basoga tribes, concluding with the observation that now, with a much bigger series of cases, he was able to see a very striking geographical distribution. At the end he acknowledges the help and encouragement he had from his surgical colleagues, and as a step towards reconciliation with the Pathology Department, he thanked Professor Davies for his help and his brother Glyn Davies for his cooperation with radiological studies.

In the audience was Prof Davies who had been warned that Denis was going to give a talk and had been asked if he would open the discussion of Denis' paper.

---

[1] Trinity College Dublin, Manuscripts & Archives Research Library IE TCD MS 11268/4/2.

© The Author(s), under exclusive license to Springer Nature Switzerland AG 2022
J. H. Cummings, *Denis Burkitt*, Springer Biographies,
https://doi.org/10.1007/978-3-030-88563-2_10

Davies was equally well prepared for the occasion with a manuscript of his reply and slides to illustrate the histology. He acknowledged that the condition, which Denis had described, was of interest but pointed out that all the meeting had heard was a clinical description and since this was a cancer, an area where he was an expert, a detailed knowledge of the pathology was essential. Working with his brother, Glyn, they had gone through the records of the Cancer Registry, which Prof Davies had helped to set up in 1951, and he presented a number of conclusions. Prof Davies thought that the cancer was primarily a jaw tumour although often starting at several sites in the body simultaneously,[2] was rarely associated with swelling of the lymph glands and had a very specific age range and geographical distribution, which was from Lake Albert east to the Elgon region. The histology suggested that it was a pleomorphic reticulum cell sarcoma.[3] In his conclusion he held out an olive branch to Denis saying 'I can only close this brief discussion by emphasising that this clinical entity to which our attention was first directed by Mr Burkitt is worthy of the closest attention. It is to be hoped that the distribution and frequency of these tumours can be worked out speedily and the possible aetiologic factors defined as they may be of worldwide significance, particularly at the present time'. Denis, always eschewing conflict and realising that he needed the pathology to fully describe his syndrome, accepted Prof Davies' gesture, and they were soon working together again. But Denis moved to find another pathologist with whom he could have a more inclusive relationship. This was initially Greg O'Conor and later Dennis Wright (Fig. 10.1).

Denis was pleased with the reception he had received from his surgical colleagues at the meeting and on returning to Mulago decided to start writing a paper for the *British Journal of Surgery* describing his syndrome. But he realised, from listening to Professor Davies' remarks, that he must include the pathology to give the paper any credibility. It was also clear to him that Davies had a legitimate interest, as a specialist cancer pathologist, in the jaw tumours and techniques and skills in understanding it that Denis did not possess. He therefore wrote to 'Dear Jack' almost immediately sending him abstracts of the case notes from 26 of the jaw tumour patients for which there was histology available and asked if Davies agreed with the views of the various pathologists who had written the reports on them. Denis suggested that if Davies agreed with all these opinions, he would write a single report summarising the pathology which he would include in his paper, as coming from one reliable source. As a quid pro quo, he offered Davies photographs of the clinical cases, which he could use at an important cancer conference in England in July at which Davies was due to speak. Jack replied very promptly to Denis' letter and enclosures sending him copies of all the pathology reports on the cases Denis had listed although not a summarising view of the condition. Denis wrote again immediately sending him some histology slides that the Pathology

---

[2] Multicentric.

[3] The tumour was eventually classified by WHO as "malignant lymphoma, undifferentiated, Burkitt's type" [1].

Fig. 10.1 Denis, Greg O'Conor and Dennis Wright at a meeting in Lyon, 1980

Department at Mulago did not have and again asking for an overview opinion of the pathology and some illustrative material, which remarkably he received.

Unbeknown to Denis, Prof Davies was starting to write a paper on 'Malignant tumors in African children. With special reference to malignant lymphoma' [2] with Dr. Greg O'Conor, a cancer pathologist newly arrived from the USA.[4] It would not be published until 2 years after Denis' initial paper, but battle had commenced to claim ownership of this cancer. The Davies/O'Conor paper would focus entirely on the pathology although acknowledged Denis' role in providing the first clinical description. From Kampala Cancer Registry data, they would collect 60 cases remarking that it was by far the commonest tumour in children, that leukaemia was rare and that other organs apart from the jaw were involved. The pathology was that of a lymphosarcoma with a characteristic 'water pot' appearance on histology.

Denis sent his completed paper off to the *British Journal of Surgery* in July 23. He named himself as sole author but quoted Davies' opinion on the histology verbatim and thanked him at the end for his 'help in endeavouring to determine the nature of this tumour'. He sent a copy of the submitted manuscript to Davies who read it whilst attending a cancer conference. Davies provided the meeting with an exhibition of material relating to the tumour, promoting extensive discussion. Those

---

[4] GT O'Conor (1924–2012) an American pathologist who had arrived in Uganda in May 1958 with his wife and six children to fulfil a calling to work in Africa under the auspices of The White Fathers Mission Services. His attempts to start a small hospital in Nyakibale a mission station in the south-west of Uganda north of Kabale had failed leading to him taking a post in Kampala.

present offered some points of agreement about the condition including that it was a new and hitherto undescribed tumour and that it was a reticulosarcoma arising in the bone marrow, which subsequently was shown not to be correct. The geographical distribution needed to be worked out and most importantly and farsighted of those attending the meeting, there was a high probability that a virus might be involved.

Denis' paper for publication was sent by the journal to Professor Ian McAdam as representing the Colonial Medical Service, who in turn sent it for refereeing to Professor Paterson-Ross [3].[5] The referee thought the paper was of great interest but asked for a fuller description of the histology. Denis was obliged to ask Professor Davies for this, promising to insert any information he sent Denis in the paper verbatim and with full acknowledgement. But he qualified his request with 'This would not of course in any way commit you to my conclusions which I know you don't endorse. I really would feel much happier to have a really expert pathological opinion expressed freely'. Thus, Denis was ensuring professional distancing.

Immediately after sending the paper to the *British Journal of Surgery*, Denis took his family on safari to the Western Province where he would do his routine visits to the hospitals in the region but now with a new motivation, to find cases of the sarcoma/lymphoma. The government paid his car expenses and hotel for himself and Olive, so converting this into a family holiday appealed to his frugal nature. Heading south-west to Masaka, Denis called at the hospital where he was met by an Indian surgeon he had known for many years. The family had lunch with him and his wife before driving a further 90 miles to Mbarara where they stayed with Dr. and Mrs. Stanley-Smith who had started the Rwanda Mission with Dr. Sharp after World War I. On the first day, Denis worked in the hospital doing clinics and ward rounds, looking for children with jaw tumours. The next day, Friday August 1, he operated all morning before heading further south to Kabale, a small town situated at 6000′ close to Lake Bunyonyi where they stayed at the White Horse Inn. In August 2 they hired a launch for a trip on the lake, taking a picnic and enjoying the more equable climate. After lunch, heading back to the shore, the boat's engine failed, and, despite Denis' engineering skills, he was unable to fix the diesel engine. He caught the attention of a passing canoe and gave them a note to take to a friend of his who lived on a nearby island, but the message came back to say he was away. Finally, they managed to get six locals to come aboard and paddle the 20′ wooden boat back to the shore.

Sunday and was taken as a day of rest with Denis reading *Livingstone's Last Journey*. The disappointments, frustrations, privations, endless setbacks and heartbreak that Livingstone had endured were a revelation to Denis that almost brought him to tears. It made him realise that his own pioneering work came at a much smaller cost to him, whilst the family saw their time in Africa as one of fun and happiness, security and adventure. That evening they visited friends at the mission

---

[5] Sir James Paterson Ross (1895–1980), distinguished neurosurgeon and President of The Royal College of Surgeons of England (1957–1960) who had operated on King George VI and Sir Winston Churchill [3].

station and heard about the revival that was happening. In his account of this safari, Denis follows his remarks about the mission with a very rare comment about Catholics and Protestants:

> In one area alone 7,000 R.C.s changed to the Anglican Church, in which there has been so much revival. The previous year it was 14,000. And this in spite of threats of poisoning etc. The R.C.s are very anxious to get political power and have highly organised political candidates for whom their adherents must vote.

Although Denis gives only passing acknowledgement in his writings to the influence his family had on him, including that of his highly talented maternal uncle Louis, he would have been hard-wired as a child and at his schools with an unquestioning Protestant faith, which very much determined his outlook on life and his opinions regarding the various sectors of Christian belief. The potential and existing conflicts between Protestants and Catholics both at home in Ireland and what he observed in East Africa never seem to have concerned him. Yet his conflict with Prof Davies exercised him considerably and had been handled by him with great care and, what he thought, compromise.

Ever conscious of the need for physical fitness, on the Bank Holiday Monday, he went for a walk in the mountains, climbing and descending several times with a policeman as a guide for much of the journey. As compensation he drove him back to Kabale and gave him 3/−. On return Denis operated all day, and then the family set off along the road that borders the Queen Elizabeth National Park heading north. After crossing the equator, they diverted to see the large copper mine at Kilembe where there was a community of over 2000 people in a self-contained village with a modern hospital around which Denis was shown. Last on the itinerary was Fort Portal, but Denis had developed nausea by the time they arrived, so he did little work on the Thursday. After operating on Friday morning, he had lunch with Olive and then returned to the hospital where an Englishman was waiting to see him. He was called Mr. Adams and was on safari but his wife, who was known to Olive, and was staying in a house adjacent to the Burkitts on the Mulago campus. He told Denis the news that dogs had broken into the cage where the Burkitts kept their rabbits and killed all but one. The next day they drove back to Kampala and were pleased to be home Denis remarking that it had been a happy week and enjoyable holiday. On only 2 out of the 10 days had he been free from either travel or hospital work. To this lifestyle Olive and the girls were already reluctantly accustomed.

Whilst awaiting publication of his paper, Denis continued his search for cases of his syndrome plotting the location of each one on a map in a similar fashion to his marking the occurrence of hydrocele of the testis in men around Lira 10 years earlier. He had also welcomed a visit from Dr. George Oettlé, a renowned cancer epidemiologist from Johannesburg, to whom Denis showed some of his patients. Oettlé told him that this tumour did not occur in South Africa. Since more than half of the children with this problem, mainly the younger ones,[6] presented with tumours of the

---

[6] Jaw involvement is age-related. All cases at age 3 have jaw tumours, this rate falling to 1 in 5 by age 10 [2, 4].

jaw, it was relatively easy to recognise, as were the large growths in the abdominal cavity arising from the adrenals, liver and kidneys. It was also the most common cancer in children under the age of 15, thus aiding the identification of cases. Oettlé was unlikely to have missed it.

Denis' paper was published in the November issue of the *British Journal of Surgery* under the title 'A sarcoma involving the jaws in African children' [5]. In it are reports on 38 cases fully identified and a further 8 without histology. He acknowledges three previously reported single cases, shows a histogram of the age distribution being between 2 and 7 years mainly and remarks that there does not seem to be any particular geographical distribution, something he had changed his mind about in the intervening 4 months since he submitted the paper. He describes the clinical picture and its relentless progress with no available effective treatment and includes Professor Davies' description of the histology with appropriate attribution. The paper is illustrated by six photographs of children taken by Denis showing how singularly unpleasant and cruel the condition is, especially for ones so young. He avoids any discussion of the organ system in the body from which the tumours might arise but justifies the paper by saying that this is a syndrome, namely, tumours occurring at multiple sites including the jaw, that had not previously been described although 'is by far the commonest malignant tumour of childhood seen at Mulago Hospital'. With this paper Denis established his undeniable priority in identifying the tumour syndrome as a single disease.

The paper produced very little response. The lack of interest by readers of the *British Journal of Surgery* in what was essentially a rare cancer of children in Africa was not surprising. Denis rationalised this by telling himself that surgeons were primarily interested in the surgical treatment of diseases, which was not mentioned in the paper because there was no surgical treatment. Denis' next paper on the subject just over 2 years later, reporting a larger series of children and the geographical distribution of the disease, proved to be a sensation and launched Denis into a new orbit. But before that there was work to do.

Undeterred by the response to his *British Journal of Surgery* paper, Denis chose to play to his strengths and follow up his preliminary observations about the distribution of the tumour cases. He decided to do a survey as widely as possible throughout East Africa to find out where the cases occurred. He records in his autobiography[7]: 'On a sheet of foolscap paper I placed photographs depicting various aspects of the facial tumours, involving either the jaw or the eye. A short, written description explaining the recognisable characteristics of the tumours in other sites was added'. He had more than 1000 copies printed and distributed them by post to the doctors working in the hundreds of government and mission hospitals as well as giving them out at conferences and to people who visited Mulago Hospital. Accompanying the information sheet was a questionnaire asking whether the doctors had seen such tumours and to please look out for them. This simple approach to the epidemiology of disease was certainly unconventional, being entirely qualitative, but Denis almost

---

[7] Wellcome Library WTI/DPB/F/1 Autobiography of Denis Burkitt Chap. 4.

took pride in this and boasted that he had never spent any time in a laboratory since he had left medical school and admitted to hardly knowing the meaning of the word epidemiology. He had received, from the Medical Headquarters Office in Uganda, two grants of £10 and £15 to cover the printing and postage expenses, hardly sums that would mark him out as a major player in the cancer world.

In 1958 he was a busy surgeon heading one of the three surgical firms at Mulago in addition to which he spent considerable time on his church duties, preaching and as Chaplain Warden at All Saints. Olive was still having short episodes of depression, despite which they entertained visitors and friends at least twice a week and made regular trips to the local cinema. Early in April Denis tore a muscle in his calf playing tennis, but this did not stop him accompanying Gerry Tewfik, the psychiatrist, on a brief but memorable journey by 'plane to Moroto a small town in the far north-east of Uganda and the only district that Denis had never visited. What surprised Denis on landing at the airport was the ground staff, which consisted of a man in a khaki drill coat, shorts and plumes of ostrich feathers protruding from mud-caked hair. His assistant was naked as were most of the men he saw as they drove from the airport into the town. The women all wore some covering but had up to 40 metal rings around their necks, which had been forged on and were never to be taken off, together with about 6 large earrings in each ear with each through a separate hole. Gerry had to give evidence in a court case allowing Denis to do a ward round in the hospital where he found men lying in bed still with their mud-caked hair and head-dresses, which could not be removed without cutting off all their hair, supported not by pillows but wooden rests or 'stools' when they lay down. Denis bought a stool from a local as a souvenir, and they were on the plane back to Entebbe at 2:40 pm taking with them a boy who needed surgery to his leg.

Whether doing his normal duties or in the quieter moments of the week, Denis' thoughts were occupied with his tumour and the pursuit of its cause. The visit of Professor Hewer, a pathologist from Bristol who had come to examine the medical students, was notable only in that Denis was able to discuss the tumour syndrome with him. What was going on in the world around him did not register or deserve any mention. This was very noticeable when Hugh Trowell and his wife Peggy suddenly left Uganda in November 1958. The Trowells had been good friends of the Burkitts ever since Denis and Hugh had met on the train going to Nairobi in March 1947. They shared a common faith, both were senior doctors at Mulago, and memorably Hugh had shown Denis the child with jaw tumours some 21 months earlier that had such an impact on Denis' future work. They had lunched together regularly since Denis arrived at Mulago in 1948. As Hugh and Peggy were about to depart, Denis called round to say goodbye but found the occasion very sad. Peggy was unwell and Hugh was having to leave without any proper farewell from his friends and colleagues. However, rioting was becoming increasingly common in Kampala as the move towards independence spread with gangs of armed thieves roaming the town. Samweri, one of the Trowells' servants, had been attacked and was saved from serious injury only by Hugh protecting him from the assailants. The Trowells' youngest daughter Jenny,[8] who was training to be a nurse in England, remembers that when on holiday with her parents in Uganda rioting had broken out in Lweza. Army trucks had arrived and smuggled the family out concealing them with their shields as they lay flat on the metal floor of the truck.

Against this background, a few weeks before their final departure date, Peggy was alone in their house when she was attacked at knife point. The attack was so brutal that the details were deliberately not described by Elizabeth, their eldest daughter who wrote about these events many years later.[9] Peggy's screams were heard by the servants who fought off the attackers, leaving her shocked and severely injured. The telephone lines had been cut so Samweri had to run the 8 or 9 miles into Kampala to summon help. She was eventually taken to hospital but the attack had been particularly savage leaving her shattered.

Hugh decided that staying in their house in Lweza, which was a few miles south of Kampala on the road to Entebbe, was no longer safe, so they moved into a small staff bungalow on Makerere Hill. Peggy's recovery was slow. She had repeated flashbacks of her ordeal, could not sleep because of nightmares and at times thought she was losing her mind.[10] A paediatric registrar who had arrived at the hospital from Great Ormond Street prescribed a new sedative which she said was very safe and was being given to babies and pregnant mums. Peggy improved initially but needed increasing doses to maintain her progress. She then became dazed, had repeated falls and lost the sensation in her feet. Her balance was upset and her vision distorted so that she could not walk unaided. She began to lapse into unconsciousness; so ill did she become that Hugh was very worried that she had developed a serious neurological disease. He sold their house, packed up their belongings and said goodbye to few close friends including Denis. Peggy was carried to the 'plane on a stretcher. On arrival at Heathrow, she was taken straight to St Thomas' Hospital where the thalidomide that had been prescribed was stopped. She never fully recovered from the ordeal.

Without regard to the prospects of the girls seeing male nudity, the Burkitt family set off for a 10-day safari to the area Denis had visited in with Gerry Tewfik in April. The excuse for the visit to this area was to see Africa virtually untouched by Western influences. They reached Mbale by the end of the first day where they stayed with the Timmises. Nkokonjeru was climbed, 2348 m/7703', but starting from 4000', the Sipi Falls viewed and a round of golf played before driving on to Moroto through the semi-desert Karamoja Region with its dry riverbeds and the mountain ranges of Kenya just over the border to the east. Passing by men with their elaborate headdresses and women with metal neck and earrings, they reached Moroto late on Saturday afternoon where they had rented a house with a swimming pool nearby.

[8] Jenny, Jennifer who would marry Thorold Masefield. https://www.royalgazette.com/other/news/article/20190612/jennifer-masefield-1940-2019/

[9] Wellcome Library PP/HCT/A.5 Hugh Trowell: Pioneer Nutritionist. Biography by Elizabeth Bray (Unpublished) 1988.

[10] As with Olive and her ordeal during the war, the consequences of this attack on Margaret Trowell would now qualify as post-traumatic stress disorder.

Leaving the family to relax, Denis and Guy decided to accompany Dr. Stuart, the station doctor, on a trip to southern Karamoja the next day to experience real bush and camping outdoors. Seventy miles south at Nabilatuk, Denis helped to see patients at the dispensary and then on to Amudat where they met Dr. and Mrs. Cox. Dr. Cox was one of the people with whom Denis had been in contact in the search for cases of his tumour syndrome, whilst his wife was from Ireland and member of the Guinness family. The next day they went off-road in Dr. Cox's Land Rover, which Denis and Guy found exhilarating with superb scenery, an almost deserted landscape and the terrain so rough that they had to stand up at the back and sway from side to side in order to keep upright. They arrived at lunchtime at the home of the Van Ryaks, an elderly couple from South Africa living in the most primitive conditions and making a living by panning for gold during the 3 months of the year when there was water in the river. A sand and water mixture was passed over matting into which the gold, being heavier, sank and was recovered at the end of the day. Employing local Karamojong labour, they were retrieving about half an ounce a day, which they were able to sell for £12 an ounce. An afternoon walk followed, guided by two boys, but without any water they returned very thirsty, and Denis drank 12 mugs of various fluids. In the evening they were invited by the old couple for drinks. They 'sat in chairs out in the open under a large tree with just endless Africa around us. A pressure lamp gave light. The table was just a huge slab of stone resting on three tree trunks that had been cut short'. Two dogs and two cats sat nearby. Two and a half hours from the nearest village and a further 60 miles from any form of centre for supplies, the couple had tried to return to South Africa, but Denis could understand why this sort of freedom could get under their skin. The three men, with Dr. Cox, set off back the next morning without having washed or shaved, transferred back into their car at Amudat where they parted from the Coxes arriving in Moroto in the early afternoon. The next day, Thursday, Denis spent the morning in the hospital and took some photographs of the men in bed resting the napes of their necks on the wooden stools, which had so fascinated Denis on his last visit. En route back to Mulago, Denis met the Rev. Robb who had been vicar of St Kevins in Dublin and had known Uncle Louis, who had so influenced Denis in his faith whilst at Trinity. He brought news of his demise and had visited his grave in Switzerland. After 10 days away, Denis had, unusually, spent only 1 day in clinical work. Family, friends and exploring the African landscape had proved to be a compelling diversion (Fig. 10.2).

The visit of the Queen Mother in February 1959 to confer degrees at the medical school, open the new library and attend a round of social events brought out the unquestioning royalist in Denis. At the degree ceremony, where the first East African woman to become a doctor graduated, Denis and family were seated in the third row from the front 'only some dozen yards from her Majesty' and later were 'standing only a few paces from her when she got into her car'. The next day the Queen Mother opened a new £140,000 sports stadium, which project Ian McAdam had instigated and chaired and was thus entitled to a seat in the Rolls-Royce with Her Majesty as they toured the site. At the Royal Tattoo in the evening, there were the massed bands of the East African Rifles, soldiers on parade, a demonstration by the

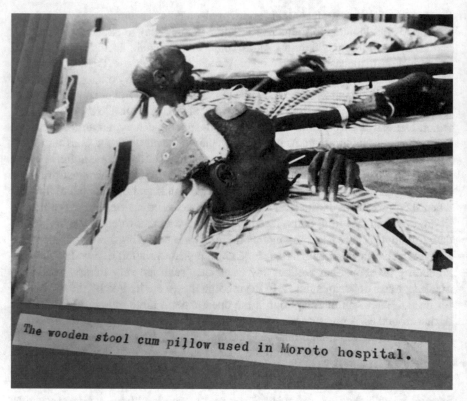

The wooden stool cum pillow used in Moroto hospital.

Fig. 10.2 Men lying in hospital in Moroto resting their heads on wooden stools cum pillows to protect their hair

Fire Brigade and an exhibition of national dancing from the surrounding regions, Scottish dancing to the skirl of the pipes and some classical Indian music. The Queen Mother arrived at the interval having had dinner with the Chief Justice and was accompanied by other royal dignitaries including the Kabaka, who attended in his own Rolls-Royce. The one in which the Queen Mother was driven around had been bought for the occasion. On Saturday there was the traditional Garden Party in the grounds of Government House in Entebbe, attended by around 4000 people. One of Olive's friends was due to be presented and the two stood close to each other as the Queen Mother approached. Denis was behind Olive but felt that as Her Majesty chatted to Olive's friend that 'she might almost have been chatting to me'. One of the ladies in waiting talked to Olive. Driving back to Mulago, Denis felt that all was right in the world. The Empire in which he served, with the Queen Mother's daughter Elizabeth at its head, was secure and bringing great benefits to Africa. On Sunday Her Majesty attended the morning service at the cathedral, admission by ticket only, together with the governor, Kabaka and their retinue. After signing the visitors' book, she was shown a former book signed by her on a visit in 1924. Such were the highlights of a royal visit. The Bishop and his wife, Leslie and Winifred

Brown, had lunch with Her Majesty after which the Royal Party left. Yet the 'wind of change' was blowing through Africa.[11] Public and political disquiet, as witnessed by the Trowell family, continued, and a negotiated settlement was eventually reached with Britain such that Uganda, a protectorate,[12] became independent on October 9, 1962, with the Queen as head of state and a year later became a republic.

The jaw tumours aside, family life was important for Denis. Now that the girls were growing up, the question of their education had to be considered. Because of the unrest and political uncertainty, Olive and Denis thought that it might be best for them to be educated in Britain, and Sherborne School was contacted with a view to taking Judy, who was now 12. Whilst waiting for Sherborne to offer Judy a place, she was sent to The Highlands School, a girls' boarding school in Eldoret north of Nairobi in Kenya. This required much less travel than going back to England every term and some of Judy's friends were already there. Although Sherborne did offer Judy a place, she preferred The Highlands School and stayed there until she was 16, then joining Cas and Rachel, who had been sent to Westonbirt School near Bristol in England. Denis was busy at home in Mulago with car maintenance, carpentry, attending classes and photography, doing his own developing and enlarging except for any colour transparencies he was taking. He had a Minolta with close-up and telephoto lenses although must have been constantly upgrading his equipment because he gave the camera to the Bishop in March and sold the lenses to 'Lewis'. And he was a good photographer winning a photography competition in June for readers of *The Times Weekly Review* with a picture of a chameleon catching an insect. Most impressive, considering he would have had none of the electronic gadgetry now available that would make the timing of the shot so much easier. The prize was 10 guineas (Fig. 10.3).

Shortly after the family had returned from their safari to the north-east, they received a letter from Robin to say that father's, James', health was declining. He died at Laragh, Ballinamallard, near Enniskillen on March 31, 1959, aged 88 and was remembered throughout Ireland for his engineering achievements as county surveyor particularly in building new roads and bridges to manage the border that was created between the two parts of Ireland in 1922. Tributes also came in from the Church of Ireland community of which he had been an outstanding figure and which had conferred on him every office open to a layman [6]. More widely, throughout Europe he was remembered for his pioneering work with birds using rings to estimate their age and plot their territories [7]. The family, including Gwen who was at James' bedside when he died, were greatly supported in their grief by their faith. Denis writes in his diary for March 31:

---

[11] From the speech given to the South African Parliament on February 3, 1960, by the Prime Minister of the UK Harold Macmillan in which he said, famously, after spending a month in Africa 'The wind of change is blowing through this continent. Whether we like it or not, this growth of national consciousness is a political fact'.

[12] Protectorate is a country protected by another but retaining its own sovereignty, whilst a colony is part of a larger country and is under its control.

Fig. 10.3 Photograph of chameleon with which Denis won first prize in a competition organised by *The Times Weekly Review*, June 1959

> Wire from Robin "Daddy passed peacefully this morning". Tears at first but I gave thanks. How appropriate to enter fuller life at Easter. I wired back to Rob and Mummy. We are much with them.

Gwen wrote in her diary:

> Darling Jim passed on peacefully at 10.10 Robin got here in afternoon Bless him – was splendid.

Jim was buried in Trory churchyard. Denis and family were clearly not able to attend, but they would soon be heading for Laragh as Denis' leave started that same day (Fig. 10.4).

The family embarked from Mombasa on a small Italian merchant vessel, the Tripolitania, in April 5 choosing a voyage that would take them almost 4 weeks just to reach Europe, rather than the now reasonably quick journey by air. Perhaps Denis felt they all needed some time together away from the challenges of Kampala and the attention they would get when they reached Ireland in the wake of James' death. Calling at Mogadishu the captain took on a large load of millet, and Denis went ashore to visit the army mess and beach where he had been stationed in 1944. Olive and the girls were not impressed. The voyage continued via Kandla, a newly built port on the west coast of India to Aden, through the Red Sea and into the Mediterranean. Denis used the time to advantage, sorting out his many clinical photographs and writing a paper on the boots and callipers he had designed for polio victims, which was published the next year [8], and one on congenital deformities,

Fig. 10.4 Yearbook
belonging to Gwendoline
Burkitt and used by her
from 1903 to 1959
'March 30
Win Langley a son. 1940
Darling Jim passed on peace-
fully at 10–10 59
Robin got here in afternoon
Bless him – was splendid
March 31
Jim retired 1940
Cable from Denis – 59'

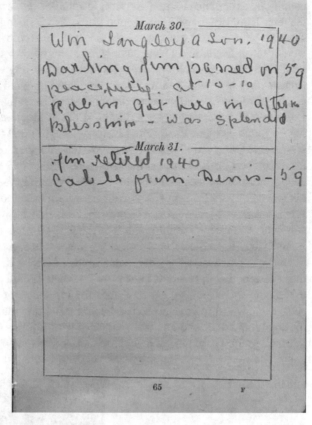

which he never sent for publication. He read widely, often with two books on the go, and shared some with Olive reading chapters to each other in the evenings. On the way up the Red Sea in April 21, the captain signalled to stop the ship at a small island with a lighthouse. A boat was lowered and went ashore with the ship's doctor. When it returned Denis was asked to see a patient they had brought as he came on board. He diagnosed tinnitus and deafness,[13] which he did not feel was worthy of such a rescue mission, but the captain thought it deserved a champagne reception to which Denis was invited. Not his favourite way to celebrate.

At Genoa they left the ship and drove to Tübingen and then the next day to the Mercedes factory near Stuttgart where they saw the museum and took delivery of their new car. They then spent 3 weeks driving back to the channel camping, sight-seeing and walking mainly in Switzerland and France. This bounty of free time, they were away from Uganda for 5 months, was an accepted part of employment in the Colonial Medical Service, thought necessary because of the hardworking

---

[13] Tinnitus, hearing sounds such as buzzing, humming, ringing, etc. that are not from coming your surroundings.

conditions experienced and disruption to family life. Arriving back in England on the ferry from Boulogne, Denis found the roads very congested and traffic dreadful. They stayed for a few days near London, visiting Robin, meeting the Tewfiks who were also on leave, and Denis gave lectures on the jaw tumours at the Middlesex Medical School and St Marys. He also met Jack Darling whom he had known since student days and who had just been appointed surgical specialist at Mulago. He would join Denis running one of the three surgical teams or 'firms' in the hospital. Denis told him that he had been touting slides of his tumour around London teaching hospitals and that 12 different histopathologists had made 11 different diagnoses. He and Olive were back at Laragh to a tearful reception at the end of May where Denis took on the jobs around the house and garden that his father would normally have done. There was much talk about James' final months and they all went to view his grave in Trory churchyard. Gwen was very pleased to show them the article by Professor David Lack, from Oxford [9], in which he placed James amongst the seven most influential ornithologists of the past 40 years (Fig. 10.5).

But James had gifted Denis something that would cause him trouble. His 50-year-old bicycle. Out for a ride with Cas on Sunday June 14, the front brake, the only one that was functional, failed as he was going down the hill into Kilkenny. Denis tried to slow down by putting his foot onto the tyre but inadvertently caught it in the spokes and was thrown over the handlebars sustaining multiple injuries including to his right humerus[14] and left wrist together with injuries to his ribs and cuts to his face and scalp. He managed to summon Olive who came with the car to pick them

Fig. 10.5 Jack Darling
1962 at Mulago

---

[14] The bone of the upper arm.

up and took him to the hospital in Enniskillen where an x-ray showed a badly com-minuted[15] fracture above the elbow and involving the joint. The junior resident on Sunday duty was persuaded by Denis to simply apply a plaster cast, and the next morning Olive drove him to the Royal Victoria Hospital in Belfast where Mr. Withers, the senior orthopaedic consultant, manipulated the arm under general anaesthetic and applied full plaster casts to both arms. Suddenly Denis was helpless. He was kept in the hospital for almost 2 weeks during which time he quickly recovered from the shock of the accident and spent the time reading, receiving visitors including Olive and watching Wimbledon. The plaster from his left arm was removed after a week and from the much more badly injured right arm after 3 weeks. At home Olive had to do everything for him because Denis could not use his right arm at first. In the absence of any physiotherapy, he built a device in the garage to help him exercise the arm, and 5 weeks after the accident, he was able to start feeding himself, drive the car and then write. He still had concerns about being able to use his camera and his ability to do the surgery he had become so skilled at and which required fine control of his fingers when he would return to Kampala in 6 weeks. Olive, who still had to help Denis dress, accompanied him to the BMA meeting in Edinburgh, which lasted a week, where he brought greetings from Uganda to the AGM. He was interviewed by the BBC along with other representatives from East Africa, saw colour television for the first time and attended the Garden Party at Holyrood Palace and the Christian Medical Fellowship (CMF) breakfast, which had been a feature of the annual BMA meeting for nearly a century. There he heard Dr. Stanley Brown a medical missionary from the Congo (DRC) and a world authority on leprosy give a talk to a packed audience.

August saw the family preparing for the return trip to Kampala, packing crates for shipping, and the 19th came what Denis had not been looking forwards to, saying goodbye to Gwen. Realising that it was unlikely he would ever see her again, given the long interval between leave periods, 'Parting with dear Mummy very difficult. I cried in car. Olive also found it very hard'.[16] Prior to departure the Colonial Office required Denis to undergo a medical examination in London to see if he was fit to practice. This went well and he was given permission to return to Uganda and was back at Mulago by the end of the month with the family following a few days later. They moved into a new house; Denis had a new office, took over ward 7 and managed an operating list successfully a few days later. Their new Mercedes did not arrive for several weeks, but Denis continued to travel widely using their old Opel, visiting hospitals, doing the surgery of difficult cases and giving talks about his jaw tumours. When the Mercedes did arrive, Denis found he was in great demand for weddings. He had not realised that that to a Bagandan family to have a Mercedes at their wedding was a great boost to prestige.

---

[15] Fracture in which the bone is broken into several pieces.
[16] Gwendoline Burkitt died on August 23, 1960.

But there was an issue that Denis found it hard to deal with. A recurring problem challenging his claim to have discovered a new disease. One which his usual approach to solving difficult issues, by talking to people, correspondence, meetings, lying awake at night and prayer, did not seem to be able to solve.

References

1. Bennett JM. Definition of Burkitt's tumor- meeting of investigators on the histopathology of Burkitt's tumor. Bull World Health Org. 1969;40:601–7.
2. O'Conor GT, Davies JNP. Malignant tumors in African children. With special reference to malignant lymphoma. J Pediatr. 1960;56:526–35.
3. https://livesonline.rcseng.ac.uk/client/en_GB/lives/search/detailnonmodal/ent:$002f$002fSD_ASSET$002f0$002fSD_ASSET:372420/one?qu=%22rcs%3A+E000233%22&rt=false%7C%7C%7CIDENTIFIER%7C%7C%7CResource+Identifier
4. Wright D. Nailing Burkitt lymphoma. Br J Haematol. 2012;156:780–2.
5. Burkitt D. A sarcoma involving the jaws in African children. Br J Surg. 1958;58:218–23.
6. Jas Parsons Burkitt. An appreciation by W.E.T. April 2nd 1959 The Impartial Reporter and Farmers' Journal.
7. C Douglas Deane First to find the age of birds. Irish Times. Circa April 1959.
8. Burkitt DP. A boot and caliper bank. S Afr Med J. 1960;57:109–12.
9. Lack D. Some pioneers in ornithological research 1859-1939. Ibis. 1959;82:299–324.

Chapter 11
Establishing Ownership of the Lymphoma

There was a strong belief in the Department of Pathology at Makerere that this tumour was a pathological entity first and foremost that only they could define and afford it an appropriate place in the classification of tumours. Denis, having written the first paper describing the childhood cancer, had no qualms about his right to follow up his interest in the syndrome, which clearly affected not just the jaws but many organs in the body. Moreover, he had different objectives and motivation. From the responses to the survey of government and mission hospitals across East Africa, he was already beginning to see clear patterns of the distribution of the tumour. These would give the first clues as to a possible viral cause. Importantly for Denis, he wanted to find a way of treating these children suffering from what was a horrendous, shocking, disfiguring and fatal condition. But there were battles ahead before it would eventually become the 'Burkitt Lymphoma'.

Professor Jack Davies, the Head of Pathology at Makerere, wrote to Denis whilst he, Denis, was on leave in June 1959, saying that he now accepted that the jaw tumours could also occur elsewhere in the body of these children. Denis had believed this from the start of his investigation, so to have Jack Davies now acknowledge it was gratifying, and Denis felt that progress was being made. However, when back at Mulago in September, Denis received a circular from Prof Davies, sent to all senior staff at Mulago, asking for information on 'lymphomatous tumours (since) he has taken this work on as a major project'. Denis felt grieved about this, thinking that Davies was trying to take credit for his research. He nevertheless offered him his clinical material once again telling himself that it was his Christian duty to go the second mile and turn the other cheek as required. Denis made contact with Greg O'Conor who told Denis that the tumours were very clearly in the category of lymphoma, although without involvement of lymph nodes, and thereafter Denis used this term to describe the condition.

Later in the year, Denis was shown, by O'Conor, a draft of the paper that Davies and he were writing about the lymphoma. Denis was very annoyed on reading it to find that it contained much of his own work without acknowledgement, writing in

© The Author(s), under exclusive license to Springer Nature Switzerland AG 2022
J. H. Cummings, *Denis Burkitt*, Springer Biographies,
https://doi.org/10.1007/978-3-030-88563-2_11
149

his diary 'This really spoilt may day'. He was incensed by what he considered to be Davies' scientific piracy. He did not know that Davies was actually working on two papers about the lymphoma, which would be published in January and April 1960, one written with his radiologist brother [1] and the other with O'Conor [2]. The papers report the pathology and histology of the African childhood lymphoma in detail and pay special attention to the recently agreed classification of lymphomas into which category fell the cases that Denis had originally described. The jaw and non-jaw tumours are considered separately at first, but when they restudied the histology, they concluded 'that they are in fact the same type of tumour….and should thus be classified as malignant lymphoma'. Denis is not a co-author on either paper but is given credit for his clinical description of the tumours and as the person who first recognised the remarkable frequency of these tumours in African children. His help in supplying case details to Davies is not acknowledged. Even the clinical photographs of the children with jaw tumours had been taken by the university photographer, Mr P Cull, when Denis had given Davies many such photographs for him to illustrate his lectures. Denis, realising that the niceties of the pathology of his children's tumours were outside his expertise, coped with all this by stepping up his search for lymphoma cases and building a map of its geographical distribution. He knew that this would give him important clues as to the possible cause of the tumour.

Early in January 1960, he met with O'Conor, and they agreed to start working on a paper together describing Denis' latest findings and the new pathology classification. Denis gave a talk on the tumour to the local Association of Surgeons in early February but was rather put out when no-one put his name forwards to become a member of the Council. He felt that neither he, now a senior surgeon in Kampala, nor his work on the lymphoma was sufficiently recognised. After the conference he had another long talk with O'Conor, who wanted the main focus of the paper to be about the pathology. This did not suit Denis at all, and after some discussion they agreed to write separate papers that they would submit jointly to a journal as Parts A and B.

Recovered from his bicycle accident, Denis was now able to swim and play tennis and started travelling again with one of his regular trips to the Northern Province. He was accompanied by Captain Anthony Bryham, whom he had met on the Tripolitania last year, and Mr Savdhan the Indian evangelist who had accompanied Denis on the Northern Province safari in November 1957. Captain Bryham was an army man with tales to tell. He had travelled extensively, lived in many countries and was on good terms with the Duke and Duchess of Windsor. He had met Mussolini, Hitler, de Gaulle and Churchill and knew lots of Italian history. But his adopted twin sons had been killed, aged 17, on the Stalingrad front fighting for the German army, and his own son also died in the war. He was an interesting man who liked the simple life of Africa.

Leaving Captain Bryham behind in Arua, Denis agreed to help Cliff Nelson with one of his sleeping sickness[1] inspections, which had been arranged months earlier.

---

[1] Trypanosomiasis.

Driving to different communities each day, about 30 miles north of Arua, they inspected between 8 and 9000 people over a period of 6 days. Denis and Cliff lived in a 'simple mud and wattle rest camp' a life that Denis found most agreeable. 'How lovely it is to sit out at dawn in the simplicity of Africa. So much that we accumulate in life is only a burden'. Whilst simplicity meant no electricity or telephone, basic sanitation and running water, they were looked after by Cliff's cook who brought them morning tea and prepared breakfast from local eggs, water for shaving in a canvas basin, a meal at lunchtime whilst they consolidated the figures from the morning's work and in the evening a bath in a canvas tub and more food. A paraffin light was all they had in the evenings, but this did not stop Denis from working on his next lymphoma paper.

They returned to Arua on Saturday after completing the sleeping sickness inspections. The next day Denis operated in the morning and then drove to Aru in the Congo (DRC) where he had arranged to see a Belgian doctor to discuss the tumour and ask if he had seen any cases. Here he found the first signs of changes that would eventually curtail his freedom to investigate the cause of the lymphoma. The hospital in the Congo was in a very poor state with refugees from the developing political crisis heading into Uganda, often arriving first at mission hospitals close to the border such as that run by Ted and Peter Williams near Kuluva. Denis realised that collecting more information on the occurrence of his tumour here was going to be increasingly difficult. Back at Mulago he decided to focus on getting the papers with Greg O'Conor finished. He felt he had a final draft ready by the middle of March, but in discussion with Greg, it became clear that there was a problem. Having worked with Prof Davies for several months, written the paper for *The Journal of Pediatrics* with him, Greg felt that this was primarily a cancer that had been described in detail, classified and brought to the attention of the world by the pathologists at Mulago. A mere surgeon like Denis had played a purely supporting role. Denis wrote in March 14: 'Discussion with Greg on our paper. He seems to think that he has done it all. This upset me and really I had collected all the material'. With a new aim of submitting clinical and pathology papers together with different authorship, they finally, by the end of April, had drafts which they were both prepared to sign off. They would co-author the first paper, but the second, about the pathology, would be authored solely by Greg. In the light of the lack of interest that Denis' first paper on the lymphoma had attracted when published in the *British Journal of Surgery*, he and Greg decided to send them to a journal in the USA, and the papers were posted to *Cancer*, a prestigious international journal published by the American Cancer Society since 1948. The papers arrived in the journal office on May 31, 1960.

When published in March 1961 [3, 4], they were to create a sensation that would launch Denis into a new orbit and bring him worldwide recognition that more than compensated for all the frustration he had experienced in Kampala. Denis would go on to write more than 40 papers about the lymphoma together with books on the subject, whilst Greg would continue to work on its pathology for the rest of his career. Prof Davies turned his attention more towards Hodgkin's lymphoma.

What was it in these two papers that would so catch the attention of the world, particularly those with an interest in cancer and its causes? Was it the very high incidence in Africa relative to other childhood cancers such as leukaemia, more details of the histology or the remarks about its cause and treatment? More likely it was the particular geography of its occurrence now becoming more defined as the results of Denis' postal survey came in. The papers were based on a much larger case series from both the Kampala Cancer Registry and Denis' large and widely canvassed group of doctors throughout Africa, 50 of whom are acknowledged. The geographical distribution, described for the first time, includes a map showing what was to become known as the lymphoma belt extending across Africa from West to East relatively close to the equator but with a tail extending down the eastern border of Tanganyika (Tanzania). The exact distribution within this area would be further refined by Denis during his long safari in 1961, but he is quite definite about one thing, that the tumour was not seen further south than the lower border of Tanganyika, that it was recognised across East, Central and West Africa[2] and 'virtually disappears north of the oblique line running through northern Kenya and northern Nigeria'. No cases had been recorded in Ethiopia, Somaliland, southern Arabia and Egypt. The clinical picture described is very much the same as in the earlier publication with now some discussion of attempts at chemotherapy [5]. The map gave the first clues as to what might be causing this tumour (Fig. 11.1).

The second paper, by Greg, was about the pathology. It is a comprehensive description of the gross appearance of the tumours, now accepted as belonging to the lymphoma category and a detailed description of the microscopic appearance of the cells, including the familiar 'starry sky' appearance. O'Conor suggests that the cell type can be divided into at least two types, lymphocytic and histiocytic, which categorisation would later be replaced by one of Dennis Wright [6]. The paper ends with the words, 'The possible relationship to a virus disease in cattle or to a parasitic infection as a priming factor is suggested as a possible explanation of the unusually high rate and limited geographical distribution through central Africa'.

Back on duty travel in the Northern Province once again in July 1960, Denis and family visited Kuluva, near to Arua, where the Wiliams had built their mission hospital. Denis spent a day and a half working but noticed several families of American missionaries who had fled the Belgian Congo. They were hoping to return once the troubles had settled, which they expected would be soon now that the country had become the fully independent Republic of the Congo (later Zaire, then the DRC) in June 30. This was not to be as the country became progressively more unstable throughout 1960. Denis also met some missionaries from the Sudan who told him that Christian work was becoming more and more difficult. No Christian schools were now allowed, all Christian medical work had been stopped, and the Church Missionary Society bookshop in Juba had to close on government orders. The Colonial powers were losing their influence, and soon this wind of change would be

---

[2] Kenya, Uganda, Tanganyika, Ruanda-Urundi, Congo, French Equatorial Africa, Cameroons, Nigeria, Ghana and French West Africa.

Fig. 11.1 First publication to show a map of tumour distribution. In print before the long safari of November/December 1961
From Burkitt D, O'Conor GT (1961). Malignant lymphoma in African children: I. A clinical syndrome. Cancer 14:258-269 [2] (With permission from John Wiley and Sons)

affecting Denis' work in Uganda. Doors of opportunity were now closing, yet Denis still had essential work to do in delineating more precisely the areas where the lymphoma of children occurred. Leaving Arua the family stayed at the home of Beth and Cliff Nelson. Cliff had done 3 years as a government medical officer and was now hoping to join the Africa Inland Mission (AIM) in Tanganyika, but between jobs Denis would invite him and Ted Williams on a great safari. The rest of the trip followed the familiar pattern with Denis sorting difficult medical problems and operating at the hospitals in Gulu and Kitgum before arriving in Lira. It was Olive's first visit since they had left 12 years earlier. Visitor facilities had been upgraded, so Denis and Olive stayed in the rest house, virtually as good as a hotel but at two thirds of the cost at 27/- (shillings) a day and no charge for meals if you ate out, which appealed to Denis' sense of frugality. Soon after they had settled in, a car drew up containing what Denis thought were American missionaries, but they were teachers including the headmaster of the school in the Congo, which they had just left leaving local guards who had been faithful to the school for years, to guard their property. Another family arrived late in the evening, missionaries with a familiar story.

The next day Olive was surprised when one of the Lira Mission ladies called round to collect her to speak at the local branch of the Uganda Council of Women. They had apparently sent a letter to Olive with the invitation to speak but it had not arrived. Denis had asked for their post to be forwarded whilst they were away, but the arrangement had not worked. Olive fulfilled her commitment, and Denis spent the next 2 days working in the hospital before they set off back to Kampala where 2 weeks' post awaited them amongst which were letters from Robin telling them that Gwen was sinking. She died 12 days later in August 23 Denis remarking 'News of dear Mummy's passing last night. We wouldn't have it otherwise. Thank the Lord for all dear dear Mummy has been to us'. A week later a letter arrived from Robin with news of Gwen's final days. Olive was sorry that she had not been able to be with Gwen at the end, but it was a cost that had to be carried for all those working abroad in the Colonial Medical Service at the time. Robin wanted to get on and sell Laragh quickly, but Olive was against this, wanting to have time to go through Gwen's personal affects. Robin concurred although they would not all be back at Laragh for a further 9 months.

The two papers in *Cancer*, whilst giving the most comprehensive description of the lymphoma, would make only suggestions as to its cause. More evidence was needed. Denis decided that to follow up his observations on the geographical distribution of the disease, obtained from his survey of hospitals, he would need to be much more precise about its limits. So was born the idea of a safari on a much grander scale than any he had hitherto undertaken, focussing particularly on the boundaries of its occurrence. This would give him the chance to determine what possible environmental factors occurred within but not beyond the limits of the lymphoma belt. To do this he planned to visit all the medical establishments along the edges of the area, questioning the medical staff about their experience of the lymphoma and asking them to look out for it and communicate with him. The argument for where to concentrate his efforts ran as follows: 'It was evident that the northern boundary was quite inappropriate because it merged into desert areas where populations were scarce. The southern limit in West Africa was the sea-coast which ruled this region out'. The Congo and Angola where there was political instability made them unwise destinations leaving the southern border on the east side the obvious choice. English was spoken except in Mozambique, the roads were good, the populations were quite dense, and there were big contrasts in climate, terrain and vegetation. For this exercise Denis needed to raise money, so he started writing to potential donors with his ideas. He was offered £100 by Dr Billington from Mengo Hospital and $300 by the pharmaceutical company Lederle. As the end of 1960 approached, they were told that all leave for government officers was cancelled in case of disturbances leading up to the national elections in March 1, part of the Britain's strategy to grant independence to Uganda in 1962. Plans were drawn up for managing any emergency and there was a sense of unease amongst the staff at Mulago.

Meanwhile, word had started to travel around the world that there was something very interesting to be discovered in Kampala. First to arrive was Sir Harold Himsworth, Secretary of the Medical Research Council of London, who visited Mulago in January 1961. Denis was invited to a party at Professor Davies' house to meet Sir Harold and Dr RJC Harris who was Head of the Department of Environmental Carcinogenesis at the Imperial Cancer Research Fund in London, Professor Marrion and Ian McAdam amongst others. The visiting scientists were clearly very interested in the lymphoma, but at a more formal meeting the next day, Denis' ego was wounded again because Davies talked about the tumour as his discovery. Ian McAdam came to the rescue and later that week Denis was given the opportunity to talk to Sir Harold on his own. Himsworth was a clinical scientist with a considerable reputation for his work on the classification of diabetes into that which was either sensitive or insensitive to insulin, now known as Type 1 and Type 2 [7].[3] He had gone on to work on liver disease, run an outstanding Medical Unit at University College Hospital in London and been knighted in 1952 and elected FRS in 1955. He became one of the longest serving and probably the most successful Secretary of the MRC, serving from 1949 to 1968, known for his wide knowledge of medical science from molecular biology through to clinical and public health medicine. His belief that research was a journey where only the beginning was clear but that outstanding people should be encouraged and helped to follow their ideas became a principal that the MRC would continue to honour for many years (Fig. 11.2).

Talking to Denis he recognised the work that Denis was doing as potentially very important and that Denis was the person to do it. At the meeting he promised immediate financial help, which amounted to £150, but later in the year, a full grant was awarded. Similarly, Dr Harris was impressed by Denis' determination to find a cause for the lymphoma and particularly excited by the idea that it might be viral in origin. Harris was a pioneer in the study of viral-induced tumours in animals and had worked on the Rous sarcoma virus for a number of years.[4] He promised to set funds aside for Denis and in March wrote to say that he had been awarded a grant of £1,200 annually for 4 years. Thus Denis, a busy surgeon working in a dilapidated hospital in Africa,[5] was able to impress people of the highest calibre in medical science in the UK by his knowledge and understanding of the lymphoma syndrome, his theory about its cause, his enthusiasm and plans for following up his ideas.

---

[3] During the 1930s he had published a series of papers amongst which was the classic description of diabetics who were insulin sensitive being 'younger and thin, and to have normal blood pressure and normal arteries, and as a rule their disease is of sudden and severe onset. The insensitive diabetics, on the other hand, tend to be elderly and obese, and to have hypertension and arteriosclerosis, and in these patients the onset of the disease is insidious.'

[4] Rous sarcoma virus was the first virus to be shown to cause cancer. Work done by Dr Rous at The Rockefeller University, New York, showed that it caused sarcoma in chickens.

[5] The new Mulago was under construction nearby and would be opened in 1962.

Fig. 11.2   Bob Harris on safari in Northern Province (1964) with Denis

The reality was that he did not have any very clear plans apart from those for the safari. He was struggling to keep up with all the lymphoma work. In addition, he had clinical duties, which included regular on-call and night-time surgery at Mulago, surgery at the European Hospital and the recently opened Nile Clinic and travel to locations in the Northern Province where the difficult cases were kept for his visits. There were also his regular church and family commitments and planning meetings for the new hospital that was being built. Meanwhile, information was coming in to Denis from his contacts throughout Africa of the occurrence of lymphoma cases, which he plotted on one of three maps in his office, of Uganda, Kenya and Tanganyika, giving more clarity to the boundaries of the lymphoma belt (Fig. 11.3).

He made a quick safari in early February 1961, the restrictions on travel having been eased, to the island of Zanzibar, off the north-east coast of Tanganyika, to find out why no-one was sending him case reports from this region when it was adjacent to the main eastern coastal extension of the lymphoma belt. The answer to his question was that no cases had been seen. This was to prove to be a highly significant observation when later in the year he and his accompanying safari venturers would start to make sense of the geography of the lymphoma. He was also working on the

Fig. 11.3 Denis in his office at Mulago with maps circa 1960

problem of phycomycosis[6] and how to manage it and wrote two papers on this with Greg O'Conor and other colleagues including Derick Jelliffe[7] [8, 9]. Meanwhile scientists were asking to come and visit Mulago and see the work that was going on, but Denis was finding it difficult to make time for them before he went on leave in March. Ian McAdam rang to say that the lymphoma situation was embarrassing, probably meaning that whilst the world understood the importance of Denis' work, the Department of Surgery at Mulago had not yet taken this on board. He asked if Denis would please take over the management of all the field studies and draw up a 3-year plan of what should be done. He agreed,[8] with the best of intentions, but for the next few months, his thoughts were occupied with the detailed plans for the long safari.

Before setting off for home leave in March, there came relief for Denis from his continuing battle to take ownership of the lymphoma but in an unexpected way and

---

[6] 'Subcutaneous phycomycosis is an infection of the subcutaneous fat by a fungus of the family Entomorphthoraceae…that results clinically in a slowly increasing painless area of subcutaneous induration' [8].

[7] D B Jelliffe (1921–1992). Pioneering American paediatrician who founded the Department of Paediatrics at Makerere. He was an advocate of breast-feeding and campaigned against the use of artificial milks, especially for developing countries.

[8] When he finally wrote a plan, it was brief, four pages, indicated the need for more safaris to define the geographical extent of the tumour and asked for a secretary, travel expenses, help for his collaborators and more office accommodation.

one that he would not have wished for. In January 27 he heard that Dilly Davies, the wife of Prof Jack Davies, had died suddenly. Two weeks later Jack rang Denis to say that he was now planning to leave Uganda and return to the UK. This would leave in Kampala only Greg O'Conor who had an interest in the pathology of the tumour, but he was planning to return to the USA and Jack's brother Glyn who was a radiologist. Denis would soon become the undisputed expert on the lymphoma although Jack had one last request that touched the sensitive spot in their relationship, offering to draft a joint paper on the lymphoma for an English publication. Denis was 'Quite absurdly hurt' by this and later talking to Olive about it writes 'This has been my battle. I, and Olive more so, have resented Davies' getting credit for what I selfishly considered my discovery'. The joint paper went ahead, amounting to only two or three pages in what would be considered a minor journal [10] and was to be the only time they would write anything about the lymphoma together. Possibly Jack had thought that this gesture of a joint paper would help to heal the relationship, which had been such a battle over the past 3 years.

Arriving by air at London Airport on Saturday March 18, 1961, with the family for home leave, which lasted almost 14 weeks, life was now different. Denis was in demand as a speaker, and the first 2 weeks were spent in the south of England where the family met Robin and they visited Westonbirt School in the Cotswolds where it was planned to send Cas. In London Denis talked to Bob Harris at the Imperial Cancer Research Fund (ICRF), whom he had met for the first time at Mulago in January, lectured at the Hammersmith Royal Postgraduate Medical School, met Drs Porter and Mannie from Lederle and Colonel Bizman at the Wellcome Museum (Fig. 11.4).

Then in March 22, Denis spoke at the Combined Medical and Surgical Staff Meeting at the Middlesex Hospital at 5:15 pm on the topic of 'The Commonest Children's Cancer in tropical Africa. A Hitherto unrecognised Syndrome'. A meeting that would become one of the pivotal moments in the story of viruses and cancer. A young research virologist, Tony Epstein[9] from the Bland-Sutton Institute of the Middlesex Hospital and Medical School, went along to the lecture. Epstein, who would become Sir Anthony Epstein FRS, was working at that time on chicken tumour viruses but had in mind the possibility that some human cancers might be virus induced. In conversation with Denis in later years,[10] he recalled that Denis 'in the past had spoken at the Institute on bizarre cases not seen in Western Europe but very common in East Africa. This time you came with a new title that caught my eye'. He went on to say that 'I listened to him, and after he had been talking for ten

---

[9] Who would become Sir Anthony Epstein (b.1921) MA, MD, PhD, FRS. Career started as Assistant Pathologist at the Middlesex Hospital Medical School, then Consultant Virologist from there to become Professor of Pathology and Head of the Department in the University of Bristol where Denis' former classmate at Trinity College, Dublin, William Gillespie would become Professor of Clinical Microbiology. Sir Anthony had a distinguished career, which including his discovery of a new herpesvirus, the Epstein-Barr virus.

[10] From videotape recording of Dr Tony Epstein in conversation with Denis in Oxford circa 1991. Family archives.

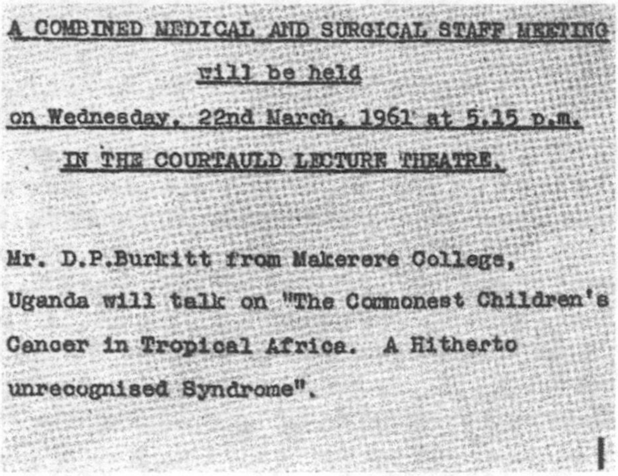

Fig. 11.4 Notice of the meeting at which Denis met Dr MA Epstein (Tony Epstein) for the first time. The talk made such an impression on Dr Epstein that he removed the notice afterwards and kept it (With thanks to Professor Owen Smith, Dublin)

A COMBINED MEDICAL AND SURGICAL STAFF MEETING

will be held

on Wednesday, 22nd March, 1961 at 5.15 p.m.

IN THE COURTAULD LECTURE THEATRE.

Mr. D.P.Burkitt from Makerere College, Uganda will talk on "The Commonest Children's Cancer in Tropical Africa. A Hitherto unrecognised Syndrome".

minutes it was absolutely clear that he had something tremendously important. It had immense theoretical implication, his idea that the geographical distribution of the disease indicated that there might be some infectious cause'. After hearing Denis talk about the possibility of a viral cause for his lymphoma, Epstein approached him afterwards and asked if it would be possible for Denis to send him some samples of the lymphoma obtained at surgery and he would pay any costs. Denis agreed, the funds were provided by the British Empire Cancer Campaign, and Dr Epstein went out to Kampala to set up a system for sending tumour samples overnight by air to London Heathrow airport where he would pick them up whatever the time of day or night they arrived. Three years later a virus would be isolated from the tumour and become known as the Epstein-Barr virus.

Useful work done in London the family headed for Laragh arriving there at the end of March. Instead of the usual warm welcome, updates on family events described around the fire at night and visits to be made to favourite haunts, there was nothing. James and Gwen were no longer around, and all that was left was for Denis and Olive to clear out their belongings, sell the furniture and then dispose of the house. They did this with heavy hearts. Without the parents there were only memories and the harsh reality of disposing of once precious items of everyday life that now looked mundane and of little value. From beyond the grave, they reached out to Denis and family knowing that they would find an envelope left for them. It contained a personal note from Gwen and £10. Gwen and James had worshiped for over 35 years at Trory Church and were buried in the churchyard. James had been an active and highly valued member, so the bishop agreed to Denis' request to have a plaque put up on the wall in the church in memory of them.

Prior to the memorial service for James and Gwen, and the unveiling of the plaque, Denis made a brief trip back to London to give a lecture at the Royal College of Surgeons of London [11]. Bob Harris from the ICRF had arranged for Denis to give the lecture after contacting Sir Cecil Wakely, a past president of the College and one of England's most distinguished surgeons. The meeting was held in the large main hall of the college and chaired by Wakely. The occasion proved to be something of a disaster. Denis describes the platform party that processed into the hall led by the mace bearer as 'fully attired in their gowns of office' with Denis in tow. There was an audience of 12. Whilst Sir Cecil had recognised the importance of Denis' work, the rest of the world had still to catch up. Back at Laragh the next day, Denis was asked to unveil the plaque to his parents on Sunday May 28, but the occasion took away his usual confidence in public, and he found it difficult to speak, instead unable to supress his tears. A few days later, Denis walked up the hill and stood by the gate where he had paused to ponder his future some 25 years earlier having been offered his first resident appointment in Chester. He closed the door of the large barn, said goodbye to the workshop where James had taught him carpentry and finally, with Olive, walked out of the front door for the last time. They drove to Holyhead and symbolically left Ireland although Denis would return on many occasions to see family and visit Trinity. Laragh was sold for £5000 (Fig. 11.5).

Whilst they had been on leave, the two papers in the journal *Cancer* had been published and in Denis' words the balloon went up. Dr Epstein arrived at Mulago for the first time on July 20, 1961, and the next day Denis found quite a crowd waiting outside the operating theatre for a lymphoma tissue biopsy 'two from New York – one from London and our own people'. *The Lancet* published an immediate editorial [12] in response to the papers in which a parallel was drawn between the cases that Burkitt and O'Conor reported and those reported by Rosenberg and colleagues from the Sloan Kettering Institute of Cancer Research in New York [13] some 3 years earlier. The editorial sought to establish that the incidence of the lymphoma in Africa was not much different from that in other parts of the world, such as the

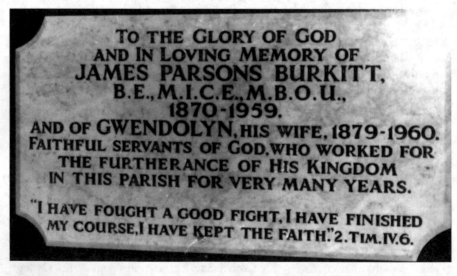

Fig. 11.5 Plaque in Trory Church to commemorate the lives of Denis' parents. Note Gwen's name should be Gwendoline and date of birth 1877

USA, that the preponderance of males to females was not significant,[11] that the meaning of the geographical distribution of cases in Africa was unclear and that all the cases were manifestations of a single syndrome. Denis didn't like this at all and wrote immediately to *The Lancet* [14] arguing that the 106 cases they had seen over an 8-year period compared with Rosenberg's 69 in 30 years at a specialised hospital 'is a rather impressive comparison'. He pointed out that the clinical and pathological characteristics of the African cases were unique and the specific limits of the tumour distribution suggested an arthropod vector. There were other clear differences between the two case series, namely, that the US children usually had lymph node enlargement whilst the African children did not, more usually presenting with jaw tumours. Denis was by now well used to defending his cause and responded by concentrating his efforts on writing about the lymphoma for various journals. He would have nine papers published in 1962 of which five would be about the lymphoma syndrome all with him as sole author. One battle was however now over. In September a final symposium was held to mark the departure of Professor Davies, whose distinguished career whilst in Kampala deserved recognition. Denis was asked to speak but kept his talk low-key talking about 'Geographical Surgery' and acknowledging Professor Davies' help with the pathology of the tumour.

Meanwhile construction of the new Mulago hospital was well advanced and Denis was able to move his unit to the new wards. He was promoted to senior specialist and received his first invitation to visit the USA from Dr Herbert F Oettgen who was one of a number of scientists from the Sloan Kettering Institute[12] now actively involved in work on the lymphoma. They were following two main lines of research, the virus story and finding a cure principally through chemotherapy. The arrival of scientists in Nairobi, Kenya, in 1960[13] from the USA with expertise in virology, immunology, chemotherapy and histopathology made Denis realise for the first time that he was not going to be able to solve all the problems relating to the cause and treatment of his lymphoma on his own. More importantly he knew that he lacked the knowledge, skills and understanding of these new sciences to make a useful contribution or form partnerships with these experts. He knew that they would be the people who would eventually identify the cause of the lymphoma and the best means of managing it. But for Denis there was still work to be done defining the geographical limits of his tumour, and as the clinician with the most experience of the condition in the world, he could collaborate with them in trying new treatments and optimising them. It was the response to treatment that would provide a further claim to Denis' ownership of the tumour.

---

[11] Male to female ratio 2:1 but Denis points out in his paper that this was unlikely to be significant, simply reflecting the ratio of male to female hospital admissions.

[12] Sloan Kettering Institute for Cancer Research was founded in 1945 in New York adjacent to the Memorial Hospital with money from General Motors (GM) named after Alfred Sloan, GM's Chairman, and Charles Kettering GM's Vice President and Director of Research.

[13] The original team included Dr Joseph H Burchenal, Bob (R D) Sullivan, two technicians and later Dr Herbert F Oettgen.

References

1. Davies AGM, Davies JNP. Tumours of the jaw in Uganda Africans. Acta Unio Internationalis Contra Cancrum. 1960;16:1320–4.
2. O'Conor GT, Davies JNP. Malignant tumors in African children – with special reference to malignant lymphoma. J Ped. 1960;56:526–35.
3. Burkitt D, O'Conor GT. Malignant lymphoma in African children: I. A clinical syndrome. Cancer. 1961;14:258–69.
4. O'Conor GT. Malignant lymphoma in African children: II. A pathological entity. Cancer. 1961;14:270–83.
5. Miles RR, Arnold S, Cairo MS. Risk factors and treatment of childhood and adolescent Burkitt lymphoma/leukaemia. Br J Haematol. 2012;156:730–43.
6. Wright DH. Cytology and histochemistry of the Burkitt lymphoma. Br J Cancer. 1963;17:50–5.
7. Black D, Gray J. Sir Harold Percival Himsworth KCB. Biogr Memoirs Fell R Soc. 1995;41:201–18.
8. Burkitt DP, Jelliffe DB, O'Conor GT. Subcutaneous phycomycosis in an East African child. J Pediatr. 1961;59:124–7.
9. Burkitt DP. Wilson AMM and Jeliffe DB 1964 Subcutaneous phycomycosis: a review of 31 cases seen in Uganda. BMJ. 1964;1:1659.
10. Burkitt DP, Davies JNP. Lymphoma syndrome in Uganda and Tropical Africa. Med Press. 1961;245:367–9.
11. Burkitt DP. A lymphoma syndrome affecting African children. Ann Roy Coll Surg Eng. 1962;30:211–9.
12. Editorial. Malignant lymphoma in African children. Lancet, 1961:1156–1157.
13. Rosenberg SA, Diamond HD, Dargeon HW, Craver LF. Lymphosarcoma in childhood. NEJM. 1958;259:504–12.
14. Burkitt D. Malignant lymphoma in African children. Lancet. 1961:1410–1.

Chapter 12
Safari

The 10,000-mile, 10-week journey that Denis made from October to December 1961 in an old Ford Station Wagon with two friends, Drs Ted Williams and Cliff Nelson, at a cost of only £678 is now the stuff of legend. From Uganda they journeyed south to Tanganyika (Tanzania), into Northern Rhodesia (Zambia), Southern Rhodesia (Zimbabwe), east to Nyasaland (Malawi), then south through Mozambique, Swaziland and finally to South Africa. The return journey took a different route through Southern and Northern Rhodesia, Tanganyika, Kenya and back to Uganda (map 1). The safari was recorded in detailed daily notes, letters to family, multiple photographs, early attempts with newly acquired movie cameras and the subsequent publication of a series of scientific papers,[1] including in the most prestigious journal at the time, *Nature*, and was the subject of a book in 1970 by the writer Bernard Glemser [1].

The safari, a timely and fruitful journey, is now regarded as a pivotal event in the search for the cause of the lymphoma. It would show the distribution of the lymphoma to be dependent on altitude, in turn interpreted as temperature. Yet as an exercise in epidemiology, it would today be considered wholly unscientific, almost naive and unlikely to attract any funding. However, given the lack of any reliable cancer registration in most of East Africa at the time, the cultural norms surrounding the rapid death of young children with a horribly disfiguring and tragic illness and the resources and infrastructure available for such a venture, the project should be seen as brilliant in conception and ground-breaking in its achievements. Sir Harold Himsworth, writing the foreword to the first book about the lymphoma, edited by Burkitt and Wright, in 1970 says:

> No explanation of cancer, or of the normal mechanism the disturbance of which underlies this, will ever be adequate unless it satisfactorily accounts for this particular tumour, its distribution, its age incidence, its response to therapy and its relation to other diseases. In a sense this tumour may well prove to be something of a Rosetta stone. [2].

---

[1] See Appendix 3, 1962.

© The Author(s), under exclusive license to Springer Nature Switzerland AG 2022   163
J. H. Cummings, *Denis Burkitt*, Springer Biographies,
https://doi.org/10.1007/978-3-030-88563-2_12

Denis could rightly claim, from his original clinical description of the lymphoma syndrome and his preliminary work on its geographical distribution, that understanding the lymphoma might well point the way to finding the cause of and treatment for other cancers, such as childhood leukaemia, which was considered to be closely related. In this context the safari was an essential and visionary step on the path to discovery. And given the rapidly growing interest by scientists around the world in the cell biology, histopathology and virology of the lymphoma, a grand safari in Africa was something to which Denis uniquely could contribute and was within his comfort zone. With David Livingstone as his hero, he set off with confidence (Fig. 12.1).

The principal record of what came to be known as 'The Long Safari' lies in the letters Denis wrote to Olive more or less daily on an Olivetti Lettera 22. They amount to about 80 A4 pages of single-spaced text detailing the events of each day, where they went, who they met and the logistical problems they encountered. These papers are now housed in the Wellcome Library in London,[2] and those covering the first half of the safari, the outward journey, are extracted and printed as a final Appendix 'Three Men on a Safari' at the end of Glemser's book. Other records include Denis' notes in his personal diary, which he kept most days; a VHS tape, lasting about 40 min, transposed from the movie that Denis and his companions filmed whilst on the tour; and a large album containing about 190 black and white photographs that Denis, mostly, had taken and then developed, printed, enlarged and annotated. In addition, there are seven hand-drawn maps of the route they followed. These records, mostly, give little information about the main focus of the journey, namely, the observation and recording of the place of occurrence of cases of childhood lymphoma. They are simply a travelogue. The essential information about the tumour is remarked upon in Denis' autobiography, written 19 years after the event, but was more reliably published in the medical and scientific press in 1962 and subsequent years (Appendix 3).

For months before they departed, Denis had been writing to the doctors in charge of all the hospitals they intended to visit informing them of the purpose of the safari and giving them clinical details of the lymphoma, including photographs of children with varying manifestations of the condition. Eventually a total of 63 hospitals, mostly run by missionary societies, were incorporated into the itinerary, and everyone was given a date on which to expect the touring party. At each visit the staff were asked whether any lymphoma cases had been seen, and hospital records were examined where possible to search for details such as age, sex, surgery, pathology and outcomes if known. The information was recorded by Denis on a simple questionnaire he had devised, but what is very revealing on inspection of it is that he was also collecting data on the occurrence of other cancers.[3] He was thinking ahead (Fig. 12.2).

---

[2] Wellcome Library. Papers of Denis Burkitt WTI/DPB/B/4 Box 4.

[3] Trinity College Dublin Manuscripts and Archives Research Library; MS 11268/1/3.

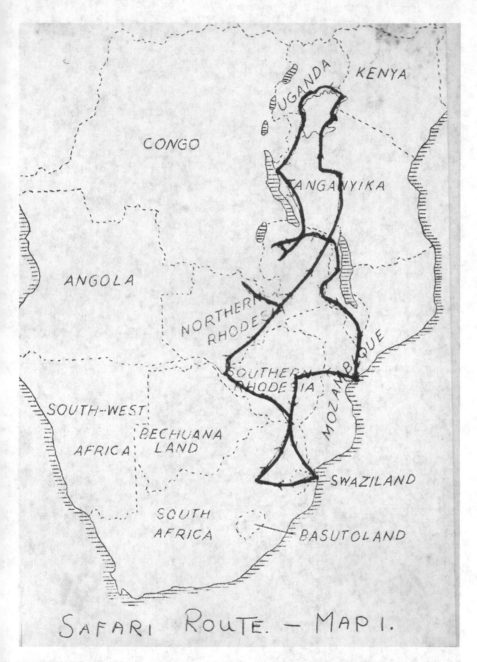

Fig. 12.1 Safari map 1. Overall route of the safari, drawn by Denis, to accompany his pictorial account of the journey

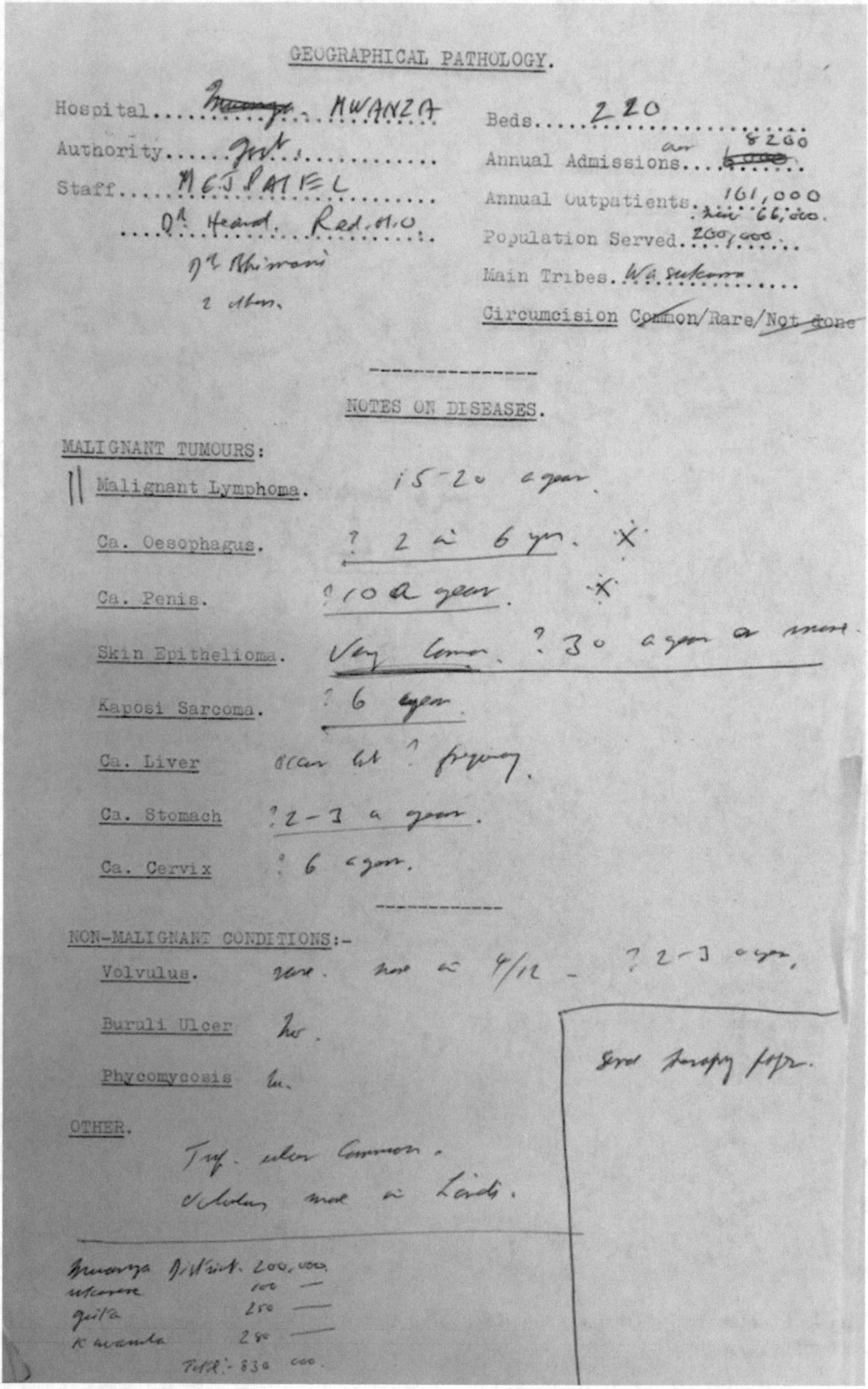

GEOGRAPHICAL PATHOLOGY.

Hospital... MWANZA   Beds... 220
Authority... Govt.   Annual Admissions... 8200
Staff... M & J PATEL   Annual Outpatients. 161,000
... Q^r Heard. Res. M.O.   Population Served. 200,000
    Q^r Bhimani   Main Tribes. Wa Sukuma
    2 others.   Circumcision Common/Rare/Not done

NOTES ON DISEASES.

MALIGNANT TUMOURS:

‖ Malignant Lymphoma.   15-20 a year.

  Ca. Oesophagus.   ? 2 à 6 yrs. X

  Ca. Penis.   ? 10 a year.   X

  Skin Epithelioma.   Very Common. ? 30 a year or more.

  Kaposi Sarcoma.   ? 6 a year.

  Ca. Liver   Occur but ? frequency.

  Ca. Stomach   ? 2-3 a year.

  Ca. Cervix   ? 6 a year.

NON-MALIGNANT CONDITIONS:-

  Volvulus.   Rare.   now à 4/12 -   ? 2-3 a year.

  Buruli Ulcer   No.

  Phycomycosis   No.   Seen frequently I figure.

OTHER.
        Trop. ulcer Common.
        Cholera more à Lands.

  Mwanza District. 200,000
  Ukerewe   100
  Geita   250
  Kwanda   280
      Total:- 830,000.

Fig. 12.2 Questionnaire about lymphoma occurrence completed by Denis at the hospital in Mwanza, just south of Lake Victoria in Tanganyika (map 7). At an altitude of 3700′, there should not have been any cases of lymphoma. Note that he is collecting information on a variety of other diseases. December 1961 (Trinity College Library, Manuscripts & Archives Research Library MS 11268/1/4)

Ted Williams and Denis left Mulago at 09.20 h on Saturday October 7 and would be joined by Cliff Nelson 2 days later. They travelled in a large 35 HP Ford V-8 American Station Wagon that Ted had obtained from an ex-Congo missionary for £250 (GBP) although he had to pay an additional £50 import duty and for insurance and registration. They would later sell the car for £125, thus defraying a significant part of their expenses, an important transaction for Denis who, with good humour, cajoled his companions into accepting a policy of minimum expenditure for the trip. Hotel rooms without a bath were always preferred, a cup of tea and biscuits instead of a proper lunch, camping where possible using equipment they had borrowed for the journey and food from the local market for their wayside picnics. From Kampala they drove down the western side of Lake Victoria to Masaka, where the tarmac road ended, and then 40 miles further south enjoyed a picnic lunch prepared by Olive, 'a perfect meal with everything labelled', which they agreed was better than a ten-course lunch at the Imperial Hotel in Kampala. Crossing the border into Tanganyika at a small town, Mutukula, there were no formalities, just a notice informing travellers of the border, and then on to the river Kagera where the ferry comprised a heavy wire cable across the fast-flowing river, attached to the boat and propelled by a team of locals walking along the deck holding the cable. Travelling east just south of the equator, they reached the town of Bukoba on the shores of the lake at 4:30 pm and went to find the district medical officer. He was Indian and a Christian, which commended him to the men. They went through the medical questionnaire that they had drawn up with him, and he brought the hospital operation registers for them to inspect. That evening a group of the DMO Christian friends gathered at his house and the two travellers had fellowship with them. The first day of the safari had ended well. There were 68 more planned (Fig. 12.3).

That night they stayed in a hotel about 17 miles south of Bukoba returning in the morning to visit the government hospital, the only one for 600,000 people, where Denis noted that heart failure was being treated with penicillin and anaemia with a form of glucose given into a vein. Neither was likely to be of any benefit but may have made the patients feel cared for. Time, and bed rest, would have healed some of them. On the way south, they stopped at a Roman Catholic mission (RCM) hospital run by two German lady doctors, and after being shown around, they were given the information about the lymphoma they were seeking following which they set off for a short drive to a Swedish Mission Hospital at Ndolage 10 miles west of the main road. Denis was well known in these hospitals, despite their being outside Uganda. He had operated on a sister from the RCM previously and removed the tonsils from the sister-in-law of the doctor at the Swedish hospital quite recently.

After staying Sunday night at the house of a doctor who was away on leave, they set off the next day shortly after 6 am continuing south to meet Cliff Nelson and his wife Beth who had left Kola Ndoto[4] at 5 am. After lunch Beth left for Mulago to stay with Olive carrying letters from Denis and Ted, whilst the three men now headed for Kibondo where they stayed in a simple thatched cottage booked in advance. The

---

[4] Kola Ndoto or Kola Ndota as written by Denis on the maps. Kolandoto today.

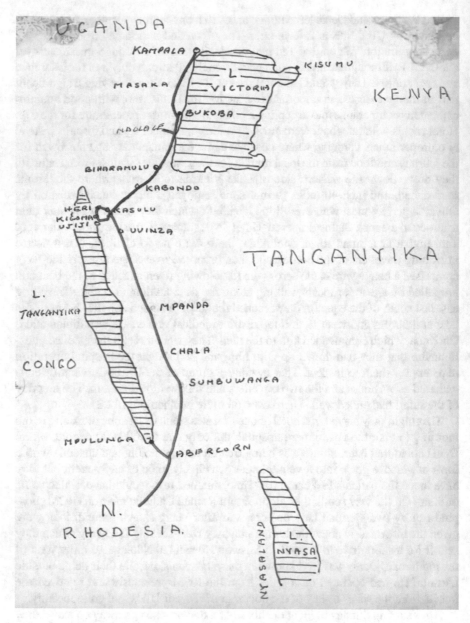

Fig. 12.3 Safari route map 2. October 7, Bukoba; 12, Mpanda; 14, Abercorn

three men shared a common faith, which clearly made for a harmonious journey because there is no record of any serious disputes or falling out during the 10 weeks they shared in each other's lives. Ted, who had been born in Nairobi and worked for the AIM for 15 years had known Denis all this time. He spoke several African languages, was a very competent engineer having built the hospital at Kuluva and was a byword for ingenuity and good humour. Cliff Nelson, a Canadian born in Alberta, had arrived in Africa in February 1958, spent 3 ½ years at a mission station and had been government medical officer of the district where Ted had worked so they were close friends. He could turn his hand to anything. He was able to join the safari because he had given up his job as DMO and was on his way to take charge of a mission hospital in Tanganyika. Both were Christians. The three wives all knew each other (Fig. 12.4).

Jobs were apportioned to each man according to his skills with Ted the engineer looking after the car and acting as treasurer, paying the bills and keeping an account. Cliff was assigned more domestic duties such as the laundry, and Denis was responsible for the main scientific purpose of the trip, gathering data on the lymphoma. Ted had an interest in bird song and had brought a tape recorder. They kept to quite a tough schedule averaging around 140 miles daily often on poor dirt roads, although traffic was light. At the end they would be defeated by conditions but meanwhile there was work to do.

Supper in Kibondo comprised baked beans, bully beef,[5] biscuits, cheese, fruit salad, tea made in a plastic jug and coffee. Declared a lovely meal, the cooking such as was necessary was done by Ted on the Primus stove they had with them. Between

Fig. 12.4 Cliff Nelson, Denis and Ted Williams. November 1961

---

[5] Corned beef, a tinned salt-cured meat which would have been very familiar to Denis following his army service.

courses Cliff wiped the plates clean with toilet paper. Water was in short supply and for bathing was heated up in an old oil drum over a wood fire. The next day, Tuesday, after more bully beef and biscuits for breakfast, they continued south-west towards Kasulu where they took a 30-mile diversion to visit a Seventh Day Adventist (SDA) hospital at which the South African missionaries who ran it were able to give Denis a lot of valuable information about the lymphoma, or lack of it. From the hospital, which was at 3000′, they had good views north to Burundi. After lunch they descended towards Kigoma on the north-eastern shore of Lake Tanganyika[6] having covered 380 miles over the 2 days, latterly on a precipitous road cut out of the mountainside. The car radiator had boiled on the ascent to the SDA hospital, the petrol tank indicator ceased to work, and there was trouble with the speedometer, but considering the age of the car, the load it was carrying, the terrain and the ambient temperature so close to the equator, it ran well under the watchful eye of Ted.

Close to Kigoma, the major port on Lake Tanganyika linked directly by a railway that ran due east to Dar es Salaam, lies Ujiji. In 1961 it was a small town of narrow dirt streets and mud and wattle houses either thatched or covered with old, flattened petrol tins and some corrugated iron. Visiting it late on Tuesday afternoon was a memorable moment for the three men. Here on October 27, 1871, Stanley had greeted Livingstone with the now immortal words 'Dr Livingstone I presume?' For Ted it was a lifetime ambition to visit the place and stand at the exact spot, under a mango tree where the meeting had taken place. Now marked with a stone monument, Denis found it an emotional experience having read most of the published accounts of Livingstone's heroic journeys of exploration. Many photographs were taken.

In Kigoma they met the DMO and had dinner with him. They were now clearly in the lymphoma area because he was able to give them a lot of information about the occurrence of the tumour locally. The next day, Wednesday, took them through an endless, monotonous landscape of dry inhospitable bush without any habitation. Stopping only to visit a Catholic mission to gather information, they pressed on to Mpanda, but en route the engine of the car started to miss and then stopped completely. Ted managed a speedy repair, wiring the battery directly to the coil, and all was well. In Mpanda they were given the keys of a house by the African officer commanding (AOC) the police where there were two rooms, a bathroom and flush toilet. Denis was allowed one room on his own by the other two in acknowledgement of his role as leader of the expedition. Supper was out of tins again, but the next day, they set off in high spirits singing hymns and watching a variety of game in more interesting country. The road was very rutted, but before leaving Kampala they had a large metal plate fixed underneath to protect, successfully, the engine and petrol tank from underlying rocks. They arrived at Sumbawanga late morning having left Mpanda at 6 am and were able to have a relaxing day writing letters and enjoying a bath and electric light in the Abercorn Hotel. They met the DMO

---

[6] Lake Tanganyika. Second largest freshwater lake in the world after Lake Baikal in Siberia. It is 420 miles (675 km) from north to south. The safari route took the men from Kigoma along the road 30 miles east of the lake to Mbala (Abercorn) on the southern tip, about 400 miles.

(district) and PMO (principal) for the usual fact-finding discussion and had the luxury of eggs for supper. Friday took them across the border into Northern Rhodesia (Zambia) and to Abercorn (Mbala) at the southern tip of the Lake. The medical officer, a South African Christian, took them all to see the Mpulungu falls and told them about Northern and Southern Rhodesia, their populations and economy. Altogether more thriving than Uganda, but with much smaller numbers of people, Northern Rhodesia is three times the size but with only one third of the population although there were 70,000 Europeans in the North compared with only 5000 in Uganda.

Thus, ended the first week. Everything had gone according to plan; they had covered about 1000 miles and visited eight or nine hospitals where useful information about the lymphoma had been gathered. They were enjoying the travel and as Denis remarked 'This must be the safest ever safari in Africa. Here we are three doctors, each with our own private stock of medicines making a beeline from one hospital to the next' [3]. The following week, October 14–21, they took a carefully planned route that had them first heading west to the border with the Congo, then returning via Abercorn to journey south-east, finishing in Livingstonia (Kondowe) in the mountains 9 miles (15 k) from the shores of Lake Nyasa in Nyasaland (Malawi) (Fig. 12.5).

Highlights of the second week included the drive on the first day to Kawambwa through 250 miles of uninhabited bush seeing only two vehicles all day. They stayed near to Kawambwa at the Mbereshi Mission Hospital where at worship in the church on the Sunday they noted the sexes sat on separate sides of the aisle although the preacher was a woman. Denis did a ward round in the hospital and they all visited the nearby leper colony. After operating on a difficult case early the next morning, they set off for Fort Rosebery (Mansa) a short drive, gathered more data on the lymphoma at the hospital and met a nun/doctor from a nearby Catholic mission hospital who had been in the year behind Denis at Trinity. Then, heading back towards Lake Tanganyika, they stayed at Kasama the provincial capital where there was a well-equipped 200-bed hospital run by a single doctor, Dr Wright. Denis operated for him in the late afternoon after which they headed for Tunduma where a small government hospital was being run by a very wide-awake lady of over 80 – Dr Trant. The reason, maybe, she gets a mention amongst the panoply of the great from the safari is that she was a graduate of Trinity. After entering Nyasaland at Chitipa, they found a rest house at Chisenga in an impressive setting looking out over the mountains and decided to spend a day there catching up on ablutions, writing and enjoying some free time. Supper was a pound of bacon shared amongst the three of them, two fried eggs each, baked beans, tinned pineapple and tea. Denis noted that the butter had melted.

From Chisenga they reached Karanga on the shore of Lake Nyasa quite quickly, visited the hospital and then set out for Livingstonia (Kondowe). The main reason was to visit the David Gordon Memorial Hospital and collect information about the lymphoma. This was an important visit because Livingstonia had been founded by missionaries of the Free Church of Scotland who had moved there in 1894 having some years earlier established a mission at Cape Maclear at the south end of the

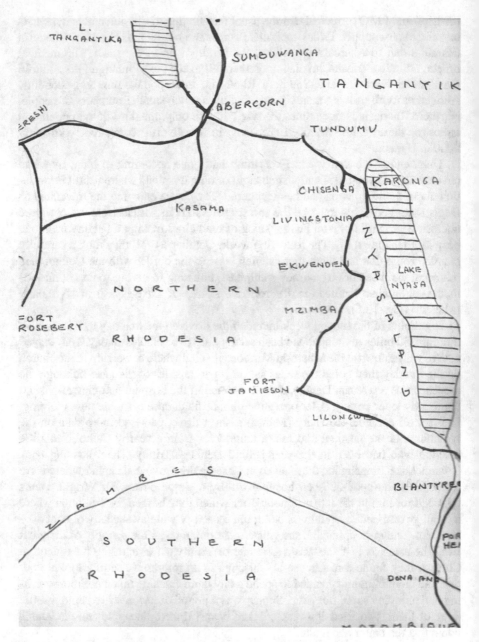

Fig. 12.5 Safari route map 3. October 16, Fort Rosebery; 20, Chisenga; 20–21, Livingstonia; 23, Fort Jamieson; 27, Blantyre

lake. They had been driven out by the mosquitos and their attendant malaria. They had travelled north and settled at Bandawe in 1881 with no more success and finally moved into the hills above the Lake. Livingstonia lies at 4247′ and was mosquito-free. Given the developing interest in a viral aetiology for the lymphoma by a vector such as the mosquito, a visit to Livingstonia was crucial to establishing the geographical and climatic limitations to the tumour belt. Turning west from the road along the shore of Lake Nyasa, about 60 miles from Karonga, Denis remarks: 'I have never ever, anywhere, been on such a road. It climbed over 2,000 feet in seven miles, which was only about two miles as the crow flies. There were 112 bends and 22 hairpin turns doubling back virtually 180°. Moreover, at these hairpin bends the turn was so steep that our car could only just make it in bottom gear. On six of the bends we just couldn't get round in one turn even with the wheels fully locked, and had to back on the corner'. Their radiator boiled repeatedly and they had to refill it and pour water over it to keep it cool. At the top their journey was rewarded with perhaps the finest view in the Federation.[7] For the night of Friday October 20, they stayed with Dr. Maclean, aged 72 years, and his wife, aged 79 years, a cultured and radiant couple who had come out of retirement to run the 150-bed hospital. They lived in a rambling old house much in need of repair, but at dinner there was a silver teapot, good china and table linen.

As they journeyed further south, it was the people they met that captured their interest, along with the changing landscapes, rather than information about the lymphoma, which was now mostly a matter of record. A short drive brought them to Ekwendeni, another hospital founded by Scottish missionaries where they met Dr. Ian Irvine who ran the 60-bed hospital without any other medical help. Continuously on call for every emergency and with responsibility for all the buildings including their mechanical and electrical facilities and with only a pit latrine in the yard, the three men realised that their own lives, ostensibly ones of sacrifice, were relatively comfortable. They stayed for Sunday at Ekwendeni, attended church and dined with Dr. Irvine whose father they discovered was a great friend of Uncle Roland in Nairobi and of Ted's father. Eighty miles further south in the town of Mzimba, they delivered a letter, that Dr. Irvine had given them, to a couple living on the outskirts of the town. Denis asked them where they came from and was told 'Enniskillen'. They asked Denis if he was the son of J.P. Burkitt. A memorable moment for them all in the middle of Africa and what WB Yeats recognised as 'the indomitable Irishry'.[8]

Fort Jamieson (Chipata) just over the border into Northern Rhodesia (Zambia) was a further 170 miles. On arrival they went straight to the hospital and asked the PMO on the 'phone whether he had seen any cases of the lymphoma. This conversation was overheard by the ward sister who went and brought a typical case for them to see, and photograph. The hospital was almost at sea level and the climate hot. Although the lymphoma studies were the province of Denis, the three men spent time discussing what they had seen, where the tumours were occurring and why. Denis realised from the time he was spending with Ted and Cliff that they were both very able people. Ted had a wide knowledge of medicine and had been an unusually

---

[7] Federation of Northern and Southern Rhodesia and Nyasaland.

[8] WB Yeats 'Under Ben Bulben'

good student, whilst Cliff had a very good brain. Denis remarks 'Although I am leader of the party, I feel in many ways intellectually behind them'. If intellect can be so simply defined, Denis' modest achievements at school and in the early years at Trinity reminded him of his limitations, but these were more than compensated for by his broad vision of life, energy, problem-solving abilities and practical skills, and he was an avid reader. It was Ted who now drew a diagram relating occurrence of the lymphoma to altitude (Fig. 12.6).

Back in Malawi they visited the hospital in Lilongwe where they met Sister Sherriden who showed them round the wards. Denis, now curious about the pervasive influence of the Irish in this part of Africa, asked her where she came from. The reply '..a town in N. Ireland called Enniskillen'. The third person he had met in 2 days from his home town. They stayed in a hotel and invited the three local doctors to dine with them. The next morning Denis felt unwell but had grown accustomed to episodes of gut infection during his many years in Africa, so they all pressed on to Blantyre, the financial capital at the southern end of the country. There they met a very experienced surgeon, Mr. Laycock, who had worked in many countries and when in Somaliland had to do a caesarean section on his theatre sister wife because no other help was available. They stayed for an extra day in Blantyre where Denis was asked to do a thyroidectomy and where he gave a talk to the local British Medical Association meeting. On Saturday October 28, they crossed the Zambezi river by the railway bridge, after many delays, having put their car onto a goods train from which they were offloaded at midnight. After crossing the bridge, they drove through the night, getting lost en route, arguing over which end of the compass needle pointed north[9] arriving at 8:30 am on Sunday in Beira on the coast of Mozambique (map 4). They went to bed, later attended a service at the European church and the following morning tried to find out about the lymphoma from the doctors at the hospital, but communication was difficult as the main language was Portuguese. Eventually one, who had worked in the USA, joined them, and they discovered that the tumour was common throughout the low-lying areas of Mozambique, and a classical case was produced for them (Fig. 12.7).

Heading west the next day to Umtali (Mutare) in Southern Rhodesia (Zimbabwe), they realised that they were in a new world. Beira, a busy port and city of over half a million founded by the Portuguese at the end of the nineteenth century, had become a cosmopolitan modern town busy with commerce, tourism, well-stocked shops, restaurants and fine buildings built in a characteristic Portuguese style, which Denis felt were unorthodox in shape and colour but exciting and pleasing. Umtali, 186 miles from Beira in the mountains at an altitude of about 3500 feet with a more benign climate, impressed the men. 'We seem to have returned to civilization completely in the past two days. There are masses of flowering shrubs, various shades of bougainvillaea, carpets of fallen blue petals under the jacaranda trees, well-kept parks and trim lawns.' A lyrical turn of phrase for Denis. In the evening they went

---

[9] From a letter by Ted Williams written to all those who gave them hospitality on the safari, dated January 25, 1962, Kaluva Hospital, containing this unique account of an argument the men had on the trip. Family archives.

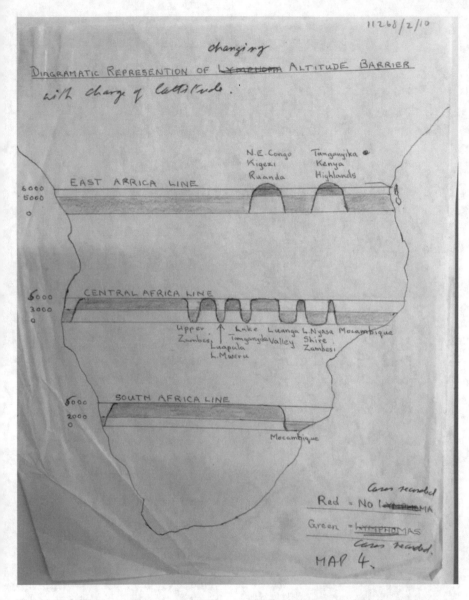

Fig. 12.6 Original map drawn by Ted Williams in 1961 to relate cases of the lymphoma to altitude. The writing is that of Dr. Williams; the amendments are in Denis' hand. Dr. Williams was almost certainly the person who first made the connection with altitude, although the three men would have had plenty of time for discussion during their time together on safari

On hearing of the death of Dr. Williams in 1992, Denis wrote 'This map was drawn by my "long safari" companion Ted Williams after the "long safari" to show the altitude dependence of the tumour which proved to be a temperature dependence' (Trinity College Dublin, Manuscripts and Archives Research Library MS 11268/2/10)

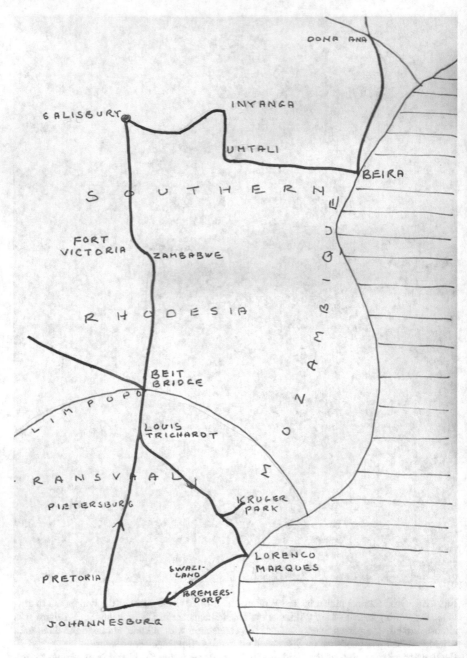

Fig. 12.7 Safari route map 4. October 31, Umtali; November 5, Zimbabwe ruins; 8, Kruger Park; 10, Lourenço Marques; 13–17, Johannesburg

to see a film about the life of Franz Liszt but were surprised to find that the cinema was reserved for Europeans only, which segregation they had not come across before. The next day they looked round the hospital. Whilst they had not been able to find any records of the lymphoma at a smaller hospital on the journey to Umtali, at 2000', there is no record of whether they were able to see any in Umtali, potentially a crucial finding in view of the altitude. Denis lectured in the evening, and the next day they drove to Salisbury (Harare), Zimbabwe's capital city where they were able to pick up post that had been waiting for them. The drive was through the most beautiful scenery of their whole trip so far causing Denis to leave the safety of recording what they saw and did on the journey to a more poetical interpretation of the landscape 'Great rocky hills, immense vistas, patterns of wattle plantations, and Paul Henry clouds.[10] Changing vistas all the time and crisp fresh air. The hilltops were wrapped with flimsy mantles of drifting cloud'. Was this the real Denis now free from the cares of hospital and university life in Kampala?

In Salisbury they were driven around the city by one of the local surgeons, Mr. Hammer, and were impressed: 'Like Manhattan in the middle of Africa', skyscrapers, a magnificent and well-equipped hospital and numerous staff with significant private practices. Lunch with a radiologist was in an elegant house furnished with antiques after which Denis met the DMO, visited the hospital and gave a lecture in the evening. He stayed with the Hammers, whilst Ted and Cliff were put up in a hotel. The next day, Sunday November 5, saw them in Fort Victoria (Masvingo) where they stayed for a day and drove out to see the ruins of the ancient city of Zimbabwe with its massive stone structures built between the eleventh and fourteenth centuries, now a world heritage site. The men were impressed by the scale but mystified by the disappearance of the civilisation and lack of development for the following 400 years. Crossing the Limpopo River at Beit Bridge, they entered South Africa and headed for the Kruger Park and some game watching. Booked in at Pretorius Kop at the southern end of the park, they spent Wednesday the eighth enjoying the huge herds of animals that were seemingly unconcerned by the presence of motor cars. Opening the windows of their station wagon to get better photographs, they were invaded by monkeys and baboons but escaped unscathed and the next day headed back to Mozambique and Lourenço Marques (Maputo) where they met Professor Prates, the best-known name in medicine in the country. Professor Prates worked at the University Hospital and was a world authority on liver cancer having recorded the highest rates in the world of this disease occurring on the Mozambique coast. He was interested in all cancers and had compiled records over many years of cases including the childhood lymphoma (Fig. 12.8).

After breakfast on Thursday morning, the three travelling doctors decided to visit the barbers where they all had haircuts, which Denis described as taking as long as a major operation. For six shillings each they were 'smothered in talcum powder, drenched in sprayed scent, and drowned in hair oil'. In addition, they had their shoes polished and walked out of the shop feeling real dandies. The rest of the day was

---

[10] Paul Henry (1877–1958) noted Irish landscape artist.

We spent most of Friday with
Dr. Prates who has done
magnificent work in his
Pathological Department. We
obtained most interesting
statistics to help with our
research programme.
The photograph below shows a
number of busts depicting
children with the tumours
we were investigating. We
found records of 40 of these
cases in his statistics over
the last five years.

Fig. 12.8 Denis' photograph of Dr. Prates and his plaster models of patients with lymphoma, from around Lourenço Marques (Maputo), made by a local man at the request of the professor of pathology

spent with Professor Prates to whom Denis had written warning him of their visit. This had allowed him to get out a series of plaster models of patients with tumours, which had been crafted, at the request of the professor of pathology, by a man who had worked in the local museum. This was the most southerly place from which Denis' tumour had been reported. They found 40 cases and plotted on their map the places where they came from because their addresses were available. Perhaps more important for Denis' future work, he was able to get detailed information about the incidence of other cancers including cancer of the oesophagus, penis, skin, liver, stomach and cervix and Kaposi's sarcoma. The absence of colorectal (bowel) cancer from this list is probably because it was almost never encountered. Denis described the time spent with Prof Prates as one of the most fruitful visits of the tour.

From Lourenço Marques they crossed the border into Swaziland and called at a fine mission hospital with 250 beds run by the Church of the Nazarene in Manzini. At lunch with the doctors, they learnt that, despite being only 100 miles from the high incidence area for the tumour in Mozambique, no lymphoma cases had ever been seen in Manzini. The city was at an altitude of 2500′, which was just above the line that Ted Williams thought divided high- and low-risk areas for the tumour, so the information they obtained here was important part to their survey. Onwards then to the most southerly point on their safari was Johannesburg where they were to spend 4 nights and meet some of the best scientists in Africa. After a drive of 260 miles on what were now excellent roads by early afternoon on Monday November 13, they were passing the spoil heaps made up from the tailings of the gold mines. They were met at the South African Institute of Medical Research (SAIMR) by Dr. George Oettlé, Director of the Cancer Research Unit at SAIMR. He was Africa's leading cancer research scientist and a pioneer in geographical or demographic pathology, in particular the occurrence of cancer according to racial origin, place of birth or residence. He had published one of the classic papers on this subject, with John Higginson in 1960 [4],[11] and was known for his fine intellect having graduated from the medical school of Witwatersrand in Johannesburg with first class honours and many prizes. Regrettably, he suffered from chronic rheumatic heart disease and was to die 6 years later at the age of 49 following surgery. Denis had met Oettlé in 1958 when he visited Kampala where he had told Denis that he had never heard of any cases of lymphoma in South Africa. This was surprising given the many low-lying coastal areas and had propelled Denis in search of the geography of his tumour. Denis lectured to the Institute staff at 5:10 pm and found the audience most receptive to his ideas about the lymphoma and its possible cause. He was informed that they had now seen two cases in South Africa one of which was in a European child, the first ever recorded. The child had spent holidays at an altitude of 2000′ much lower than Johannesburg at 5750′, which Denis thought significant. In the evening they were invited to dinner by a drug company representative from Johannesburg called Mr. Thomas, whom they had met earlier with Dr. Prates, and his wife who was an American scientist working for the company. After dinner they repaired to the hotel where they were all staying, considerably more luxurious than any other they had encountered, or would encounter, on the whole safari. There they began to talk business. The company had a new drug for the treatment of schistosomiasis which Ted and Cliff had agreed to trial, the infection being

---

[11] John Higginson (1922–2013) was head of the Geographical Pathology Unit and Cancer Registry at SAIMR. An Irishman he went in 1966 to be the first Director of the WHO International Agency for Research on Cancer (IARC) in Lyon.

endemic[12] in the areas where they worked. Denis felt that the expensive dinner and hotel were a small price that the drug company was paying for the opportunity to test their product in the field. But always trying to see the best in everyone, Denis was won over by the help he had received from Mr. Thomas and his kindness so offered to put him in touch with key people in Kampala who could test the drug for him.

The next day they were taken to Soweto to visit one of the world's largest and most famous hospitals, Baragwanath. It had started life as a military hospital in 1942 but after the war was designated specifically to serve the indigenous Black and Coloured populations where it attracted outstanding medical staff of all backgrounds who did not favour the racial policies of the then South African government. The many wards for the 3500 patients, operating theatres and other facilities were in huts spread over a vast area. Denis, Ted and Cliff were warmly welcomed and shown round parts of the hospital. They picked up many ideas that they would take back to their respective units. The following day, Wednesday, was even fuller and more interesting at the Virus Research Unit 7 miles from town. During the many hours of discussion that the three men had during the safari, it was becoming apparent to them that the pattern of occurrence of the lymphoma indicated an insect-borne virus as the cause. At the Institute they saw tissue cultures of living viruses and maps of their distribution. At the SAIMR in the afternoon, they met experts on maps, the distribution of insects, animals and vegetation. This was priceless information as they sought to find out what factors might be responsible for the distribution of the tumour. They were able to obtain a number of useful maps to take back to Kampala. That a virus might be involved was now firmly on their agenda and was a proposal of the highest importance for cancer research. The evenings were spent with the Oettlés and their six children. They were Plymouth brethren, and before going to bed, each child would read a couple of verses from the Bible to the assembled family and guests.

The final day they were due to meet three more authorities in different subjects at SAIMR and then have lunch with the professor of surgery and senior surgeons at Coronation Hospital. Of the meeting with one of these scientists, Denis records that he had some time with a nutrition expert who told him that it was better to be undernourished than over nourished and that coronary heart disease was common amongst richer Europeans in Johannesburg. This authority was almost certainly Dr. A.R.P. Walker who would become a close friend and in whose wisdom and judgement Denis would have the greatest confidence as Denis broadened his research beyond the lymphoma into other cancers (Fig. 12.9).

Leaving Johannesburg on Friday November 17, the return journey followed a different route through most of the countries they had already visited except those on the coast and would take only 4 weeks. They firstly headed north to Pietersburg where a routine that had been developed was now in place. They would visit the local mission hospital and meet the doctor in charge, who at the 500-bed hospital in

---

[12] An identifiable area where a disease is frequently found.

Fig. 12.9 Alec Walker in
1986 in Cambridge

Pietersburg was Dr. Naude, and ask him or her whether they had seen any cases of
the jaw tumours or lymphoma and fill in the questionnaire about other conditions,
and in the evening Denis would address the local medical society. Dinner was usually with their hosts and accommodation in a modest hotel and then on to the next
town staying longer in the major centres for sightseeing, shopping, collecting their
post and discussions with medical staff. Sundays were always reserved for attendance at the local church, meeting people and rest from travel (Fig. 12.10).

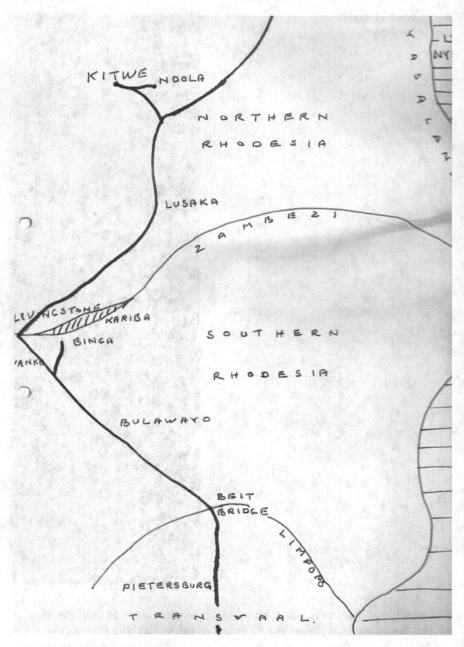

Fig. 12.10 Safari route map 5. Return journey. November 18–22, Bulawayo; 24, Victoria Falls; 25–27, Livingstone; 28, Lusaka; 29–December 1, Kitwe

From Pietersburg to Bulawayo in Southern Rhodesia was a journey of 360 miles, where they arrived on Saturday November 18 the day on which a new tunnel[13] was being opened by Dr. Verwoerd[14] the Prime Minister of South Africa. They stayed there for 4 nights during which their visit attracted the interest of the press. A reporter from *The Chronicle*, a national newspaper published in Bulawayo, interviewed Denis, and on Tuesday 21st an article was carried by the paper which showed how far Denis' ideas about the cause of the lymphoma had developed during the 6 weeks of the safari. Under the headline 'Do insects carry a cancer virus?', the article was given prominence on the front page alongside reports of severe flooding in Kenya and the impending visit of the new Colonial Secretary Mr. Reginald Maudling who was attempting to manage the move towards independence. The report summarised neatly what might be considered to be the state of the art at that time with regard to the tumour.

'A cancer affecting children in tropical Africa may be caused by a virus, probably carried by an insect like a mosquito, Mr. Denis Burkitt, a Uganda Government surgeon, said in Bulawayo yesterday.

'If this can be established it will be a major breakthrough in cancer research.

'Mr Burkitt is leading a three-man team on a 10,000 mile trip through Central and South Africa to locate the disease geographically.

'He explained that the cancer in question is the most common among children in tropical Africa, attacking them between the ages of two to 14 – usually in the jaw.

'The team has found that the cancer occurs only in areas below an altitude of 5000 ft. near the equator and in areas in the region of Southern Rhodesia, below 2000 ft.

'British, French and American research groups are coming to Africa to study the disease.

'Mr Burkitt and his colleagues at the Surgical and Pathological departments of Makerere College, Kampala, have worked on the disease for the past 5 years.

'The party leaves for Northern Rhodesia tomorrow.'

This was a story of growing importance although it would be another 3 years before the virus was seen, cultured and identified (Fig. 12.11).

They pressed on north to Livingstone calling at three mission hospitals on the way. The men were very conscious that much of the information they were gathering, and on which ideas about the cause of the tumour were based, came from medical staff who were denying themselves all the comforts and amenities of Western culture because of their commitment to care for the poor and sick. They were particularly impressed when they called in on one RC hospital about halfway from Bulawayo. It was run by a German lady for whom they had the greatest admiration. She had started off 13 years ago with a mud hut and had built up a well-equipped

---

[13] The only tunnel on the railway is located between Thomson Junction and Hwange.

[14] Dr. H F Verwoerd (1901–1966) leader of the Nationalist Party and Prime Minister from 1958 to 1966 when he was assassinated. Declared South Africa a republic on May 31, 1961, after a referendum in 1960 when only Whites were allowed to vote.

Fig. 12.11 Press cutting from *The Chronicle*, Bulawayo, November 21, 1961. The first public indication of viral involvement with the tumour

MR. D. BURKITT

DO INSECTS CARRY CANCER VIRUS?

A cancer affecting children in tropical Africa may be caused by a virus, probably carried by an insect like a mosquito, Mr. Denis Burkitt, a Uganda Government surgeon, said in Bulawayo yesterday.

If this can be established it will be a major breakthrough in cancer research.

Mr. Burkitt is leading a three-man team on a 10,000-mile trip through Central and South Africa to locate the disease geographically.

He explained that the cancer in question is the most common among children in tropical Africa, attacking them between the ages of two to 14 — usually in the jaw.

The team has found that the cancer occurs only in areas below an altitude of 5,000ft. near the equator and in areas in the region of Southern Rhodesia, below 2,000ft.

British, French and American research groups are coming to Africa to study the disease.

Mr. Burkitt and his colleagues at the Surgical and Pathological departments of Makerere College, Kampala, have worked on the disease for the past five years.

The party leaves for Northern Rhodesia tomorrow.

80-bed hospital with a very fine building. She was tired because she had only had 4 days' leave in the last 6 years.

Livingstone was a favourite place where they stayed for 3 nights in a rest camp near the Victoria Falls. Apart from the natural splendour of the falls, although with river levels low at the end of the dry season they were not quite so turbulent and threatening as during the rainy season, it was the association with David Livingstone that made the place so special for them. Denis was reading a book in which Livingstone described his first view of the falls, almost exactly 106 years earlier on November 16, 1855, painting a vivid picture in words. The men had their cameras both still and cine but realised that they could not do justice to the great expanse of water cascading over the cliff with unstoppable force. In Livingstone they visited the museum where there were many artefacts from his life whilst outside a bronze statue of him striding forwards with a telescope over one shoulder and his satchel containing his notebooks over the other. In his hand a Bible signalled to the three men a reassurance that their voyage of discovery and the Christian fellowship they enjoyed would have the approval of the great man.

Back across the Zambezi and into Northern Rhodesia, they were able to obtain valuable information about the areas where the tumour was found. It occurred in the valleys of the great rivers, the Zambezi and Limpopo, but not in the country between the two rivers, which was much higher. They noted that 'The distribution pattern was getting clearer and clearer and was indisputably related to altitude'. In Kitwe, the second largest city in the country, they met a pathologist, Maurice King, who would later join Denis in Makerere. He would become a household name in mission and other small hospitals throughout the world for his book *Medical Care in Developing Countries* [5]. Ten years later Judy would marry one of his cousins, Philip Howard. Denis' lecture in Kitwe to a BMA meeting was very well received, a now anticipated response to his talks. With his considerable experience of public speaking and with a good story to tell, he was a charismatic figure who would engage his audience with illustrations from his extensive collection of photographs and good stories about the children with lymphoma some of whom he could now count as being cured with chemotherapy. After the lecture Denis was uncomfortable with his success and wrote 'Save me Lord from Pride'. He would come to terms with his success and in later years enjoy engaging with audiences who would fill any lecture venue where he was billed to speak (Fig. 12.12).

After a long drive north of Kitwe, they arrived in Mpika and stayed in a small hotel. Visiting an RCM hospital the next day, again staffed by Irish sisters, they then went on to the estate of an Englishman Sir Stuart Gore-Brown who had built a castle in the middle of his 23,000-acre land. He was away, but they were able to stay in a guest house on the estate and dined in the evening in the castle with the Harveys.[15] Pushing on into southern Tanganyika (Tanzania), they passed through Tunduma, which they had visited on the earlier part of the safari, and to Mbeya north of Lake Nyasa and then a short drive to the small town of Tukuyu situated at 5000′ where they found no cases of the tumour in contrast to the situation by the shores of the lake. Tuesday December 5 saw them driving 345 miles north to Itigi through the most beautiful country they had seen on the whole trip.

Finally, their luck ran out. In Itigi the car developed a flat tyre, the first of the whole journey. They carried a spare, so Ted was able to change it requiring them to stay the night in the railway rest house. Itigi had a main line railway station which gave them the chance to put the car onto the train and head the 135 miles west to Tabora. There they stayed again in a railway hotel but were unable to get the car off the train until 4 pm. They took the time to visit the Tabora Hospital but were depressed by the standard of surgery there. Leaving late in the afternoon for Kola Ndota near Shinyanga, where Cliff worked at the Africa Inland Mission Hospital, a bolt on the chassis sheared, but Ted came to the rescue again and repaired the fault with the benefit of the headlights of another car. They drove on into a violent storm so bad that they could see only a few feet in front of them. Arriving at what was normally a dry riverbed, they found a powerful torrent of swirling water. They had

---

[15] John Harvey managed the estate at Shiwa Ngandu for Sir Stuart Gore-Brown who was a distinguished soldier and pioneer White settler in Northern Rhodesia (Zambia).

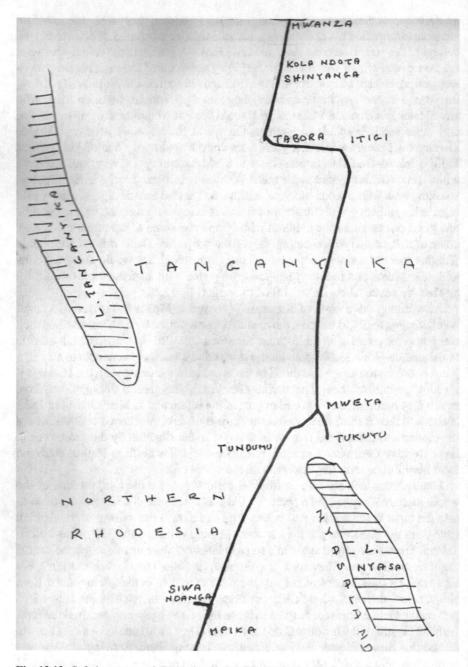

Fig. 12.12   Safari route map 6. December 1– 2, Mpika; 6–7, Tabora; 8, Kola Ndota

no option other than to stay in the car overnight, Thursday December 7, trying to sleep stretched out on top of their luggage either sweating with the windows closed or 'mercilessly bitten by mosquitos with the windows open'. The next day they commenced their journey soon after 6 am and managed to cross the river expecting the 65 miles to Cliff's hospital to be more straightforward. That was not to be. A few miles further on, they came to a far larger river 30–40 yards across which they decided to try and cross. 'The car stuck completely about a third of the way over, and we couldn't get an inch further. With the river swirling all round and up to the car floorboards, we tried to dig out the sand with a spade but this was to no avail'. They managed to get back to the bank which they had left where a dozen locals appeared and with Ted at the wheel pushed the car to the other side. Passing other vehicles stuck in the mud, they eventually arrived at Kola Ndota at 8:30 pm tired and dirty. There was worse to come, and eventually they would have to abandon their plans to continue the safari into Ruanda and head for home (Figs. 12.13 and 12.14).

They spent 3 days with Cliff at the mission recovering from their ordeal and found records of 21 cases of the tumour, more than in the whole of Northern and Southern Rhodesia. On Saturday December 9 1961 the day on which Tanganyika became independent, they visited the Williamson diamond mines where Cliff normally operated at the hospital and, as was their custom, attended church on Sunday including a service for lepers, which impressed them. They had arranged for post to be forwarded to the mission amongst which was a letter from Sir Harold Himsworth

Fig. 12.13 Ford Station Wagon marooned in the mud on the final leg of the journey north of Kisumu. Ted Williams at the back and Cliff Nelson in front of the car. December 14, 1961

Fig. 12.14 Cliff and Beth Nelson at Kola Ndota

informing Denis unofficially that the MRC would be giving him a grant to continue his work and appoint a part-time secretary. News was reaching them that the parts of Africa through which they intended to travel were experiencing the worst rains in living memory, so they decided to take the most direct route home. The plan was to drive north to Mwanza on the southern shore of Lake Victoria where they would get a steamer up the eastern shores of the lake, but the road was impassable, deep in water with whole sections washed away. They returned to Shinyanga close to Cliff's mission and found that a goods train was leaving for Mwanza that evening. An added complication to their journey however was the national holiday associated with Independence Day, and so there were no railway staff to help them. They found an empty wagon which they moved into a siding where there was a ramp that enabled the three of them to push the car onto the train, hoping that someone would help them to hitch it up to the train. They were lucky, and when the train reached Mwanza, a crane was available to lift the car onto the boat, and they headed overnight up the lake to Kisumu (Fig. 12.15).

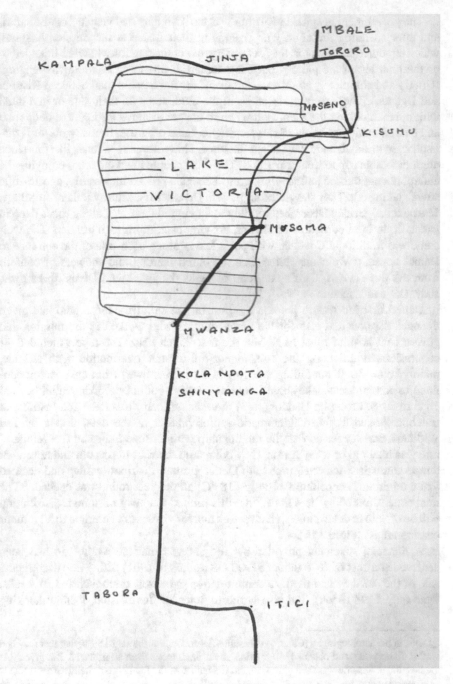

Fig. 12.15 Safari route map 7. December 11, left Kola Ndota; 12, on Lake Victoria; 13, Kisumu; 14, Tororo; 15, home

The government hospital in Kisumu was the 45th they had visited[16] on the safari, and after spending the night with friends in their house at the mission hospital, which straddled the equator, they set out the next morning for what was intended to be the final leg of the journey back to Mulago. It proved to be quite challenging. 'Utterly and completely stuck on a detour off the impassable road between Kisumu and Uganda'. There were buses stuck in the mud, some on their side or in a ditch with others blocking the road. Lorries were unable to move through the deep mud and behind them was a row of cars hoping to make progress. Denis wrote 'It is difficult even to stand, and I have fallen flat once. I have never seen roads like this since our memorable trip to Kenya in 1952'. The local people lined the roads enjoying the entertainment, but the journey through to Kampala proved impossible, one 15-mile stretch taking 3 h. They stayed the night with Dr. and Mrs. Murray Baker[17] in Mbale. The next day, Friday December 15, they manoeuvred their way along roads deep in water, up to the floorboards of the car, arriving back at Mulago at noon. The welcome was traditional Ugandan with palm leaves along each side of the entrance to Denis' house, a Welcome Home arch, coloured bunting and showers of confetti from the three wives, Cliff's son, one of Ted's daughters and Denis' three girls, Judy, Cas and Rachel.

Denis' first job was to prepare a report on the safari for those who had given financial support and also for the many doctors who worked at the mission and government hospitals that had given them so much help. The report included 11 conclusions highlighting the coincidence of tumour distribution with altitude, mainly below 5000′ but falling to below 3000′ further away from the equator, confined to coastal areas, lake shores and river valleys but not occurring at all beyond 18°S to 20°S. Professor Haddow from the East African Virus Research Institute in Entebbe looked at the diagrammatic representation of the data, drawn by Ted Williams, and suggested that the relationship with altitude indicated that temperature was likely to be a key factor. They compared maps of maximum and minimum temperatures in Southern Africa with Denis' tumour distribution map and decided that a minimum temperature of 64 °F (18 °C) all the year round was needed.[18] The final conclusion being 'It is hoped that this limited mapping of tumour distribution will be of help in determining what vectors and, or viruses are common to all tumour bearing areas' (Fig. 12.16).

A financial statement provided by Ted showed that the whole trip had been achieved at a cost of £678 (about $1900 US dollars in 1961) taking into account the sale of the car they had used. Accommodation expenses, three men for 10 weeks, were only £198 ($560). Ted also wrote to some of the mission doctors thanking

---

[16] Denis in his autobiography (Chap. 5) concludes his account of the safari by saying that they had visited 48 hospitals and obtained information about the tumour from a further 12. But given that the hospital at Kisumu was their 45th visit and the nature of the final part of their journey, there is a discrepancy here.

[17] Murray Baker would subsequently become administrative Dean of Makerere Medical School and later work for the Medical Research Council in London, for which Denis would also work.

[18] Later revised to 60 °F or 15 °C.

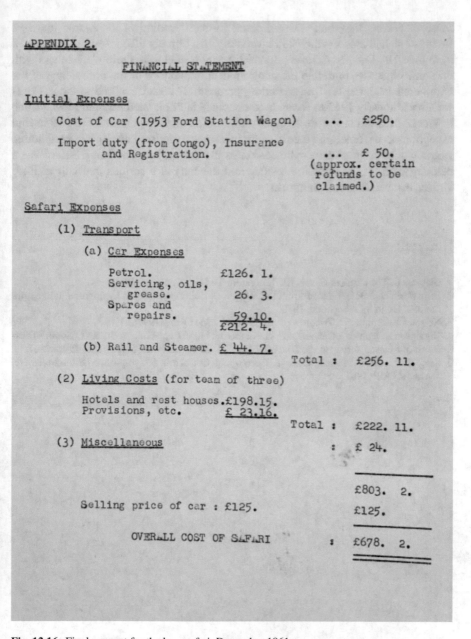

APPENDIX 2.

FINANCIAL STATEMENT

Initial Expenses

    Cost of Car (1953 Ford Station Wagon)   ...   £250.

    Import duty (from Congo), Insurance
        and Registration.   ...   £ 50.
                                 (approx. certain
                                 refunds to be
                                 claimed.)

Safari Expenses

    (1) Transport

        (a) Car Expenses

               Petrol.   £126. 1.
               Servicing, oils,
                  grease.   26. 3.
               Spares and
                  repairs.   59.10.
                           £212. 4.

        (b) Rail and Steamer. £ 44. 7.
                              Total :   £256. 11.

    (2) Living Costs (for team of three)

        Hotels and rest houses.£198.15.
        Provisions, etc.   £ 23.16.
                          Total :   £222. 11.

    (3) Miscellaneous   :   £ 24.

                                £803. 2.

    Selling price of car : £125.   £125.

        OVERALL COST OF SAFARI   :   £678. 2.

Fig. 12.16 Final account for the long safari, December 1961

them for their kindness and highlighting aspects of the trip the three men had enjoyed. Important for all of them had been the visit to Ujiji where Livingstone had met Stanley and the Victoria Falls where he had stood some 106 years earlier. Any sense of destiny they had was tempered by their reading of the heroic journeys Livingstone had made, which had ultimately cost him his life.

Whilst The Long Safari has attracted the most attention, there were others with the same objective, to define the geographical boundaries of the occurrence of the lymphoma. With the 1961 safari ending prematurely because of bad weather, Denis went back in early 1962 to Rwanda to conclude their planned trip and in May/June to West Africa to clarify the relation with rainfall and temperature. He was to find exceptions to the emerging theory of an insect-carried virus causing the lymphoma, which would bring a further dimension to the theory. But The Long Safari was a defining point in the lymphoma story and the start of a serious quest to identify viruses that cause cancer in humans.

References

1. Glemser B. The long safari, vol. 1970. London: The Bodley Head; 1908–1990.
2. Himsworth H. Foreword. In: Burkitt DP, Wright DH, editors. Burkitt's Lymphoma. Edinburgh/London: E and S Livingstone; 1970.
3. Nelson CL, Temple NJ. Tribute to Denis Burkitt. J Med Biog. 1994;2:180–3.
4. Higginson J, Oettlé AG. Cancer incidence in the Bantu and Cape-Coloured races of South Africa-report of a cancer survey in the Transvaal (1953-1955). J Natl Cancer Inst. 1960;24:589–67.
5. King MH, editor. Medical Care in Developing Countries. A symposium from Makerere. Tanzania: OUP; 1967.

Chapter 13
'We Have Been to Africa and Met Dr Burkitt'

Life was never to be the same again for Denis following the long safari. Exceptionally busy with correspondence, surgical duties, church activities, developing the photographs he had taken so that he could tell the story of the safari, he also had a lot of writing to do. There were reports to be sent to the sponsors and papers for medical and scientific journals. Over the next 2 years, he would be the lead author on 15 papers giving evidence for the progression of ideas about the lymphoma.[1] The publication of the results made Kampala a honeypot attracting the best in cancer research and virology from all over the world, and many famous figures visited Mulago. Denis made lasting friendships with several of them,[2] which would be important to his future research. The media were very anxious to catch the story, yet whilst the staff of the medical school were pleased when the *Reader's Digest* and *Time Magazine* carried features about Denis' work, he learned an important lesson with his first exposure to television journalism. He spent many hours one day with a team from the USA only to find that his contribution was confined to 3.5 min in the final programme. There was also his family who felt they were due a piece of him now. By Christmas Day he was exhausted having been up the previous 2 nights operating on emergencies. There were 18 to lunch, all managed by Olive, but he was not sure that the present he had bought for her was quite right. Never one for social occasions he was glad when Christmas was over (Fig. 13.1).

Visitors arrived daily as news spread that a cancer of African children might be caused by a virus. Dr Frank Horsfall,[3] recently appointed as Director of the Sloan Kettering Institute in New York, who had contributed funds towards the cost of the safari, came with his wife and over lunch with Olive and Denis invited them to New York promising to pay all their expenses. Suddenly Denis was being launched

---

[1] Appendix 3.

[2] Jack Darling. Letter to Ethel Nelson giving an account of life at Mulago in 1960 dated June 20, 1991.

[3] Dr Frank Horsfall (1906–1971), eminent virologist and immunologist.

© The Author(s), under exclusive license to Springer Nature Switzerland AG 2022
J. H. Cummings, *Denis Burkitt*, Springer Biographies,
https://doi.org/10.1007/978-3-030-88563-2_13

Fig. 13.1 Church duties. Saturday February 3, 1962. Outside All Saints Church, Kampala Induction of new Chaplain by Bishop of Uganda. L to R, Dr KL Batten The People's Warden, Archdeacon Lubwama The Rev Asa Byama Assistant Chaplain, Archbishop of Uganda Dr Leslie Brown, New Chaplain Dr Derek Matten, Mr Victor Ravensdale Lay Reader and Mr Denis Burkitt, Chaplain's Warden, looking unusually smart (*Uganda Argus*, Wednesday February 7, 1962, Kampala (original photograph in family archives))

into a new orbit in which he would remain for the rest of his life – one of international recognition for his work, worldwide travel and more confidence in his own judgement. To help in the management of his evolving status and workload, he rapidly moved to appoint a research secretary, Miss Mary Ward, and bought a typewriter for the office. He was fortunate in all this by having a good income from his government appointment at Mulago and his private practice at the European and Nile hospitals together with the loyalty, love and support of Olive and the family. Denis did not seek to acquire wealth for its status or utility but used his money for what he felt at the time to be in keeping with his Christian ideals. Their major expenditure was on their two, and soon to be three, children at private schools, but the cost of living in Kampala was not high, and both Denis and Olive had been brought up to be careful with their spending and to 'make and mend' for their everyday requirements (Fig. 13.2).

The social side of his new status came with responsibilities to meet people and entertain them. Always impatient with social rituals and gossip, Denis was able to leave Olive to manage this on his behalf with a well-run household and intelligent understanding of what was required. Despite relatively frequent episodes when she was 'low', they were able to spend time each day together, usually late in the

Fig. 13.2 Guy Timmis
1962 at Mulago before
departure to New Zealand

evening, to read a book or the *Daily Light* and share their prayers. It was a relationship that would survive with the changes that lay ahead for them all. Uganda would become independent of the UK on October 9, 1962, and a republic in 1963. Thereafter politics divided along tribal lines with an uneasy relationship between Milton Obote the Prime Minister, who was from the north, and the Baganda people under the leadership of the Kabaka in the region around Kampala in the south. Denis would soon have to relinquish his job at Mulago and eventually return to England where he would turn his mind to solving, with renewed success, the problem of other chronic, non-infective, diseases. Looking ahead in February 1962, Denis felt there was a definite case for investing in a house either in Ireland or England.

Working hard, Denis nevertheless allowed some time for relaxation with the family and went regularly to the cinema. During the week of February 19, he saw 'I'm All Right Jack', described as excellent and then went with Olive and Rachel to see 'Whistle Down the Wind' and on the Saturday with Jack Darling[4] to see 'Canaris' a film about Admiral Wilhelm Canaris of the German Navy, who was executed by the Nazis at the end of the war (April 9, 1945) along with Dietrich Bonhoeffer,

---

[4] Jack Darling had known Denis since student days in Ireland when both were members of the Christian Union and had met at No 40 meetings where Jack once spoke. Jack attended Queen's University, Belfast. Jack's father was a keen ornithologist and was invited by Denis' father James to accompany him, with Jack and his brother, to visit some of his favourite bird haunts. But Denis was away at school when this occurred. After army service Jack was posted to Tanganyika as surgeon to the groundnuts scheme but got to know Denis again on his travels. Eventually, 1960, he took up the post of specialist surgeon at Mulago and would run a surgical 'firm' in the hospital with Denis. He and his wife Beryl were very great friends of Olive and Denis. Jack would speak at Denis' memorial service in 1993.

General Hans Oster, General Kar Sack and Ludwig Gehre. In February 22 Olive records 'Voting day for the Buganda Lukiko[5] – the first ever direct elections to the Lukiko. Barbee & Yokana both went off to vote & didn't return until lunchtime – so I had all the cleaning, cooking etc single handed & V.I.P's to lunch in the shape of Mr and Mrs Kettering – American millionaires of the Sloan-Kettering Research Institute & Dr & Mrs Horsfall. Dr Horsfall is the Director of the Research Institute. All very easy and pleasant. Took ladies round hospital after lunch and Ian (McAdam) came around to lunch too'. Denis, in addition to spending time with the visitors, drove the Horsfalls to Entebbe in the late afternoon and then went on to evening Bible study. Not surprisingly Denis developed symptoms of a duodenal ulcer but was buoyed up by the receipt of several letters of commendation from people to whom he had sent an account of the safari. That month he was also visited by Bob Harris from ICRF, which was taking a great interest in his work, and Dr Laithwaite from the UK Colonial Office, who was conscious of the impending changes in Uganda.

Of greater significance, in February 28, Tony Epstein arrived for the first time in Uganda with a plan to inject tumour tissue into monkeys with the hope that any virus contained in the tissue might survive and induce the cancer. They went out early one morning in a motor launch to an island about 10 miles into Lake Victoria where they met the local headman and walked with a guide and interpreter through the forest enlisting their help to catch monkeys. On the way back to the shore, their boat engine failed and they were marooned on the Lake in the dark. Tony Epstein remembers that 'we came to be adrift in a dug-out canoe on the waters of Lake Victoria for several hours in the middle of a rather dark night'.[6] This was potentially dangerous given that fierce storms could blow up with very little warning on one of the world's largest lakes. Their African navigators managed to get them back to the shore, but the monkey experiment did not work although many years later the tumour was induced in South American primates with tissue-culture extracts of the lymphomas.

The heavy workload began to take its toll on Denis. In addition to his ulcer symptoms, he developed a bad cold complicated by a 'Stye in eye & boil in nose'. The family doctor prescribed the antibiotic Aureomycin, but Denis felt tired, and when Olive caught his cold, precipitating an episode where she felt low again, it left Denis without his underpinning family support. After taking Judy back to school in Eldoret and staying there for the night, he returned to meet Sir Robert Mackintosh, Professor of Anaesthetics in Oxford, whom he entertained at the club in Kampala. In return for this hospitality, Prof Mackintosh gave the anaesthetics for Denis' Friday operating list at the European Hospital. Professor Mackenzie from Alberta, Canada, arrived the same week with his wife and wanted to discuss the tumour syndrome with Denis. A party for the guests at Prof McAdam's on the Wednesday was

[5] Uganda Lukiiko the legislative assembly for the Buganda region of which Kampala was the capital.
[6] Letter from Professor Epstein to Ethel Nelson, February 28, 1991. Family Archives. Dug-out canoe or motor launch?

followed by dinner for 16 at the Burkitts on Saturday managed by Olive. Denis, unusually for him, was too tired to attend evening service on the Sunday, but as the antibiotic began to work, Denis felt better. In the middle of the next week, he heard that Dennis Wright,[7] who had become Denis' principal partner in the Pathology Department following the exodus of Prof Jack Davies and Greg O'Conor's return to the USA, was to give a lecture in Moscow about what Denis now regarded as his tumour. He found himself jealous about this, but on reflection he asked himself what harm it could do to him and what a selfish attitude he was taking. Operating later that week, he drained a hydronephrosis[8] containing an enormous 15 ½ pints (9.4 litres) of fluid. His surgical skills encompassed a range of surgery that would not be acquired today by surgeons in most developed countries, but he was a man with inborn practical capabilities who enjoyed operating and would miss greatly the opportunity to help people in this way when he returned to England in 1966.

Olive was low for the next 2 weeks having caught Denis' cold, and the family mood was not helped by the departure of their very good friends Guy and Dawn Timmis. Denis and Guy had done their first jobs together at Chester Hospital and later in 1938 the FRCS course in Edinburgh. They had been as close as brothers over the years. On their final evening, the Drowns, Billingtons and Matters came to supper where Guy and Dawn, who were to spend 16 months in a mission hospital in Tanganyika (Tanzania) before settling in Australia, gave Denis a briefcase. Their departure made him feel very sad thinking that he would probably never see them again, although he would on his 'world tour' in 1980. As March drew to a close, Denis lectured on his safari to a general audience of the Uganda Medical Association in the Pathology Lecture Theatre, now the Davis Theatre after the departed Professor Jack Davis, and met Professor Wells from Liverpool who was very impressed by the tumour work.

A much-needed family holiday followed, but Denis spent most of the 2 weeks reading and writing. After a full Friday operating list, they headed for Eldoret to collect Judy from school and then to Kaptagat to stay with the Drowns. They drove north to the Aberdare Forest Farm, on the edge of the Aberdare National Park where they stayed in a guest house and enjoyed morning walks – what Olive described as restful days. After they had walked about 3 miles one day, Olive fell behind for no special reason but writes in her diary:[9] 'I was alone for a few minutes while others walked on – & as I stood drinking in the magnificent view, mountains – plains – forests – sky cloud – I suddenly had the most NEAR feeling of God – & found tears trickling down my face – I could only lift my head to Him & and thank Him & praise Him. A most wonderful experience'. Each day after breakfast, they continued their usual practice of family prayers. On Wednesday April 11, they drove to Nairobi for shopping where Denis was introduced to the Dalldorfs and then worked on a

---

[7] Dennis H Wright (1931–2020). Qualified from Bristol Medical School, England. Pathologist and expert in lymph node pathology who defined most clearly the characteristic microscopic appearances of the Burkitt's lymphoma.

[8] Hydronephrosis, collection of fluid/urine in the channels draining the kidney into the ureter.

[9] April 10, 1962.

paper with Dr Oettgen [1] before returning to their forest house where they received their first delivery of mail. This included an invitation from Professor du Plessiss to Denis to give a lecture to the South African Surgical Congress in September. By the time they returned home for Easter, Denis had written the first drafts of papers about the safari for the *British Journal of Cancer*, about congenital limb deformities for the *Journal of Bone and Joint Surgery* and about the lymphoma for the *International Review of Experimental Pathology*. Whilst the girls had been unsuccessful with their fishing in the river, all felt it had been a very good family holiday.

Visitors now came to Kampala not only to learn more about the tumour syndrome but often brought their wives, husbands or partners and were able to say on returning home 'We have been to Africa and met Dr Burkitt'. Yet there was still much work to do. The question of the role for a virus was far from being answered, and work was ongoing to find suitable treatment regimens for the children. A week later a visit from Dr Gilbert Dalldorf[10] also from the Sloan Kettering Institute, a well-known pathologist and virologist who had discovered the Coxsackie virus, made Denis realise that the final part of the long safari, which they had to cancel because of the weather, to Rwanda and Burundi was now of crucial importance. These two small countries, lying between Uganda to the north-east, Tanganyika (Tanzania) to the south-east and Zaire (DRC) to the west, are mostly mountainous, 6–11,000 feet (1800–3300 m), but are densely populated and according to Denis' sources were almost completely free of the lymphoma.

The plan for the safari, with Dr Dalldorf, was to visit the two capitals Bujumbura in Burundi (2,500') and Kigali in Rwanda (5,000'). Bujumbura had a well-established medical school and Kigali one in development. Both places were staffed by Belgian doctors with long experience in the two countries. The safari, which could have taken 7–10 days, was accomplished in only three courtesy of the use of a Piper Aztec aeroplane piloted by Paul Nones of the Flying Doctor Service run by the East African Medical and Research Foundation. This had been set up by Mr Michael Wood in Nairobi, whom Denis had met through the East African Association of Surgeons (Fig. 13.3).

On Tuesday April 24, Denis left Kampala at 07:15 and met Paul Nones with Dr Dalldorf, who had flown over from Nairobi, at Entebbe. They embarked on board at 9:00 and were in Bujumbura, the largest city in Burundi, 2 hours later where they spent time talking to the doctors, looking through operation records and pathology reports, but there was no evidence of the tumour having been seen. The next day they met the Minister of Health and several other doctors and were taken on a sightseeing trip. On to Kigali, capital city of Rwanda, where some of the doctors had been working for 25 years in the hospital and at a large mission hospital 100 miles away, again, there was no evidence of the tumour. They returned to Uganda that evening, Thursday. Denis had enjoyed the trip especially having the small aircraft at

---

[10] Gilbert Dalldorf (1900–1979) a graduate of New York University and Bellevue Hospital Medical School was a distinguished virologist who discovered the Coxsackie viruses and showed the ability of infection with one virus to modify the course of another. https://www.nap.edu/read/4548/chapter/5

Fig. 13.3 In Kigali on trip to Rwanda and Burundi. Thursday April 26, 1962, about 5 pm. Piper Aztec aeroplane piloted by Paul Nones of the Flying Doctor Service run by the East African Medical and Research Foundation. Dr Gibert Dalldorf is looking at the 'plane with his back to Denis, the photographer

their disposal. He felt that they were in a different class to the ordinary passengers waiting in the airport departure lounges.

The Rwanda–Burundi trip further underlined for Denis that whatever was causing the tumour, it was altitude and temperature dependent. But there were anomalies in the data, particularly the absence of tumour cases in some of the low-lying but arid regions they had visited on the long safari. The next logical step was to explore West Africa, which Denis decided to do alone. It was to add another important dimension to the story. The landscape of the West was very different from that which they had driven through in the East, with few mountains, and there were many reports of children with the lymphoma. He set off in May 15 for what was a journey by train and aeroplane that lasted 3 weeks and took him to the Congo,[11] Nigeria and Ghana, having made preliminary postal contact with as many hospitals and medical authorities as possible. Arriving first in Leopoldville (DRC Kinshasa), Denis immediately felt quite insecure because of the lack of any competent authority particularly when his passport was taken at the airport and not returned. He spent almost the whole of day 2, with the help of the senior administration officer at the university, going from office to office, often distances of many miles until he finally traced a bundle of passports that included his own. At the University Hospital, Denis

---

[11] Congo. In 1962 there were two independent states known as the Republic of the Congo to the north-west of the Congo River, capital Brazzaville, and a much larger state to the south and east now known as the Democratic Republic of the Congo, capital Kinshasa. The DRC is the second largest country in Africa, whilst its Congolese neighbour is much smaller.

showed the staff a well-illustrated album depicting the clinical, pathological and epidemiological features of his syndrome, but neither the paediatricians, surgeons nor pathologists had seen any cases from the city area. Denis thought that the most likely explanation for this was the energetic and successful methods employed by the Belgian authorities to rid the area of mosquitos.[12] A quick trip across the Congo River to Brazzaville also failed to find any reported cases, which surprised Denis. There was something else in the equation that he was missing.

With a sense of relief, Denis flew out of the Congo to Lagos in Nigeria and the next morning to Port Harcourt in the Niger Delta. There he had discussions with the doctors at the government hospital but because they were relatively new to the area were not able to give him reliable information. However, his personal contacts inland at Enugu and Okigwe told him that they had seen cases of the tumour. Returning to Lagos he was met by Will Davey who drove him to Ibadan, about 75 miles north-east of Lagos, where he was professor of surgery. There in the Pathology Department were Professors Eddington and Osukoya who were familiar with the tumour and would publish extensively about it in the future along with Prof Davey's successor Victor Ngu. With Denis, Professor Ngu would receive the Albert Lasker Clinical Medical Research Award in 1972 for their contributions to the drug treatment of the lymphoma. Driving round other hospitals in the area, Denis was left in no doubt that the lymphoma was one of the commonest tumours in children in Southern Nigeria. At University College Ibadan, Denis personally examined seven children with different manifestations of the lymphoma and learnt that intestinal obstruction due to ileal tumours was common in Ilesha (Ilesa), about 60 miles east of Ibadan. After lecturing late in the afternoon, Denis caught the train to Kano in the northern region arriving at 2 am to be met by a driver from the medical department who took him to what Denis thought was a very luxurious hotel where he stayed for 2 nights (Fig. 13.4).

Kano immediately fascinated Denis and was the greatest possible contrast to Lagos and Ibadan. In 1962 this was Africa's largest city to be populated solely by people of the Muslim faith. Only believers in Islam were allowed to live within the 13 miles of city walls and women were rarely seen out in the open. The houses had mud walls and open sewers ran down the middle of the streets. Denis describes the local market as 'like being back in Biblical times'. The main buildings were the Emir's Palace, the Mosque and a new orthopaedic hospital built with funds raised by Frank Bryson the orthopaedic surgeon. The region was almost a desert with rainfall of about 30" confined to the months of June, July and August, yet despite complying with the altitude and temperature requirements that Denis and colleagues had come to believe were essential to finding the cause of the lymphoma, there were no records of the tumour having been seen in Kano. The large city hospital served a population of over three million and was run by medical staff including Dr Bryson, who did all the jaw surgery and had not seen a case in 10 years; similarly Dr Doyle

---

[12] Report by DPB on the safari written on his return and completed by June 18, 1962. In album of photographs of the safari 'W. Afr 1962', family archives.

Fig. 13.4 Groundnut (peanut) pyramids, Kano, Northern Nigeria, a city Denis found to be one of the 'most fascinating in all of Africa'

the physician had never seen it, whilst the hospital surgeon Dr Craddock had seen a few cases in 7 years. The distinguished radiologist, Professor Howard Middlemass from Bristol whom Denis knew well, was visiting the hospital at the same time. After a further night in the luxurious hotel, Denis flew south to the city of Jos, which he had planned to visit because at over 4,000' on the Bukuru Plateau it is one of the few places in Nigeria at altitude. Dr Parker who had been the surgeon to the 280-bed hospital in Jos for the past 10 years saw one or two cases of the tumour each year but was not sure whether they came from the plateau or the plains below. Driving next to the Sudan United Mission Hospital in Vom, a similar sized hospital to the one in Jos, six to eight jaw tumour cases were seen every year one of which Denis saw personally whilst he was there. The hospital is situated on the south-west edge of the plateau with most patients coming from the plains below. Rainfall was much higher than at Kano and occurred more evenly throughout the year. Denis drew a map marking areas over 3000' and those with rainfall of under 40" per annum. Suddenly it all made sense. A new climatic condition had emerged from his study of the occurrence of the lymphoma in Nigeria. In addition to an altitude of less than 3000' and average daily temperature never falling below 60 °F, the agent causing the tumour required rain.

Retracing his steps by air, Denis flew to Ibadan and then Lagos where he stayed overnight with the Maguires, whose marriage was in total disarray and where there was great unhappiness in the house. This upset Denis who did his best to heal the rift by talking to them about the Christian view of marriage before giving his usual lecture to the medical staff in the evening. Following this he had to attend a sund-owner given in his honour with invited officials and doctors that went on to 11 pm, which as usual he hated. Not a good day but he was able to get more time talking to

Alison Maguire the next morning as she showed him round Lagos before visiting the medical research unit and then flying to Accra in Ghana. There he was met by Dr Cofie George, who had been in his year at Trinity, but they had not met since 1939. Cofie was a Ghanaian who had struggled at Trinity but had been helped financially by Denis and other members of his Church. Denis had taken Cofie to his home in Enniskillen during one holiday and taken him into the village school. Africans were a great rarity in Ireland, and the children were surprised when Cofie pointed out on a map where in Africa he came from. Cofie had married well, his wife Sylvia being the daughter of Sir Emmanuel Quist the speaker of the first parliament following Ghana's independence in 1957. Now he was running the Salem Nursing Home in Accra, and when Denis went to dinner with Cofie and Sylvia in the evening, he was given traditional Kente cloth and a model of an Ashanti stool to take back for Olive (Fig. 13.5).

In pursuit of information about his tumour, Denis visited both Accra and Kumasi about 150 miles to the north-west where his host was Dr Charles Bowesman, another Trinity graduate and Ghana's leading surgeon. In Accra the jaw tumours had been seen, but there were perhaps only two to three cases a year, whilst in Kumasi Dr Bowesman, who had worked for over 20 years in various hospitals including both Accra and Kumasi, was convinced the tumour was more prevalent with at least six to eight cases admitted annually, possibly more. Both cities were at less than 1000', but rainfall in Accra was recorded by Denis as being only 27" per year, whilst in Kumasi it was over 60". Population density was similar in the two areas. Denis lectured in both places, met Lady Quist, Cofie's mother-in-law, spent time going through the hospital operation notes and turning for home met Dr Dalldorf for lunch in Accra. Finally before leaving West Africa, he purchased several school atlases that showed climate, vegetation, population densities and other characteristics of the countries he had visited.

Arriving back at Entebbe after an overnight flight, Denis was met by Olive with all the local news including a request from Bill Davies, Director of Medical Services at Mulago, asking if Denis would be the surgeon in charge of the Royal Party during the Independence Day celebrations. Tired and with a mountain of jobs waiting for him, Denis went early to bed. His priority the next day was to get started on his report of the West African safari for the MRC who had financed it. He had, over the 11 years since he had written his first paper, on hydroceles, for the scientific literature, developed an ability to write and with an easy style that would be the envy of any nascent medical scientist today. And now, in addition he had Mary Ward to do the typing. The report 'Children's Lymphoma Syndrome West Africa Safari, May 15th – June 7th, 1962' was ready by June 18 and comprised six faultless pages of professionally set out text followed by three maps that Denis had drawn. The main conclusions were that whilst previous safaris had shown that the incidence of the tumour was related to altitude, now interpreted as temperature of over 60°F throughout the year, in West Africa he had found that at low altitude where temperature was over 60°F there were great variations in tumour rates. This Denis ascribed to differences in humidity, by which he meant rainfall, with a lower limit of 40"/year required for tumour risk, according to his maps. His suggestion for further research

Fig. 13.5 Cofie George and his mother-in-law Lady Quist in Accra, Ghana, 1962 West African safari

was to follow up on a much wider scale the contacts he had made in over 50 mission and other hospitals in the Congo and to compile a list of government and mission medical units throughout Nigeria and the north of Ghana in order to substantiate the geography of the tumour. There was another motive. The questionnaires, which Denis was sending to all these contacts, asked for information not only about his lymphoma but also a range of other cancers and conditions.

Did the idea that rainfall was important emerge from the West African safari and was Denis the originator of the idea? There is no mention of rainfall following either the long safari at the end of 1961, nor was it apparently discussed with Gilbert Dalldorf on the brief expedition to Rwanda and Burundi. On the long safari, Denis had been accompanied by two colleagues both of whom were acknowledged by him to be very able and with whom during the many days they spent together would have discussed the precise geographical limits of the tumour distribution. Whilst Ted Williams had drawn the first maps showing the relationship with altitude and Ted was clearly much better at drawing maps than Denis, rainfall is not mentioned. But on the safari to West Africa, Denis travelled alone and drew the maps for his report. These maps clearly identify rainfall as significant factor in explaining the variation in tumour distribution at low altitudes where the temperature was more than 60°F year-round. However, before he set off for the safari, Denis had written and had published three papers about the lymphoma in which the idea of rainfall being important is mentioned. Significantly these papers include a map showing the coincidence of the tumour syndrome with areas in Central Africa where altitude was less than 5,000', mean temperature more than 60°F and rainfall greater than 20" (Fig. 13.6).

The first paper of the 3 was one written to set down for a wider audience the essence of a lecture that Denis had given at the Royal College of Surgeons in London on May 24, 1961 [2], where, despite much formality, there was an audience of only 12. Denis started writing this paper on returning to Africa following home leave in June 1961 and had sent it off to the journal before he went on the long safari in October. It was published in April 1962 and included discussion of the distribution of the tumour and the accompanying geographical features of the lymphoma belt. In the paper is a map attributed to Dr A J Haddow, Director of the East African Virus Research Institute at Entebbe whom Denis knew well and with whom he had frequent discussions about the lymphoma.[13] Denis had shown Haddow his own maps of the tumour distribution, acquired from his responses to the questionnaires he had sent out to all his mission hospital contacts, displayed in his office on the wall before the long safari. Looking at them with great interest, Haddow said 'I have seen something like that'. He went back to his office and sent Denis a map of the distribution of an insect disease vector, a mosquito, which was virtually a replica of those of Denis – covering the same warm humid tropical and subtropical terrain. The possibility of a causative factor carried by an insect became an exciting vision. No such factor had ever been proved before in any human cancer.

---

[13] Professor Alexander John Haddow (1912–1978), CMG, FRS, FRSE, FRCPSG FRES graduated from Glasgow University with first class honours in zoology followed by a degree in medicine in 1938. He became a world-renowned entomologist known for his work on viral infection especially yellow fever, the Zika virus and Burkitt's lymphoma. Elected FRS in 1972, the same year as Denis. PCC Garnham (1980), Biographical Memoirs of Fellows of the Royal Society 225–254 and other sources.

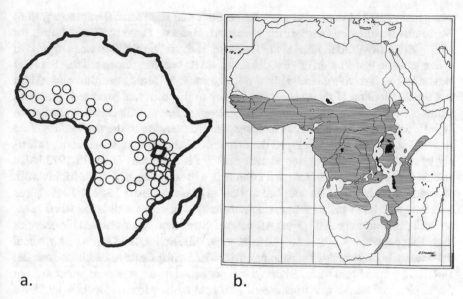

a.   b.

Fig. 13.6 '(a) Map showing known distribution of the "tumour syndrome" in Africa. Each circle represents an area where the condition has been recognised. (b) Map of Africa from which have been eliminated: (i) Areas over 5000 ft. and areas where seasonal mean temperature may fall below 60 °F.; (ii) areas with less than 20 inches of rainfall per year'. Map **b** is credited to Professor Haddow by Denis in the paper and is probably signed by him. Figure 6 from Ann Roy Coll Surgeons 1962 [2]. Lecture delivered at the Royal College of Surgeons of England on May 24, 1961. (By kind permission of the Royal College of Surgeons of England)

Haddow had drawn for Denis a map of Africa from which had been eliminated areas over 5,000', where temperature might fall below 60°F and where rainfall was less than 20" annually. The remaining area corresponded closely with Denis' tumour maps, and Haddow concluded that there was a strong prima facie case that an insect vector such as a mosquito must be involved. This same map also appears in Denis' paper in *Nature* [3] published the same month as the Annals and earlier in the *Postgraduate Medical Journal* [4] with acknowledgement of the Annals paper as the source and with discussion of Haddow's ideas. Denis gives full credit to Haddow for taking notice of the maps he had drawn of tumour occurrence across Africa and relating the patterns of distribution to climate and the associated possible insect vectors.[14] Before the long safari, the idea of altitude, temperature and rainfall was known to Denis, and his safaris in East and West Africa and Ruanda were intended to gather data to help explain the anomalies in the geography and confirm or refute the idea that the climatic conditions favoured a mosquito-borne virus causing the tumour.

Haddow was one of the world's leading entomologists and virologists and was Director of the East African Virus Research Institute in Entebbe from 1953 to 1965.

---

[14] DP Burkitt Autobiography Chapter 7. Wellcome Library WTI/DPB/F/1.

Denis had visited him many times at the Institute and soon after Denis' return from West Africa; whilst he was writing his report, Haddow came to Denis' house for lunch. Humidity/rainfall was always part of Haddow's thinking when trying to understand the distribution of viral diseases, and he became convinced that the lymphoma had an infective cause [5]. The information that Denis gave him showed him that the distribution of the tumour was similar to that of trypanosomiasis (sleeping sickness), yellow fever and o'nyong nyong fever. Haddow, with all his vast entomological experience had been able to bring a new dimension to the studies. Together Haddow and Denis had developed the essence of the mosquito virus theory probably by mid 1961. Haddow was elected FRS at same time as Denis in 1972. Also elected the same year was Hermann Lehmann, a haematologist notable for his work on abnormal haemoglobins who had worked at Makerere from 1947 to 1949. Three scientists who had worked at the same university all elected FRS in the same year.

With the report about the West African safari finished, Denis turned his energies to writing papers, but life was about to change. Bill Davies, the Director of Medical Services at Mulago and a longstanding friend of Denis, came to lunch on June 22, 1962, and warned him that, following Independence, there was now pressure from the Ministry of Health to Africanise all jobs held by ex-patriots. Denis would have to consider resigning from his hospital appointment in the not too distant future. Denis was quite shocked by this because he had never really thought of doing any other job. Moreover, his chances of obtaining a surgical appointment in England would be slim given his age, now 51, and lack of specialist experience. Shortly after this meeting, Ian McAdam, the professor of surgery and family friend, talked to Denis and suggested that he should consider going full-time into the study of geographical medicine continuing and expanding the work he had done on the lymphoma. It had never occurred to Denis that he could get sufficient support to follow his research ideas full-time, but after discussions with Olive, they both felt that all would turn out for the best and that they should take one step at a time on this new road. That week they received copies of two American magazines, *Life* and *Image*, both of which ran articles on Denis' research. Perhaps after all this would be a good way forwards. More discussions took place in the ensuing weeks with Ian and those affected by the new staffing policy, which proved difficult for all. Denis was to a certain extent in a stronger position than the others given that he now had an international reputation for his tumour research and there were still major issues to be followed up. Ian wrote to Sir Harold Himsworth at the MRC in London, suggesting that Denis might be taken on as an MRC scientist. Himsworth replied quite promptly saying that the MRC would try to help Denis if he was no longer employed by the government. Eventually this would prove to be a fruitful opportunity not least because the new Professor of Pathology at Makerere, who had taken over from Jack Davies, was Michael Hutt one of whose interests was geographical pathology. They would work together both in Uganda and after they both returned to England writing papers and books [6], enlarging the whole study of the geography of non-infectious diseases (Fig. 13.7).

Meanwhile the New Mulago Hospital had been completed and areas A and B were open to patients. Denis moved into his new office, attended his first patients in

Fig. 13.7 New Mulago Hospital 1962. (By kind permission of Richard Price)

August 8 and did his first full operating list on the 15th. Soon after the Presidents of both the Edinburgh and London Colleges of Surgeons, Sir Arthur Porritt and Sir John Bruce, visited Kampala to see the new hospital and hear about the work of Denis and the Department of Surgery at Makerere. It was decided by Ian McAdam to show the distinguished guests examples of cases that they would not normally encounter in the UK. Sitting on the front row of a small lecture theatre, they were firstly shown, by one of the African surgeons Alec Odonga, a man covered in bandages with the remark 'Sirs, this is a rather unusually severe case of hippopotamus bite. You may not be acquainted with such injuries'. There followed a case of an enormous inguinal (groin) hernia in a man presented by Joe Shepherd who went on to be Professor of Surgery in Tasmania. A classification of such hernias was described as reaching the knee or beyond, perhaps something of an exaggeration to emphasise the severity of hernias seen in Uganda and impress his audience. The meeting finished with Denis showing four children each of whom had lymphoma affecting both sides and both upper and lower jaws. Presentation of the tumour in all four quadrants of the jaws was still very unusual. To have four cases to show highlighted the world-leading role of the Department of Surgery at Mulago (Fig. 13.8).

After the visit of the Royal Colleges' Presidents, Denis left for Johannesburg to give the opening lecture of the biannual meeting of the South African Association of Surgeons. It was a very formal occasion with Denis in his dinner jacket talking about the events surrounding the discovery of what was at this time known as the African lymphoma although in a few months' time this designation would change. The lecture was very well received being the combination of a good story, told to an audience pleased to hear about an African success, narrated by the very person who had discovered the tumour syndrome and in an engaging style practiced now over many public performances that kept their attention until the end. Denis was a star in Africa and soon would be a sought-after speaker all over the world.

Fig. 13.8 From a 12-page supplement to *Uganda Argus*, September 17, 1962, published to mark the opening of the New Mulago Hospital in Kampala. After an introduction by Dr Emmanuel Lumu, Uganda's Minister of Health, Denis' work is the first to be highlighted. Shows the first map of tumour distribution and Denis at his desk surrounded by other maps

On his return to Kampala, it was clear that arrangements for Independence Day, Tuesday October 9, and the associated royal visit by the Duke and Duchess of Kent were starting to dominate life.[15] Denis' duties as surgeon to the Royal Party were to accompany them, at a proper distance, at all times throughout the 10 days of their visit. Olive was struck down with one of her bouts of tiredness and 'the old arch enemy' depression at the beginning of the week before the visit and feared she would not be able to take her place beside Denis for the formal occasions. But she kept busy and gradually improved during the week finally getting her confidence back when listening to the challenging sermon of Rev Dr Donald Coggan, Archbishop of York at this time, who was in Kampala to represent the Church for the celebrations. The first event took place on Saturday 6th when Denis and Jack Darling attended an international athletics meeting at which were competing athletes from other African states, India, Europe including the UK and the USA. Almost all Uganda's open records were beaten and some winning times were close to world records. The royal couple arrived at Entebbe the next day, and in the library of Government House waited Denis and Olive, amongst others, to be presented to their Royal Highnesses. Olive thought the Duchess was 'radiant in vivid turquoise' [7], whilst Denis described

---

[15] Wellcome Library WTI/DPB/D/1/9. Itineraries of Denis Burkitt's lecture tours and safaris, 1963 and 1967–1979.

the encounter as 'Very brief'. There was a picnic lunch in the Botanic Gardens followed by a canoe regatta. The next day was a public holiday with the Independence Tattoo at Kololo Stadium followed at midnight, Monday October 8, 1962, by the lowering of the Union Jack, replaced by the new flag of Uganda.

Back in the stadium the next morning at 10:30, the ceremony of swearing-in of the Governor-General, Sir Walter Fleming Coutts, and the new Prime Minister Milton Obote of the Uganda People's Congress (UPC) took place. The Duke of Kent then read a message from the Queen, made his own address on behalf of the Queen and handed the documents of Independence Day celebrations to the new prime minister. Now officially in office, Milton Obote spoke appreciatively of what the British Government had done for Uganda and paid special tribute to the work of the missionaries, which pleased Denis. In the evening there was a State Ball enjoyed by Olive, endured by Denis who was no dancer and hoped he might escape the occasion by being called to do a difficult operation early in the evening but did pluck up courage to dance with one of Olive's African friends. After a cold buffet supper, Olive and Denis left at 1 am to catch a few hours' sleep before returning for the State Opening of Parliament by the Duke of Kent on Wednesday morning. What impressed Denis most was the fleet of Rolls-Royce and Bentleys that arrived bringing dignitaries from all over Uganda and beyond. There followed a Royal Garden Party at Government House attended by over 5,000 guests, a tradition that would not survive the changeover of power.

The Royal Party, with Denis and other functionaries accompanying, was due to fly out of Entebbe on Thursday to visit Jinja, Tororo and Mbale, but bad weather caused some disruption to the programme. By mid-morning the schedule was underway and in each town children lined the streets to welcome the Royal Party. Eventually arriving in Mbale, Denis had lunch, comprising soup, trout, steak or cold buffet, pineapple and cream, which Denis enjoyed in the company of a senior police officer and the Duke's valet, out of England for the first time. Next stop was Gulu where the Independence Day celebrations were repeated in the local stadium with speeches, exchange of gifts and then local dancing. On Saturday the party set off to see the Murchison Falls and Queen Elizabeth Game Park where they stayed for 2 nights. Denis spent a lot of time just sitting around and talking to members of the Royal Party, but things looked up on Sunday evening when the Private Secretary to the Duke, Lt Commander Buckley, told Denis that their Royal Highnesses would like Denis to join them at their table for supper. He sat to the right of the Duchess and found her very easy to talk to. Although several wines were offered during the meal, the Duchess declined them all, which Denis noted with interest in the light of his own aversion to anything containing alcohol.

The safari over, the party moved to Fort Portal for the final Independence Day celebrations of the tour before returning to Kampala to open the New Mulago Hospital in October 16. In the speech, which the Duchess read,[16] she made generous

---

[16] According to Denis' autobiography, but Olive's diary says the speech was given by the Minister of Health.

reference to Denis' lymphoma work and the international interest it had created, which was applauded by the audience. Later Denis was called to Government House where he had a private audience with their Royal Highnesses and was given a signed portrait of them. The couple left for Nairobi later in the week to inaugurate a new broadcasting and television studio. 'Independence' with Queen Elizabeth as Head of State lasted 1 year after which time Uganda declared itself a republic but retained its membership of the Commonwealth of Nations. Milton Obote was deposed in a military coup led by Idi Amin in January 1971. The sum of Denis' medical work during the visit had been one nosebleed, which had already stopped, a maid with a rash on her lip, a police officer who needed a plaster to cover a spot on his neck and another who asked for a laxative, one swollen ankle and a request for painkillers for gout by the manager of the lodge in the game park.

Relieved of official duties, Denis was able to return to his comfort zone which at this time included preparing three papers for a meeting of the UICC.[17] This was the leading international organisation working to reduce cancer worldwide, scheduled to meet in Dakar, Senegal, early in 1963 to be attended by scientists and doctors from all over the world. Amongst the latter were Dr Ralph Blocksma and his wife Ruth. Ralph was a plastic surgeon from Grand Rapids, Michigan, USA, who made an immediate and lasting impression on Denis. He and his wife had been medical missionaries in North India and were practising Christians looking for ways to help others rather than building a comfortable life for themselves. Visiting Mulago Ralph spoke at the Thursday Bible study which was held in the Burkitts' home, and realising the common ground they shared, he invited Denis and Olive to visit them in the USA, which they would do on many occasions in the ensuing years. It was Blocksma who encouraged Denis to write down the story of his life, which he did eventually in 1980 (Fig. 13.9).

Meanwhile the social programme at Mulago was gearing up for Christmas and before that the first casualty of the government's Africanisation policy, the departure of the hospital's Medical Director Bill Davies and his wife Joan. To Olive fell the responsibility and task of organising the farewell. There were 18 for supper on Tuesday November 6 the menu for which was grapefruit, chicken with almonds, rice, mixed vegetables, meringues and strawberries. Although Olive had help in the kitchen, entertaining 18 to a meal does not happen without a lot of planning. She was fully occupied running the family home, keeping in touch with those back in England, church activities, flower arranging, shopping, tennis and swimming, helping in the Mengo library and entertaining often at short notice the continuous import of visitors who were delighted to be able to say that they had dined at the Burkitt home. Denis was more than happy for Olive to be in control of these activities; in fact he was totally unsuited as a manager of social events. The occasion of Bill and Jean Davies' departure is described by Denis as 'Olive's Party'. Neither was Christmas his favourite time of the year. Cajoled into attending the local pantomime he, felt it was 'Late – long – and mostly just a vanity show. Did not enjoy it'.

---

[17] UICC, Union Internationale Contre le Cancer.

Fig. 13.9 Dr Ralph
Blocksma, 1963

Christmas day there were 15 to lunch including five Burkitts, the four MacAdams, three Drowns, Alex Alderdice the superintendent of the hospital at Mulago and Pooh Blanding and Sue Graham, nurses on secondment from Great Ormond Street. The visitors mostly had left by 3:30 pm, but the Drowns stayed all day leaving at 11:30 pm. This was not Denis' idea of a good day. 'I don't really enjoy these more or less extravagant Christmases. All the joy and meaning of it is entirely due to Olive. I fear I'm rather a wet rag on these occasions'. The family joined the Drowns at Budo for Boxing Day, but Denis managed to plead work commitments and spent the day writing a paper for the *Ethiopian Medical Journal* [8]. More enjoyable was a short family trip up country to Masindi and the Murchison Falls Game Park where the sight of large herds of elephant and buffalo brought in the New Year. But Denis, now working at the limits of his time and abilities, was feeling tired, so Olive had to drive the family back to Mulago. Within a few weeks, an event would occur which would admit Denis to the pantheon of great names in cancer biology.

References

1. Burkitt DP, Oettgen HF, Clifford P. Malignant lymphoma involving the jaw in African children. Treatment with alkylating agents and actinomycin D. Cancer Chemother Rep. 1963;28:25–34.
2. Burkitt DP. A lymphoma syndrome affecting African children. Ann Roy Coll Surg Eng. 1962;30:211–9.
3. Burkitt DP. A children's cancer dependent on climatic factors. Nature. 1962;194:232–4.
4. Burkitt DP. A tumour syndrome affecting children in tropical Africa. Postgrad Med J. 1962;38:71–9.

5. Haddow AJ. Epidemiological evidence suggesting an infective element in the aetiology. In: Burkitt DP, Wright DH, editors. Burkitt's lymphoma. Edinburgh: Livingstone; 1970.
6. Hutt MSR, Burkitt DP. The geography of non-infectious disease. Oxford: Oxford University Press; 1986.
7. https://www.youtube.com/watch?v=ylDMrkHrWcg, http://royalwatcherblog.com/2018/10/09/ugandan-independence-celebrations-1962/
8. Burkitt DP. A climatic dependent children's cancer. Ethiop Med J. 1963;1:254–8.

Chapter 14
A Cure for Burkitt's Lymphoma?

The UICC meeting was to be about 'Tumours of the Lympho-reticular System in Africa' to which category Denis' jaw tumour syndrome clearly belonged. He and two colleagues from Makerere made arrangements to travel together taking the opportunity to gather more data on tumour incidence on the way. However, it was clear that the South African delegates would not be able to attend a meeting in Senegal; in fact, there was no possible conference venue in Africa at which every-one could attend,[1] so the meeting was moved to Paris and took place in February 1963. It proved to be entirely about Denis' childhood lymphoma.

The journey to Paris was less than direct, but the outcome of the meeting would immortalise the Burkitt name and be a defining moment in Denis' life and career as a surgeon. Olive drove Denis to Entebbe from where he flew to Nairobi, then on to Rome where he took a train to Milan and finally through the snow-covered moun-tains of the Alps to Paris. Arriving on Saturday, February 9, he noticed the change in temperature and spent the next day acclimatising and enjoying Paris with Dennis Wright, George Oettlé, Professor Murray and Ralph Dorfman. The conference, held in the UNESCO building, lasted 5 days, Monday to Friday, and was mainly about Denis' childhood tumour. He gave lectures on the Monday and Tuesday in which he laid emphasis on the clinical presentation and probable environmental cause of his lymphoma but had to sit through a lot of talks about the histology of the tumour, which Denis described as 'entirely beyond my understanding'.[2] In the evenings, he went out exploring Paris with Dennis Wright, eating together and visiting the Louvre and Notre Dame.

On the final day, there was a lot of discussion around what to call the tumour. The UICC, being the world's leading cancer organisation, had the right to decide on such issues and with the distinguished group of participants in the meeting had

---

[1] Many delegates from S. Africa would be denied entry to newly independent African countries, and delegates from these would have refused to attend a conference in S. Africa.

[2] Wellcome Library Papers of Denis Burkitt WTI/WPB/D/1/11.

© The Author(s), under exclusive license to Springer Nature Switzerland AG 2022
J. H. Cummings, *Denis Burkitt*, Springer Biographies,
https://doi.org/10.1007/978-3-030-88563-2_14

access to a wide spectrum of opinion and knowledge. In East Africa, where the tumour had been first described and where it was the most commonly occurring tumour in children, it was known colloquially as 'Burkitt's tumour' because of Denis' known interest and work. In Paris, a sub-committee had been set up to decide the matter and reported back on the Friday with the suggestion that it be called 'Reticulo-endothelial Tumour of Burkitt'. This was rejected on the grounds that there was insufficient evidence for this histology and the suggestion was made that it be called 'Burkitt's sarcoma'. In the final discussion on Friday, Dr. Stewart, one the most senior participants, said that there was so little known about the tumour and its histology 'That perhaps it would be better to call it "Burkitt's tumour" until more is known about its histogenesis'. Dennis Wright immediately supported this suggestion. A vote was taken to decide between the two names, but the result was tied. The chairman cast his vote in favour of 'Burkitt's tumour' with Denis demurring and saying that much of the credit should go to his colleagues in Kampala [1]. But Denis had finally and irrevocably taken ownership of the tumour syndrome and wrote immediately to Olive to tell her the news, which reached her on February 20[3] and was received by her with delight.

Not long after the conference the term tumour was changed to lymphoma, more by common consent than any ruling from UICC or other body, and Burkitt's lymphoma, a more precise descriptive term, it has remained ever since. In 1967, as part of a series of meetings to agree an 'International Histological Classification of Tumours', the WHO organised a consultation amongst the world's most experienced haematopathologists and cytologists with a view to deciding on 'an authoritative definition which would be internationally acceptable' [2].

Discussions were detailed, thorough and driven in part by the need to draw a clear distinction between Burkitt's tumour, the leukaemias of childhood and the other lymphomas known at the time. The final agreement was to classify the tumour as 'malignant lymphoma, undifferentiated, Burkitt's type', but retaining the name 'Burkitt's lymphoma'. The group noted that it was not limited to Africa but occurred in many parts of the world. The meeting was attended by 18 leaders in the field of pathology and included Ralph Dorfman, Greg O'Conor and Dennis Wright.

In their 2016 revised classification of lymphoid tumours, the WHO [3] placed Burkitt's lymphoma in the category arising from B-cells.[4] There are around 50 different types in this category now identified of which only two are personalised in recognition of the work done to describe them: Waldenström and Burkitt. Of the other lymphoid tumours, a further 50 varieties, there are only three ascribed to individual scientists, Hodgkin, Langerhans and Erdheim-Chester. Denis had joined a distinguished group whose names have become bywords in cancer biology. Much research has since been done on the pathology and genomics of Burkitt's lymphoma such that there are now three distinct subtypes recognised: endemic, being the most

---

[3] The 20th anniversary of their engagement.

[4] B-cells. A type of white blood cell known as the lymphocyte responsible for antibody production and other immune functions.

common paediatric malignancy in sub-Saharan Africa; sporadic, occurring sponta-
neously around the world; and the third subtype is seen in immunodeficient patients
such as those with HIV infection [4].

With renewed confidence and an accolade afforded to only a very select few living doctors, a disease that carried their name, Denis left Paris for New York and his
first visit to the USA. The primary purpose of his visit was to attend and speak at a
conference in Houston on viruses and nucleic acids following which he had been
invited to visit the Sloan Kettering Institute in New York. The Institute staff and its
director, Dr. Frank Horsfall, had taken an early interest in Denis' work on the lymphoma, providing funds and expertise to help him. Arriving in New York in the
evening, the temperature was around freezing, but Denis now more acclimatised
was impressed by the enormous size of the cars, and the attractive house architec-
ture, which gave way to the skyscrapers of the city centre. He was put up for the
night at the Chemists' Club courtesy of a small cancer organisation, the directors of
which, Drs. Denues and Clarke, invited him out to dinner that evening where he met
Dr. Dalldorf again. A long discussion about the tumour ensued with Denis more
than tired when he finally got to bed at around 9 pm. From New York, Denis took
the train to Washington, then to New Orleans, giving himself 2 days to adapt to the
time and temperature and to see something of the countryside. His plan was to fly
from New Orleans to Houston for the conference, but he received a telegram asking
him to fly to Dallas to give his opinion on a possible lymphoma patient.[5] The patient
did indeed have the lymphoma, the first patient he had seen outside the tropics and
a piece of the jigsaw, a sporadic case, that did not readily fit into his developing
theory of the cause of the disease.

Denis gave a lecture in Dallas, then flew to Houston for the conference on viruses.
He was put up at the Hilton Hotel and, not yet used to the hospitality afforded to
distinguished speakers at meetings in the USA, he was amazed by the luxury in the
Shamrock Hilton. He asked for the cheapest room at $8/night but was given in one
at $15/night with thick carpets, two double beds, four standard lamps, two arm-
chairs, an electric cooker, separate room for his clothes and 'artificial heating'. The
towels and bed linen were changed daily. An enormous television dominated the
room, although the programmes were interrupted frequently by adverts, even the
news, which he found disconcerting. He was warned not to eat in the hotel because
of the cost, having a daily allowance of only $45, so he and Bob Harris, from ICRF
London, had breakfast at the local drug store. The conference, with 914 registered
participants, was in the ballroom of the hotel and lasted 2 days with Denis giving the
penultimate talk. He understood almost nothing of the lectures, remarking 'Someone
is talking an unknown language on cell chemistry' and later 'Another lecture, even
if possible, less intelligible, has started. By looking up occasionally at the illustrations (strange graphs meaning nothing to me), I can pretend I am intensely interested

---

[5] Mentioned by Nelson (p. 130) but not in Denis' diary nor his account of the trip. An alternative
source suggests 'Denis then flew to Dallas where he was met at the airport by two lady doctors who
drove him to where he gave a lecture to about 25 people all having their lunch, a new experience
for him'.

and taking profuse notes!'.[6] He was worried that his talk would be too clinical and simplistic for the scientists, but it was very well received; he was given a standing ovation and, presumably planned in advance, made an Honorary Citizen of the State of Texas. Press interest was overwhelming, and with so many journalists wanting to interview him, he agreed to hold a joint press conference with Bob Harris. He gave an interview to the local television network and was told that he had made the subject very easy for people to understand. The interview was featured on the news that evening, but Denis was again disappointed at the brief coverage it was given, brief by his standards. He attended little over half the lectures but was interested to see an eminent scientist on the front row chewing gum and another wearing red socks. This was not Africa where the, by now fragile, social order was maintained by a series of conventions born out of the playing fields and parade grounds of England.

After the conference, he was driven to a medical centre on the other side of the city where Dr. Hay, a former colleague from Kampala, had invited him to give a lecture. He had supper with about 15 doctors; then Dr. Hay showed a film about his time there. It was mostly about hunting, which Denis found quite repugnant. 'It quite sickened me to see these beautiful animals lying dead and Hay holding up their lifeless faces by the horns for photography; to see a prolonged shot of boys hacking off an elephant's feet with an axe and gouging out the blood-stained tusks'. He had to turn away. After a day of sightseeing in Houston, he was driven to Galveston by Jim Ouler, a medical student, where he gave the first of what would be many talks at a meeting of the Christian Medical Society (CMS), which on this occasion was largely students. There followed two very busy days meeting people, being entertained, giving talks on the lymphoma and the long safari, now always illustrated with his colour slides, visiting the hospital and talking to the doctors, all of which he found tiring.

6 pm, Monday, February 25

> his has been a full day talking to group after group of people. I had lunch in the hospital cafeteria. I started with haematologists, then over to paediatricians, then to pathologists. I lectured to full hall on the "lymphoma story" at 4.30. At 7 I am lecturing again.

But his hotel was up to standard with lights operated by buttons beside his bed. Finally, 4 days after the Houston Conference, he journeyed to New York where at Idlewild Airport the plane drew up beside a corridor projecting from the terminal building and Denis was able to walk inside without descending to the ground. A new experience for him. At the Sloan Kettering Institute, he renewed his acquaintance with Joe Burchenal, Gilbert Dalldorf and Frank Horsfall, the director, and gave a lecture for an hour, which stimulated one of the longest discussions at the Institute. That evening, he had dinner in an exclusive club in New York where even Frank Horsfall was not a member but gained access for the occasion through Mr. Kettering, a director of General Motors. His room for the night was more luxurious than ever with antique furniture, original paintings and a frame containing oval portraits with

---

[6] Wellcome Library Personal, Biographical and Autobiographical Material of Denis Burkitt WTI/DPB/D/1/11.

the names of the 19 members of the Thursday Night Whist Club in 1864. He was overwhelmed and felt undeserving of the attention he was getting. The next morning was spent at the Institute before he took the subway 10 miles north to a 1400-bed Veterans Hospital where he was met by Dr. Ludwig Gross who had arranged for him to lecture to the staff. This was becoming routine for Denis, but this time in the audience was one of the great names in cancer research, Dr. Peyton Rous whom Denis realised was a privilege to meet. Dr. Rous was the first person to show that a virus can cause cancer, the Rous sarcoma in chickens, and he would be awarded the Nobel Prize in Physiology or Medicine in 1966, 3 years after meeting Denis but 55 years after his discovery.

Back in London, he was met by Robin who took him home for an exchange of news and brief chance to recuperate, but Denis felt cold and was unwell. Soldiering on he travelled back to London for meetings at the Department of Technical Co-operation, which was now funding his work in Uganda, then Sir Harold Himsworth at the MRC and finally with a senior MRC statistician, Richard Doll, with whom Denis would develop a long and fruitful working relationship once permanently back in England. He left for Uganda on March 4 after almost a month away during which he had experienced for the first time the international dimension to science, the high esteem in which he was held and the rigours of travel across the world. Arriving at Entebbe, he was met as always by Olive and although now clearly suffering with 'flu went straight into a busy schedule, which included dinner with the visiting external examiner for surgery, Mr. Selwyn Taylor from London who had travelled on the same flight as Denis. But life was changing in Uganda, long-established staff at Mulago were leaving and Denis was in active discussion with the new Chief Medical Officer at the hospital and the MRC in London about moving to full-time research.

The Paris meeting and his trip to the USA had come at a good time for him because it made him realise that he could no longer control nor even contribute to the ever-expanding research into the lymphoma. From the outset during discussions with Professor Davies, Greg O'Conor and Dennis Wright, he had to rely on their expertise in characterising the tumour pathology, and this become even more evident in Paris when he listened to the arguments that took place about the definition of the lymphoma. Equally problematic was the increasing complexity of the epidemiology of the lymphoma. The rainfall and temperature maps no longer were able to explain the distribution of the syndrome, which was now being reported sporadically from countries across the world, and in adults as well as children. The search for a virus was beyond him, and as the Houston meeting brought home to him, there were realms of science relevant to the tumour story that he could not even understand. Laboratory medicine had never been an interest. But it was now his tumour, named after him, and wherever he went in the world, he would be greeted by people as the person who had originated the story and set in train the hunt for a possible virus that was causing cancer, with all the implications that had for cancer research. Denis was not demoralised by losing control, as he might have been 6 years earlier, far from it. On his visit to the USA, he had enjoyed the reception given to his talks, even by the most erudite and knowledgeable scientists working on his tumour. And there was one part of the narrative account he gave that had everyone sitting up and paying attention. He was now talking about a cure.

From his initial encounter with children with lymphoma, he knew that the only option for treatment was chemotherapy. Because the nature of the lymphoma was to arise in several sites in the body simultaneously, surgery was not realistic, and there was no radiotherapy equipment available anywhere in Uganda in 1961. In his first talk about the jaw tumours in February 1958 to the East African Association of Surgeons, he had mentioned treating two children, unsuccessfully, with nitrogen mustard.[7] The arrival in 1960 in Nairobi of scientists from the Sloan Kettering Institute, particularly Joseph Burchenal and Herbert Oettgen, changed that (Fig. 14.1).

The team from New York had based themselves in Nairobi at the Kenyatta National Hospital with Peter Clifford, a distinguished head and neck surgeon, who had recognised a case of the lymphoma in 1955 and realised that this was a new disease after reading Denis' 1958 *British Journal of Cancer* paper. He was Kenya's most eminent surgeon who had a lot of experience treating cancer with chemotherapy and was a pioneer in regional perfusion with drugs [5]. The Americans were all specifically expert in the development and use of chemotherapeutic agents to treat cancer and had wide experience of childhood cancers. One of the team, Joe Burchenal, visited Denis at Mulago and gave him some methotrexate to try. It was the first chemotherapeutic agent that had been used successfully to treat cancer[8] and had the advantage that it could be given either by mouth or into the bloodstream. Peter

Fig. 14.1 Dr. Joseph H Burchenal (1912–2006), pioneer in developing chemotherapy for Burkitt's lymphoma who was part of the Sloan Kettering group. (Photograph kindly provided the Lasker Foundation, New York)

---

[7] Nitrogen mustard. A close relation of mustard gas used in chemical warfare in the Second World War.

[8] Choriocarcinoma, a tumour of the placenta. Methotrexate remains a first-line treatment for this cancer.

Clifford in Nairobi believed that chemotherapy of cancer should involve high doses of the agent being used given repeatedly until signs of toxicity occurred. Denis was not able to do this. His patients were reluctant to stay in hospital, and local resources demanded the treatment be as cheap and simple as possible. But methotrexate worked. Even given by mouth, tumour regression was seen with just a few doses of methotrexate spread over a week. This, at times quite dramatic response to chemotherapy, was thought to be because the lymphoma was the fastest-growing cancer known, the rapidly dividing cells in the tumour providing an ideal target for the drugs. Working out the best dose and regimen for his patients, Denis was helped by Dr. Oettgen who together with Dr. Burchenal treated a series of 31 children with jaw tumours and reported that in early stage, a disease survival of up to 21 months was seen [6]. Follow-up of the children was difficult, but as time went by, Denis was able to recount anecdotes of much longer survival. One 5-year-old girl named Namusisi who presented with jaw tumours was scheduled to be given three courses of methotrexate each comprising eight pills a day for 5 days. However, after only one course, Namusisi's mother came to Mulago one night and took her home. Very often, people were frightened of hospitals and doctors. A year later, one of Denis' African workers came across Namusisi in her home in the bush. She was well and appeared completely cured. Denis subsequently managed to see her frequently after that, and she remained well for many years. Cured with one course of pills [7] (Fig. 14.2).

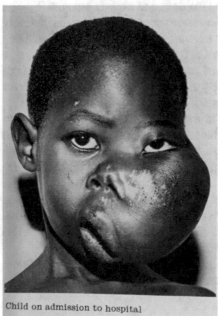
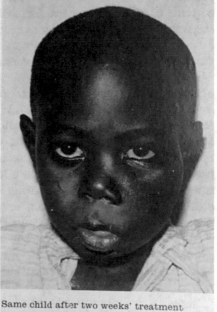

Child on admission to hospital

Same child after two weeks' treatment

Fig. 14.2 Child with jaw tumours. On admission to Mulago and then 2 weeks after start of treatment. (Wellcome Images; WTI/DPB/F/8 and with permission from the *Journal of Laryngology & Otology*. 1965 vol LXXIX p. 933)

It was the response to treatment that provided the next claim to Denis' ownership of the tumour. Caught early and with a simple course of oral methotrexate, the tumour would start to regress within days. This was a regimen developed very much by Denis of necessity. The implications for cancer chemotherapy worldwide were quickly recognised. Denis would write extensively about it, but he had two problems. He was no expert in understanding the cellular biology of cancer chemotherapy, and Mulago Hospital could not afford such expensive drugs.

Obtaining cheap supplies of the drugs was a challenge that Denis relished. His innate love of a bargain, passionate belief in what he was doing for these patients and a touch of Irishness made him a difficult person to refuse. He started by approaching the Nairobi representative of Lederle Laboratories, Mr. Innes, who manufactured methotrexate, and Denis suggested that giving him supplies of the drug would provide them with a very good opportunity to observe its effectiveness in patients who had not had either surgery or radiotherapy. They gave free supplies of the drug to Mulago for years. Later, whilst on safari in South Africa, he learnt of another chemotherapeutic agent, cyclophosphamide, a recently approved anti-cancer agent, and on his return to Kampala, he persuaded the German manufacturer of the drug, Asta-Werke, to give him 'liberal amounts'. It worked well for the lymphoma. Subsequently, he obtained an expensive drug, vincristine, from Eli Lilly. A review of the use of these drugs and their effectiveness is given in Denis' chapter on the subject in a book comprehensively addressing all aspects of the lymphoma, published in 1970, co-authored with Dr. D H Wright, the histopathologist [8]. The best outlook for his patients was for those who were young and presented at an early stage of the disease. Lymph node or central nervous system involvement meant that the outlook was much less favourable. Peter Clifford had access to better funding, a steady supply of patients with lymphoma and was able to try 17 drugs, a number of which proved beneficial, although again early diagnosis was key [9]. The search for a cure for these children was well underway.

On his return from his Paris/USA trip in early March 1963, Denis decided to focus more on treatment and gave his first talk solely on this subject in Kampala in May. More chemotherapeutic agents were becoming available, now including cyclophosphamide and vincristine, such that every child diagnosed with the lymphoma was offered treatment. The tumour proved to be very sensitive to these agents largely because of its unusually rapid growth, which coupled with a strong immunological response to the tumour gave remarkable results to treatment. Denis started adding the outcomes of treatment to all his papers about the lymphoma and was now in search of nothing less than a cure.

A letter written on August 8, 1963, to him by the mother of one of the children he had treated must have offered encouragement (Fig. 14.3).

Nabyajwe
Buddu
9.8.63

Mr Sir,
Mr Bukitt,
How are you Mr? Any news over there?
Thank you for all what you do. Sir, our people you are with, are they well?

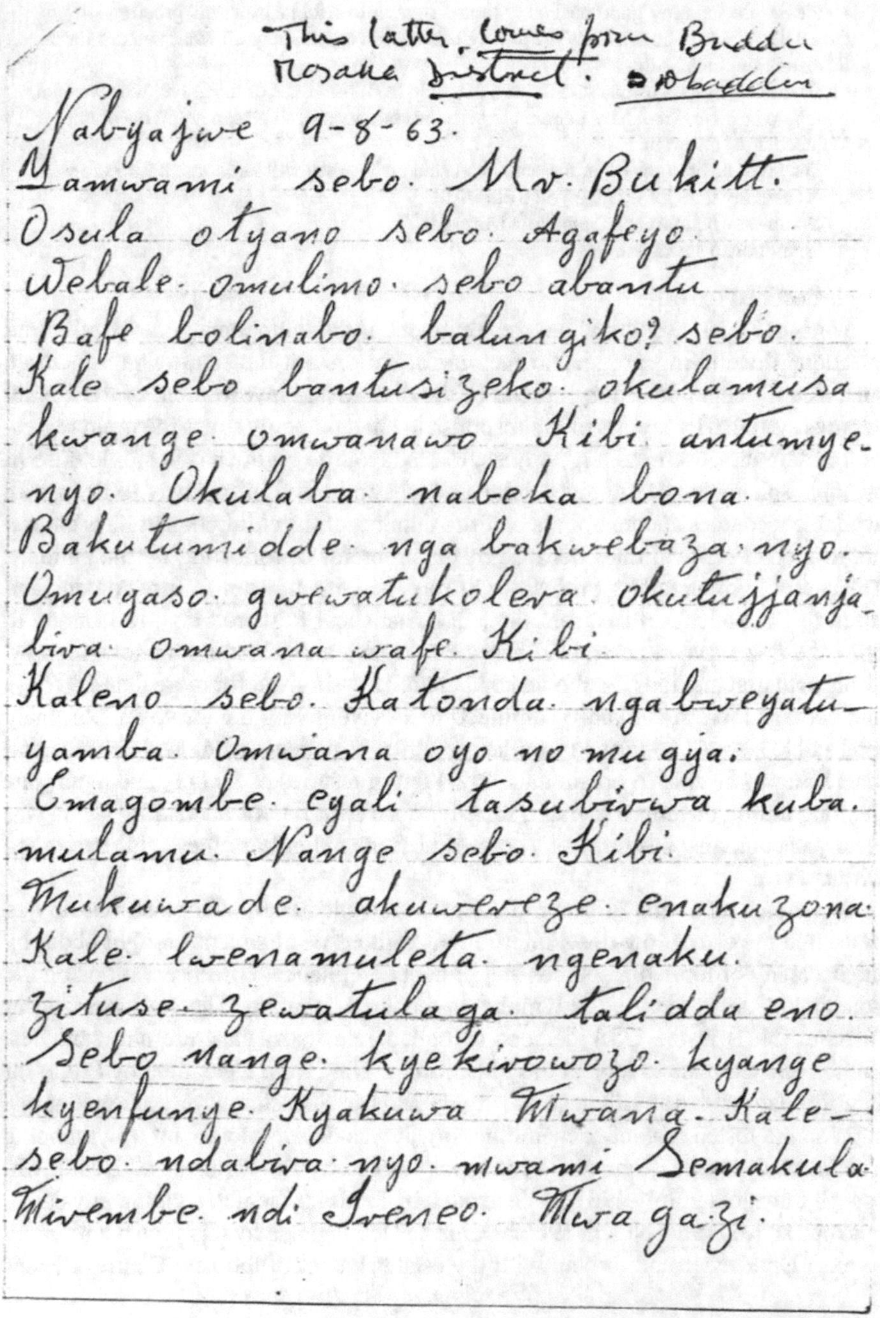

This latter, comes from Buddu
Rosaka District. Obaddu

Nabyajwe 9-8-63.
Tamwami sebo. Mr. Bukitt.
Osula otyano sebo. Agafeyo.
Webale. omulimo. sebo abantu.
Bafe bolinabo. balungiko? sebo.
Kale sebo bantusizeko okulamusa
kwange omwanawo Kibi antumye.
nyo. Okulaba nabeka bona.
Bakutumidde nga bakwebaza nyo.
Omugaso gwewatukolera okutujjanja
bira omwana wafe Kibi.
Kaleno sebo. Katonda nga bweyatu
yamba. Omwana oyo no mugya.
Emagombe egali tasubirwa kuba.
mulamu. Nange sebo. Kibi.
Mukuwade. akuwereze enakuzona.
Kale lwenamuleta ngenaku.
zituse zewatulaga taliddu eno.
Sebo Nange. kyekiwowozo kyange
kyenfunye. Kyakuwa Mwana. Kale
sebo. ndabira nyo. mwami Semakula
Mivembe ndi Ireneo. Mwagazi.

Fig. 14.3 Letter about the patient Kibi in which his parents tell Denis that they wish to give the child to him. The language is Luganda. (With thanks to Richard Price of Homerton College Cambridge and Aisha Shuamazzi for help with translation from Luganda and with cultural aspects)

Please- do send my good wishes to them. Your child Kibi has sent his greetings to you and all the people at home. They are saying thank you very much to all what you did for us, of treating our child Kibi.

Sir, as God helped us that you treated the child from the edges of the grave, who was not thought to be alive, Sir- I have decided to give the child to you, so he can help you whenever you need him (for ever).

They will be brought when the dates you said arrive, they will not come back here with me. That is my thought, to give you a child.

Ok Sir- say hello to Mr Ssemakula Mirembe.

Yours Ireneo Mwagazi.

Denis declined the offer.

Much of Denis' writing about the treatment of the lymphoma is descriptive and anecdotal, identifying prognostic factors for children at the time of presentation, such as age, length of history, spread of the disease and involvement of the central nervous system. His reports of the response to chemotherapy stress the rapid regression of tumours, even the largest ones, to his favourite regimen of a single dose of cyclophosphamide, 40 mg given intravenously optimised in this way to fit in with available resources and compliance of the families of the children. But he is careful not to predict survival rates because of the problems of following up the patients. The Nairobi group, led by Professor Clifford, had much better follow-up data, and it fell to Dr. Malcolm Pike from the MRC Statistical Research Unit in London to draw the first survival curve [10] based on 42 patients treated with chemotherapy. Using actuarial methods, he showed that about 15% survived for more than 250 days and perhaps more remarkably continued to survive giving the curve an extremely long tail. This was confirmed in a larger series of patients from Mulago Hospital where they were able to obtain data on 74 out of a series of 80 [11] and again saw the long tail on the curve giving a long-term survival rate of about 21%. This was clearly encouraging but was not a cure and left the majority of these children dying within a year.

It was not for Denis to become a world-renowned chemotherapist, but he had pioneered the use of low-dose, intermittent regimes of treatment brought about by the necessity of limited resources and patient compliance. His observations on the potential of drugs to cure the lymphoma attracted attention. The National Cancer Institute (NCI) in the USA decided to collaborate with Makerere and sent Drs. Zubrod and Carbone to look at the possibilities. They were welcomed by Professor Ian McAdam who was still head of surgery and Mr. Sebastian Kyalwazi who was a surgeon interested in cancer chemotherapy. It was decided to set up a Lymphoma Treatment Centre in Kampala, which was opened by the Minister of Health Mr. Joseph Lutwama in July 1967, to be headed by Dr. John Ziegler, a young physician, who had arrived at the NCI in 1966.[9] Denis had left Uganda by this time but returned to meet Dr. Ziegler and to help with the establishment of the new Centre. Ziegler

---

[9] Dr. John L Ziegler b. 1938, MD Cornell 1960. Joined the NCI in 1966. Uganda 1967–1972. Returned to the NCI as head of clinical oncology, then in 1981 to the University of California, San Francisco, where he became founding director of the Graduate Sciences Programme. Awarded Lasker Prize in 1972, with DPB and Burkitt Medal of TCD in 2014.

worked hard to build up the new unit with the support of Richard Morrow, an epidemiologist from the WHO; Professors Michael Hutt and William Parsons, heads of pathology and medicine, respectively; and Sebastian Kyalwazi, who became professor of surgery. The new unit was very successful and by 1969 had become the Uganda Cancer Institute (UCI) comprising 40 research beds and attracting staff from all over the world. Following the military coup led by Idi Amin in 1971, there was a period of unrest and civil war in Uganda. Many of the staff left the Institute including Ziegler who had been director from 1967 to 1972. By 1988, peace had returned to the country and the epidemic of HIV/AIDS presented new challenges to the world of cancer research and the UCI thrived once more (Fig. 14.4).

Stimulated by Denis' early suggestion that the sometimes spontaneous [12] and long-term remission of his lymphoma with treatment in some patients [13] was due to host immunity, early work on cancer and immunity was pioneered at the UCI [14]. Ziegler's success with cyclophosphamide chemotherapy for the lymphoma and work on immunity in cancer resulted in him sharing the Lasker Prize, with Denis and others, in 1972. In 2015, thanks to the University of Washington, Seattle, the Institute moved to a new building now known as the Uganda Cancer Institute/Fred Hutchinson Cancer Research Center Clinic and Training Institute and is a fine example of collaborative research across the world that has led to important discoveries. Its origins go back to Denis and his work on the lymphoma and the Institute stands as a legacy to his work [15]. Today, thanks to worldwide research into the chemotherapy of the lymphomas and leukaemia, survival in the paediatric (endemic) form of Burkitt's lymphoma now approaches 90% [16] in countries where resources are available. In sub-Saharan Africa, the children, who have contributed so much to

Fig. 14.4 Dr. John Ziegler. Director, Uganda Cancer Institute Kampala, Uganda. 1967–1972. (Photograph kindly provided the Lasker Foundation, New York)

the understanding of haematological cancers all over the world, have not benefited from these advances in therapy. Because of a lack of access to good medical care and significant socioeconomic challenges, cure rates are less than 50% [17], although they are improving [18].

The year 1963 saw Denis writing in his every spare moment to fulfil requests for papers to be published in connection with meetings where he was speaking, from journal editors who saw the lymphoma story attracting the attention of clinicians and scientists interested in cancer the world over. To report his most recent clinical observations and ideas about the cause of the lymphoma, he had seven papers published in 1963, all in reputable journals or as chapters in books and for five of which he was the sole author. In 1964, there were two publications about the lymphoma (in 1965, four; 1966, five), and then from 1967 onwards, he would author or co-author around a dozen papers on various related topics each year for the next 10 years. For someone who never ran a large department or gathered a supporting team around him, this reflected the huge amount of work that arose out of his enthusiasm, dedication to serve those around him and perhaps ambition. His principal hobby, photography, which continued throughout his life, led him into moviemaking when affordable cameras became available. The film of the long safari that he had made with the help of Ted Williams and Cliff Nelson is best described as a home movie of an unusual holiday. But it was good practice for his next and more important venture, a film about Burkitt's lymphoma. Made in part to save him from having repeatedly to give lectures on the subject, the finished version had its premiere on July 18 when Gilbert Dalldorf and the Nairobi team visited Mulago.

Home leave started soon after the movie was finished and saw Olive and Denis journey by air first to London where he did some more work on the soundtrack for the film and talked to the MRC about his future. Progress seemed to be slow. Then to North Devon for a week of rest and recuperation at the Christian Retreat and Holiday Centre, Lee Abbey interrupted with a drive to Westonbirt near Bristol to collect Cas and Rachel from school. With all the family together, they made their way to Fishguard on the west coast of Wales where they embarked on a boat for Cork. Now without the focus of Laragh, James or Gwen, they settled for a week in the south renting a house, then heading back to Cork and to Coolmore where they were enthusiastically received by the Newenhams and other members of Gwen's family. After a further week, they drove to Dublin where they showed the children Trinity College and some of their favourite haunts, but the cold and wet weather did not agree with them, and Denis developed 'flu so they decided to return to Coolmore where he went to bed. Olive also became unwell, and the girls took over managing the household. By the middle of September, they were all recovered but were back on the Cork to Fishguard ferry on Friday the 13th en route to Gatwick by train in time for Judy to catch her flight back to Nairobi for the start of the school term.

Denis then travelled to London by train where he made a brief visit to Kodak to talk about copying some x-rays, then to Slough where he had lunch with Robin and Vi. He borrowed their car to drive to Oxford where he lectured at the Radcliffe

Infirmary on his 'cancer hobby' at the invitation of Mr. Macbeth, an ear, nose and throat (ENT) surgeon, who had visited Denis in Kampala. After the now customary dinner party in college,[10] Denis stayed overnight, returning to meet the family and drove to the village of Nettlebed in the Chiltern Hills a few miles from Henley on Thames. There they took over a house, Cedarwood, described by Olive as 'Charming and most suitable little house', which a friend had kindly agreed to let them rent so that they would have a base in England for the winter. The plan was for Olive, Cas and Rachel to be based in England for a few months whilst Denis sorted out their future in Uganda with the MRC and University. They all returned to Worthing where Olive was organising Denis' imminent trip to the USA on which, for the first time, she would accompany him.

Back again in London, Denis spent time with Professor Cohen at the Royal College of Surgeons who was particularly interested in jaw tumours, then to the Middlesex Hospital where he met one of the pathologists, to whom he had given tumour tissue earlier and gave his customary lecture to the staff at 4 pm. Two days later, he again visited Kodak, then after lunch at the RSM with the director of one of the drug firms providing chemotherapeutic agents to Denis in Uganda, he went to Film Services, a commercial company, who were copying the lymphoma film. It was due to be shown as a premiere, with soundtrack, to a distinguished audience at the ICRF, which included Sir Cecil Wakeley of the Royal College of Surgeons, Sir Harold Himsworth, Sir Miles Clifford of the Leverhulme Trust who had a distinguished career in the Colonial Service including in East Africa and was a Governor of the ICRF, Professor Cohen, Dr. Lewthwaite, and others including Jack Darling and Will Davey from Ibadan. It was a considerable tribute to Denis' rapidly growing reputation that such an audience should be gathered to see him and his film. It was only 5 years since his first paper on the childhood cancer had been published. But the film was not ready, so Denis had to show the original, without the soundtrack, and give a commentary off the cuff.

After the meeting at the ICRF, Denis had discussions with the Colonial Office and with the MRC about his future, once again. It was becoming clear that the MRC would take him on to their staff and that he could continue his work at Mulago. He spent the evening with Robin and Vi, together with Olive and Aunt May, at Farnham Common before leaving the next morning with Olive for Harrogate via London. There he met Dr. Jim Wilson, a good friend from the days when Denis was a junior doctor and who now as president of the Harrogate Medical Society had invited Denis to talk. The members of the Society were luckier than they might have known. On the way through London, Denis had picked up the finished version of his lymphoma film, now including the soundtrack, which they were to be the first of many to view. It was September 1963 and Denis was about to meet the greatest admirers of his work.

---

[10] Oriel.

References

1. Roulet FC, editor. Symposium on Lymphoma-reticular tumours in Africa. Basel, New York: Karger; 1964.
2. Histopathological definition of Burkitt's tumour. Bull World Health Organisation. 1969;40:601–7.
3. Swerdlow SH, Campo E, Pileri SA, Harris NL, Stein H, Siebert R, et al. The 2016 revision of the World Health Organisation classification of lymphoid neoplasms. Blood. 2016;127:2375–90.
4. Panea RI, Love CL, Shingleton JR, Reddy A, Bailey JA, Moormann AM, et al. The whole-genome landscape of Burkitt lymphoma subtypes. Blood. 2019;134:1598–607.
5. Clifford P. Further studies in the treatment of Burkitt's lymphoma. East Afr J Med. 1966:195–9.
6. Oettgen HF, Burkitt D, Burchenal JH. Malignant lymphoma involving the jaw in African children: treatment with methotrexate. Cancer. 1963;16:616–23.
7. Glemser B. The long safari. London: The Bodley Head; 1970.
8. Burkitt DP, Wright DH, editors. Burkitt's lymphoma. Edinburgh and London: E and S Livingstone; 1970.
9. Clifford P. Treatment. In: Burkitt DP, Wright DH, editors. Burkitt's lymphoma. Edinburgh and London: E and S Livingstone; 1970. p. 52–63.
10. Pike MC. Chemotherapy in Burkitt's tumour. Lancet. 1966;2(7468):856.
11. Morrow RH, Pike MC, Kisuule A. Survival of Burkitt's lymphoma patients in Mulago Hospital. Uganda. BMJ. 1967;4:323–7.
12. Burkitt DP, Kyalwazi SK. Spontaneous remission of African lymphoma. Brit J Cancer. 1967;21:14–6.
13. Burkitt DP. Clinical evidence suggesting the development of an immunological response against African Lymphoma. UICC Monograph Series, Vol. 8. In: Burchenal JH, Burkitt DP, editors. 1967. p. 197–203.
14. Ziegler JL, Cohen MH, Morrow RH, Kyalwazi SK, Carbone PP. Immunological studies in Burkitts Lymphoma. Cancer. 1970;25:734–9.
15. Ziegler JL. Into and out of Africa – taking over from Denis Burkitt. Br J Haematol. 2012;156:766–9.
16. Dozzo M, Carobolante F, Donisi PM, Scattolin PN, Maino E, Sancetta R, et al. Burkitt lymphoma in adolescents and young adults: management challenges. Adolesc Health Med Ther. 2017;8:11–29.
17. Esau D. Denis Burkitt: a legacy of global health. J Med Biogr. 2019;27:4–8. https://doi.org/10.1177/0967772016658785.
18. Scanlan T, Kaijage J. From Denis Burkitt to Dar es Salaam. What happened next in East Africa? – Tanzania's story. B J Haematol. 2012;156:704–8.

Chapter 15
America

Having had a brief taste of the USA and Canada in February, Denis was looking forward to returning, this time with Olive and an exceptionally busy schedule to fulfil. He would make many visits to North America over the next 30 years, more than to any other part of the world. He was always received most warmly and generously and with respect bordering at times on adulation for his unique achievements with the lymphoma and later with dietary fibre. America loved him. His genuine humility, amusing yet self-deprecating talks, interest in people and how they lived, lack of interest in the material things of life, limited tolerance of social conventions and for his Christian faith. And he loved America. Its varied and monumental geography, the 'can do' attitude of the people he worked with whom he described as 'charming and interesting', the amazing contrasts with where he worked in Uganda and, probably without acknowledging it to himself, he liked being the star attraction wherever he went.

The next morning, September 28, 1963, Jim Wilson drove them to Liverpool on the first step of a journey that was to prove both busy and historically eventful. They embarked on the RMS Sylvania and found themselves in first class, but a rough crossing prevented Olive from enjoying the exceptional cuisine on board. Denis was fascinated by their fellow first-class passengers who were mostly Americans over 60. He noticed that 'money seemed to flow from them' and was rather fearful of the expectations of the crew for tips. In the library, he found a copy of Winston Churchill's 1908 account of his travels in East Africa and read it so that he could understand better the changes that had gone on in the intervening 55 years. But not wishing to be completely idle, he decided to write an article for a Swiss drug firm, Sandoz, with the title they had requested of 'Can Cancer Be Transmitted by Insects?' An additional motivation, other than his wish to tell the world about his tumour and its treatment, was the 1200 Swiss francs the company offered him, a substantial sum

Wellcome Library Papers of Denis Burkitt WTI/DPB/D/1/12.

© The Author(s), under exclusive license to Springer Nature Switzerland AG 2022
J. H. Cummings, *Denis Burkitt*, Springer Biographies,
https://doi.org/10.1007/978-3-030-88563-2_15

for an article of this type and far more than he had ever had for writing a paper. The article was printed in the Sandoz company journal in 1964.

After almost a week at sea, they arrived in New York on Saturday, October 5, and headed for the Sloan Kettering Institute where they stopped briefly to leave surplus luggage before catching a bus to Washington. They both enjoyed the 4.5-h journey because of the lack of traffic compared with London and it was the fall with the endless shades of brown, yellow, gold and red of the trees along the early part of the route through Philadelphia and Baltimore. They took a taxi to the house of Dr. Nicholson, a general practitioner, and his wife with whom they had been put in touch by the Christian Medical Society. They were of a similar age to Denis and Olive but earning £10,000–£15,000 annually compared with Denis £2,000–£3,000 pa. Denis noted their luxurious, by his standards, house with carpets even in the children's bedrooms, en suite bathrooms, the largest television he had ever seen, central heating for the winter, air-conditioning for the summer and two enormous cars on the drive outside. Sunday was spent sightseeing in Washington and going to church, after which life became more hectic. On day 1 proper of the trip, Denis was collected from the Nicholsons by Major Price, who had spent 18 months at Makerere, and driven to the house of their next hosts, Dr. Albert and Mrs. Louise Bright. After a brief chance to get his slides in order, Denis was taken to the Armed Forces Institute of Pathology (AFIP) to lecture, followed by a formal lunch at the Officers Club. He was presented with a medal that had his name engraved on it, after which a secretary was put at his disposal so that he could deal with all the correspondence that had been forwarded to the hospital. Next, he was taken to the Medical Museum where at 4:30 pm he lectured to the Washington Society of Pathologists, then dinner where everyone ate steak. Denis remarked 'They were nearly an inch thick' and chose chicken. Returning to see Olive at 11 pm, he was tired. Two lectures were one thing, two formal meals another, but it was the constant flow of people who wanted to talk to him, coupled with endless travel, that proved so tiring. Back at the AFIP the next morning, Denis was booked to talk to a number of staff members, who presumably presented their research to him. As with the meetings he had attended in Paris and Houston earlier in the year, the subjects for discussion with him were largely unintelligible. At lunch time, he was fascinated by a machine where on inserting a coin dispensed a glass of 'Coke' with ice. At the National Institutes of Health (NIH) the next day, he was amazed by what he felt was a gigantic set-up with a staff complement of around 10,000 and a budget of close to a billion dollars, more than 20 times that for the whole of Uganda. At the NIH, he talked to various members of staff and was looked after by Greg O'Conor with whom he had worked on the early pathology of the lymphoma at Makerere. Denis showed his film in the director's office and was promptly asked if they could have a copy. Along with a similar request from the AFIP, Denis now needed to make two copies, which he was happy to do. A lecture to the Dental faculty followed, where a number of his slides were broken by a faulty projector, then home for dinner with Greg and family. Denis had always believed Greg had very little money apart from his salary in Uganda, but his huge and lavishly furnished house told another story. Greg's father, Daniel J O'Conor with Herbert Faber, had invented Formica in 1912.

Arriving in Buffalo, they were met by an old army friend of Denis, Dr. Adrian Kanaar, who was a driving force in the local CMS. An early start the next day took them to Roswell Park Memorial Institute, one of the oldest and best cancer research and treatment institutes in the world. Denis lectured at 9 am, showed them his film, then moved on to the children's hospital for a repeat performance at midday. Lunch followed, after which Dr. Kanaar drove him to the Edward Meyer Hospital where he was on the staff and he had the kudos of being able to introduce the great Dr. Burkitt, his friend from army days and now world-renowned cancer specialist, to lecture. Denis was back at the Kanaar's house with just enough time to wash, change and shave before leaving with Olive for the CMS banquet where his role was to speak first on the work in Africa, then 'apply the message of the Gospel and point to the obvious limitations of medicine alone in treating whole men and making men whole'. Sharing credit for his work with God was not a problem for Denis. There was no danger he would open the *The Lancet* one morning and find a report of a new treatment for the lymphoma by Him. His faith opened many doors in his travels around the world and brought Olive and himself many friends (Fig. 15.1).

The next day, Saturday, was for sightseeing. They were driven around Buffalo, at that time one of the largest centres for flour milling in the world, by Dr. Kanaar's son David. Denis was impressed by the modernity of the city especially the elevated highway that crossed over above the houses and the intricate flyovers that allowed people to continue their journeys uninterrupted. The next event was a picnic at Letchworth State Park, about 80 miles out of town, organised by the CMS where there were a lot of international students. Denis spoke about relationships, probably using a few scribbled notes put together during the event. It was the end of their first

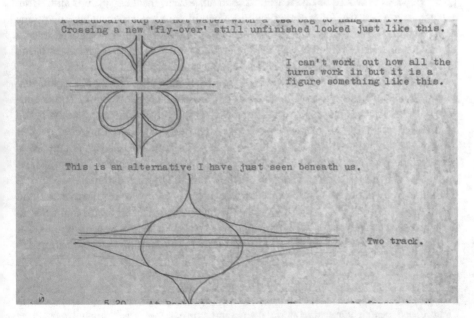

Fig. 15.1 Denis' observations from an aircraft window on seeing a major road junction below

week in the USA and Denis' eighth talk of the trip. His audiences had included high-end scientists in leading national institutions, expert groups of pathologists, dentists, hospital doctors and members of the CMS, each requiring a different approach and emphasis. It was very demanding, and it would be over 6 weeks before they were back at home.

Sunday was for church, then a trip to see the Niagara Falls before heading to Toronto and the home of Donald and Elizabeth Gibson who had spent 7 years at the Mengo Mission Hospital in Kampala. Denis was not impressed by the Falls, being too commercialised for him, instead preferring the Victoria Falls, which in 1963 remained largely unexploited. What did impress him was an aerosol can which, on depressing the nozzle, exuded what was said to be cream. He was sure that nothing relating to any cow was involved in its production. Waking the next morning, he walked down to the shores of Lake Ontario, surrounded by trees clothed in their autumn colours. It was Thanksgiving Day, a day to be with Olive whose company on the trip made life more enjoyable and to whom he could happily leave managing social activities and keeping up correspondence with the family. However, the next 3 days were spent with Olive at the Gibsons because she had a cold, whilst Denis visited hospitals in Toronto, Kingston and London, Ontario. A pattern was now emerging for these visits. Denis would be driven from place to place, often by doctors or academic staff from the various hospitals and institutes, some of whom had visited him in Kampala but now welcomed him as the great cancer specialist from Africa. He lectured at least once at every venue his talk now comprising half an hour of slides, the illustrations for which were now drawn by eldest daughter Judy, with appropriate commentary, followed by the showing of his lymphoma film. This format seemed to go down best. He would be taken around each hospital to meet key groups and, in the evening, there was usually a dinner, called a banquet, in his honour where everyone would want to be introduced to him and shake his hand. He quickly realised that he would not be able to remember all these people or even get chance to talk to them. But they would always be able to say that they had met the great Dr. Burkitt.

Journeying to Grand Rapids via Chicago, they met the Blocksmas again and joined in some of the activities of the Christian Medical Society there. The friendship with the Blocksmas, which would grow over the years, was not so much based on a common interest in surgery, missionary work in Africa nor the geography of disease but their shared Christian faith, which had established an immediate bond between the two men when they first met in Kampala [1].[1] Arriving at 4 pm, Ralph and his wife Ruth gave the Burkitts tea, after which they proceeded to the CMS dinner attended by about 60 where Denis looked at the programme for the evening and found the title of his talk. After the meal, there were two singing items and then, to their surprise, Olive was asked to speak. Denis thought that she 'was quite magnificent and won all hearts' then, loosely interpreting his given topic he talked about his

---

[1] Ralph Blocksma 1914–2001 of Dutch descent was born in Grand Rapids, where he remained for the whole of his career as a plastic surgeon. He was a committed member of the local Christian Churches and spent significant time abroad doing short-term missionary work helping to found the Medical Assistance Programme (MAP) and chairing the Board of Affordable Medicines for Africa (AMFA).

work in Africa and the limitations when treating a person's illness to have no regard for what Denis would call their soul. The next morning at the Christian Reformed Church, which had seats for over 1000 and was packed to the doors, Denis discovered that he was due to speak at an after-service meeting on 'Christian Aspects of Medical Missions in Uganda', but again and without warning, Olive was asked to speak first. By Monday morning, Olive was beginning to feel unwell, but their commitments required them to move on.

Denis' eyes were opened to a different way of life in the USA. Visiting Ralph's office and hospital the next day, to his amazement, he was able to get a cup of coffee from a machine and saw that much of the hospital equipment such as needles, syringes and surgical supplies was disposable. He gave a short, illustrated talk to the resident staff before returning to the Blocksma's house. Although less ostentatious than those occupied by his medical colleagues, it had an automatic garage door, which could be opened from the kitchen when Ruth saw Ralph approaching in his car, and where downstairs the kitchen, dining room and living area were open plan, separated by only small partitions. The idea he was told was so that mother could keep an eye on the whole family. The next day, they were shown around a furniture factory that was turning out good quality pieces by mass production and in addition 'antiques' suitably distressed to make them look old. They were also making Chinese furniture. On the way back to the city centre, Denis enquired of Ralph what might be the cost of an electric razor he had seen, which could be charged and then used for several days – very useful when travelling. Ralph insisted on buying him one as a gift from the CMS. After lunch at an exclusive club for doctors and their wives, Denis lectured at the hospital and received a standing ovation. That evening, he gave the Susan Lowe Memorial Lecture before departing the next day by train for Chicago. There they were met by Dr. Crown who took them to a Salvation Army Centre where he was treating alcoholics. Olive was by now feeling continually tired and stressed, so they headed to the house of their hosts for the visit, the Knightons, where she was able to get an early night. Even Denis was feeling tired, but the Knightons had arranged for a number of people to meet Denis including Cliff Nelson who had been on the long safari. After a lecture to the Department of Surgery the next day, Olive joined Denis at the railway station, where they had to move their own luggage using a trolley, before boarding the California Zephyr for the 48-h, 2532-mile journey to San Francisco.

Olive found the journey to San Francisco quite stressful, but the train was comfortable, they had a two-berth coach, and the next morning, they both went up to the observation car to enjoy the views as the train passed through the Rockies. For Denis, every minute of the journey was an experience to be enjoyed and recounted at length,[2] but as the day wore on, Olive felt ill and exhausted. Denis gave her a sedative, after which she slept well but still felt rotten the next morning and 'Feel so

---

[2] 10:30 Sunday morning, 'We are passing through a galaxy of glory. Evergreen pines mixed with fall colours in mountainous country. Great river gorges flanked with green pines and set aflame with the sun shining on the golden aspen leaves. The extent and profusion of ever changing vistas is almost intoxicating'.

desperately sorry for my Deny being encumbered with a wife like me'. Arriving in San Francisco late on Sunday afternoon, they were met by their hosts for this part of the trip, Dr. and Mrs. Johnson who lived in San Jose where he was a general surgeon. The 50-mile journey between San Jose and San Francisco was really too far for Denis' convenience. He would have to make the return trip every day by bus because he was attending the Annual Meeting of the American College of Surgeons, and he was also beginning to worry about Olive.

For the next 2 weeks, he had a very busy schedule with travel and lectures during which he did not think Olive would be able to accompany him because he felt she needed to rest. He could not cancel these commitments because the honoraria from the lectures were paying for the journey. Moreover, he felt he had an obligation to the people that had invited and who awaited his visits with high expectations. Good Christian that he was, he nevertheless enjoyed the welcome he received at every meeting and the respect he was shown for his work with the lymphoma. Fortunately, Olive had felt very much at home with the Blocksmas, so Denis rang them and asked if she could return and stay in Grand Rapids. They readily agreed, and a couple of days later, Ruth Blocksma met Olive at Chicago airport, and after a night in a hotel, they drove to Grand Rapids. Ralph, finding Olive to be quite unwell, suggested she see a Christian psychiatrist, Dr. Bergsma, who said to her that 'depression often a sign of inferiority – not being able to measure up to someone or something'. Olive thought that summed up her situation very well. A few days of quiet domesticity followed coupled with rest and the kindness of the Blocksmas, but her mood did not improve, so she decided to go and see Dr. Bergsma again. This time he changed her medication, and reading the next day Catherine Marshall's 'Beyond Ourselves', Olive felt compelled to do what she had done and to be more accepting of God's will for her. On November 18, after 2.5 weeks with the Blocksmas, Olive flew to Pittsburgh to meet Denis but was too tearful to tell him all that had gone on since they last met. Olive had been unwell for 4 weeks when events overtook them and the world on November 22.

Having dispatched Olive to the Blocksmas, Denis headed north to stay with the Iversons in Eureka before catching a Greyhound bus the next morning for the 455-mile journey along the west coast to Portland. He was due to speak at a dinner that evening but took the bus in order to see the redwood forests, which despite it raining all day allowed him to relax. Changing in the Gentlemen's toilet on arrival at the Hilton Hotel, he made it to the dinner as the guests were finishing drinks, something he was not unhappy to miss, and gave his lymphoma talk and film show. His hosts that evening were the family of the Iversons who delivered Denis to the local hospital for a seminar at 9:15 the next morning where he sat between two very distinguished surgeons,[3] gave his talk and two media interviews, then headed for Vancouver. Sunday was less busy with morning service at a Mennonite Brethren Church, a new experience for Denis, sightseeing around Vancouver, then a CMS

---

[3] Sir James Patterson Ross, past president of the Royal College of Surgeons of England, and Professor Dumfry, president of the American College of Surgeons.

group in the evening where Denis spoke. After a somewhat inconsequential next day, Denis was surprised to be greeted at the door of the lecture theatre where he was to speak in the evening by Professor Ian McAdam, which gave them both the chance to catch up with news from Kampala long into the evening. After giving the Simmons and MacBride lecture on Tuesday evening, Denis boarded a train which climbed very slowly up through the Rockies where the view of the snow-covered countryside was largely obscured by mist. After his favourite lunch in the snack bar, eggs and bacon, tea and brown bread, he was engaged in conversation by a fellow passenger who asked him where he came from and what his work was in Africa. Shortly afterwards, she returned with a page of the current *Vancouver Sun*, which carried a photo of Denis and an account of his work in Africa.

After Banff, Denis took a plane to Edmonton where, on Thursday morning, he met Professor Patterson Ross again and was given a tour of a hospital devoted entirely to the care of Indian and Eskimo people. Denis lectured at 1 pm and again to students later in the afternoon after Professor Ross had spoken. There was the usual media interest this time shared with Professor Ross. Denis pointed out that Sir Patterson Ross was the 'big fish and I very small fry' but found the great professor easy to engage with and was able to talk to him, thanks to him going out of his way to make Denis feel an equal. On to Saskatoon flying over deeply snow-covered landscapes, Denis marvelled at 'the wonderful privilege that has been afforded to me to see this fascinating country and to meet so many charming and interesting people. When I think that until a little over two years ago, I had never left Uganda on medical work, expenses paid'. In Saskatoon, he joined up with another visitor, Dr. Paul Brand, professor of orthopaedic surgery at Vellore who was known throughout the world for his revolutionary approach to treating leprosy deformities of the hand. At the CMS dinner that evening for once, Denis was not the main speaker, but he was allowed a few minutes first. Somewhat in the shadow of Professor Brand, who was trying to raise money for his Mission to Lepers, Denis spoke to a group of students and nurses then became engaged in discussions about their local hospital problems. He did not leave for his hotel until 11:20 pm, which for someone who liked to get to bed early was not to his liking.

Monday morning, November 11, saw Denis looking around the hospital in Saskatoon where to his amazement he saw for the first time two patients in whom cardiac pacemakers had been inserted. Whilst not a cardiologist, he was fascinated by the technology, the way the gadget could monitor the heartbeat and correct it with an embedded battery that would last 5 years. He gave a talk to surgical staff after lunch; then, following a pleasant evening with his hosts, Dr. Abe Dick and family, he had to go to another doctors' house at 9 pm where more medical staff wanted to meet him. Asking to be excused at 11:30 pm, he was fearful that he would be too tired to give of his best at the series of lectures he was due to give during the week. Each day now followed a familiar pattern with a visit to the hospital and ward round in the morning where Denis would be asked to comment on special cases, followed by lunch to meet the staff, lecture then off to the airport. His next destination was Winnipeg, arriving just in time to wash and change before a CMS dinner where he had to speak. A similar day followed including interviews with the local paper, the *Winnipeg Free Press*, who reported that 'Dr. Burkitt disclaimed credit for discovering the (virus and environment) theories but gave praise to other men who have been active in the field including Dr. J.N.P. Davies, Dr. G. O'Conor, Dr. Gilbert Dalldorf and Prof. Alexander Haddow, who is director of the Virus Research Institute in Uganda'.

Denis was now clearly confident with his name now attached to the tumour to share the credit, a little. Back in the air heading for St Louis via Chicago, he arrived in time for an evening lecture to the St Louis Paediatric Society and was in bed by 10:30. An early start the following day found Denis listening to a lecture about lymphoma in cattle at the children's hospital, then to the Homer G Phillips Hospital where he lectured at 10 am and met endless staff wanting to talk to him. He was scheduled to speak again at 2 pm in the medical school but asked his hosts if he could just sit and rest for an hour saying to himself, 'Please don't introduce me to anyone and let me just stop talking for a bit'. He was reaching the limit of his physical, mental and emotional resources. His lecture was well received, which made him feel the effort had been worthwhile, so he felt more confident going into a meeting of senior staff at the home of Dr. Cowdry, then, after what Denis described as a sumptuous dinner in a hotel, early to bed for a change.

Arriving in Pittsburgh on Sunday, November 17, at the start of a momentous week for the world, Denis was met by Dr. Rogers of the Department of Preventive Medicine. The next morning, he lectured to the staff, had lunch in a private room with some of them and then talked to a group of pathologists at 3:30 pm. Denis managed to get to the airport for 5 pm in time to meet an improved Olive. During their 18 days apart he had visited seven major cities in Canada and the USA, given 18 lectures and taken part in many ward rounds and seminars, attended dinners in his honour most evenings and met countless people. Olive, who had battled her demon depression for the whole time, was tearful at their reunion and noted 'absolutely blissful to see my Deny'. Neither of them could sleep that night, so Denis booked a flight to New York for the next morning where they arrived soon after midday. He went straight to the Sloan Kettering Institute (SKI) to collect his post and find out what the programme was for the week and was then able to spend the evening with Olive. After lecturing at the Institute and talking to Dr. Oettgen, Denis and Olive had dinner at a restaurant where several of the waitresses were Irish, one from Ballybunion where they had spent 6 weeks in the summer. All seemed to be right with the world. The next day, they were driven to one of the satellite laboratories of SKI in Bronx where they met Dr. Herbert Dalldorf, the virologist, and his wife with whom they had lunch. Returning to SKI, Denis showed his film at a conference; then he and Olive were guests of honour, which Denis thought to be absurd, at a special dinner at the Institute attended by the Dalldorfs, the director Dr. Frank Horsfall and his wife, Dr. and Mrs. Kettering, a director of General Motors and major funder of SKI, and Mr. and Mrs. Bruce, another director of the company.

Friday, November 22, 1963, found Denis in the hospital attached to SKI and Olive shopping in New York. He and Olive remember exactly what they were doing soon after 12:30[4] that day when the news of John F Kennedy's assassination broke.

_____

[4] Central Standard Time USA.

Denis was talking to the husband of a patient who had been operated on that morning and found to have incurable cancer. They were discussing whether or not to tell her. A secretary came into the room with the news. Denis left the hospital and met Olive at the hotel. She had returned from her shopping trip to be met by someone coming up the stairs from the basement saying, 'Kennedy's been assassinated'. They spent the rest of the day watching events unfold on the television as gradually everything came to a halt in a country trying to come to terms with the tragedy. Saturday morning, Denis caught the subway to Columbia University where he was due to speak. On arrival, he found a notice on the door saying that the meeting had been cancelled along with all other University activities. A number of doctors arrived, and they all started talking together. They were still anxious to see the film and hear Denis speak and perhaps being more accustomed to dealing with death, its circumstances and consequences felt it would not be in contravention of any regulations if they were to have an informal meeting. They had promised Denis an honorarium, which pleased him and afterwards he was driven back to the hotel by one of the doctors who had been a classmate of Kennedy at Harvard. Sunday was spent with church friends, Laurence and Margaret Wynne, whom Denis had known in Enniskillen but was disrupted by the news of the shooting of Lee Harvey Oswald, Kennedy's assassin, the incident covered in full on television.

Denis and Olive watched coverage of the funeral on TV on Monday 24th and were unexpectedly moved by everything.

> Mrs. Kennedy has just arrived, and all the other dignitaries walking before us. Often tears come to one's own eyes. Mrs. Kennedy and brother now kneeling at the coffin. How real it all is. The motionless guard of honour. She slowly walks backwards, in deep black, holding her brother's hand. A last farewell to her husband. How composed she looks. The military guard now slowly carrying out the casket. Six white horses are lined up to draw the bier. A riderless but saddled black horse stands behind. The guard of honour present arms. "Hail to the Chief" is played by the band. The casket comes down the steps, draped in the U.S. flag. The sun is shining, the Service chiefs are saluting. Some of the pall bearers are coloured men. The coffin is raised on to the carriage. The family, Service chiefs and other dignitaries fall in behind the cortege. The procession from the Capitol to the R.C. Cathedral starts.

Olive and Denis left New York the next day arriving in London in the evening to be met as usual by Robin. It was just over 8 weeks since they had left Liverpool on RMS Sylvania. Denis reckoned he had travelled over 17,000 miles, visited many US States and Canadian Provinces, fulfilled a very busy schedule that included 44 talks often combined with the showing of his film and had observed history in the making. The next day, they were at Nettlebed where Olive would spend the winter whilst Denis returned to Uganda. The house and garden were in good order, thanks to the housekeeper, Mrs. Sadler, and gardener. Olive immediately felt at home, more settled and was able to stop all her drugs apart from a sleeping tablet. She and Denis enjoyed a few days together free from work commitments, although she accompanied him to London where he was interviewed by people from the MRC and brought up to date about affairs in Uganda by Foreign Office staff. He was to become full-time employed by the MRC on April 1, 1964.

Denis flew back to Entebbe a few days later. Leaving Olive in Nettlebed was a planned decision by the two of them but for reasons that are not set down anywhere, even in the detailed diary in which Olive describes her everyday life, thoughts and emotions. There is little doubt that the two of them were very much attached to each other, interdependent and in love in a way that transcends its physical dimension. The day on which Denis flew back to Africa, he describes as 'A Terrible day. Left Olive. Hardest parting ever' whilst Olive writes 'Wanting to put off the goodbyes yet wanting to get them over. Parting is such agony for us both – but we wouldn't have it otherwise'. Denis had left a letter for her. During the afternoon of his departure, she was 'travelling with my Deny all the time. Sat down at about 5.15 to read his letter to me – lovely. Lump in throat and wet eyes'. The following morning, 'Missing Deny dreadfully – worse than any other parting. Lovely note from him from Airport also one from Rachel – bless her'. The parting left her feeling sad and lonely soon to be followed by a relapse of her 'old arch enemy' depression. Maybe aggravated by stopping all her tablets but somewhat ameliorated by the arrival of Judy from school and Rachel and Cas from Westonbirt to spend Christmas together. Her letters to Denis were sad, but he was pleased that the girls were there to look after her. It was an unequal partnership, born out of the necessities of war, shaped by the formalities of the doctor-nurse relationship, bonded by family and their shared faith but strained by long periods spent apart. Soon life would change completely for the family, but before that, there was the question of the cause of the lymphoma to be solved.

Reference

1. https://www.legacy.com/obituaries/timesunion/obituary.aspx?n=ralph-blocksma&pid=108474

Chapter 16

A Virus Causing Cancer? [1–3]

Ex Africa semper aliquid novi
   (From Africa always something new)
   Pliny 23–79 C.E.

That a virus might be the causal agent of Burkitt's lymphoma had been a matter for discussion from the earliest days of the lymphoma story. After seeing the first two children with the condition, Denis had immediately started to examine hospital records for earlier cases and had soon realised that there was a distinct geography to the disease. At his talk to the East African Association of Surgeons in February 1958, his first major presentation on the topic, he showed a slide of the districts in which the cases were found, which gave a clear indication that they came mainly from only 3 of the 13 districts around Kampala.

At what stage did Denis discuss these findings with Professor Haddow? He was head of the East African Virus Research Institute and had been studying mosquito-borne viral diseases such as yellow fever for many years and was aware of geographical and climatic factors affecting the prevalence of the lymphoma. In 1961, before the long safari, Denis had written, in reply to a letter from a Dr. Engel, saying that 'The evidence is, in my opinion, mounting that this syndrome is due to a virus'. In the letter, he describes how he and Professor Haddow had discussed maps of tumour distribution, climate and the geography of other viral diseases. By the time of the long safari at the end of 1961, the media were reporting Denis' findings on climatic factors and were highlighting the possibility of a viral cause. It was big news. On Tuesday, November 21, 1961, an article had been carried by *The Chronicle*, a national newspaper published in Bulawayo, under the headline 'Do Insects Carry a Cancer Virus?' The article was given prominence on the front page. Earlier, in July 1958, Professor Jack Davies, with whom Denis' relationship had at times been fraught, had been at a cancer conference in London where he had presented the clinical and pathological features of the tumour syndrome. Whilst at the meeting he had received from Denis the final pre-publication draft of his first paper describing the cases and the early geography of the tumour distribution, which intrigued the assembled pathologists and promoted much discussion. Amongst a number of

© The Author(s), under exclusive license to Springer Nature Switzerland AG 2022
J. H. Cummings, *Denis Burkitt*, Springer Biographies,
https://doi.org/10.1007/978-3-030-88563-2_16

suggestions was the possibility that a virus might be involved, based on the knowledge that some cancers in animals were due to a virus. Greg O'Conor, the pathologist with whom Denis worked most closely in 1959/1960, was aware that a condition in cattle known as lymphocytic bovine leukaemia had a distinct geographical distribution in Europe thought to be due to a virus [4] (Fig. 16.1).

The belief that a virus was responsible in some way for this cancer was clearly on the agenda from very soon after the first cases were described by Denis in 1958. However, without his key observations on the characteristic geographical and climatic limits of the lymphoma, such a theory might have taken much longer to be recognised. After the publication of his first paper on the tumour in November 1958, Denis sent out a questionnaire to all the mission hospitals in Africa and elsewhere asking them if they had seen children with this problem. Within 2 or 3 months, he had gathered enough replies to plot the distribution of the cancer on a map. It is likely therefore that he would have sought Professor Haddow's views by early in 1959, before he left for home leave in April 1959. On seeing Denis' map of tumour distribution, Haddow responded immediately with evidence of the geographical distribution of viral infections in Africa, in which matter he was already a world-renowned expert. The virus theory proper was probably born by the second half of 1959. At this time, no virus had been shown to be causal in any human cancer, but by 1964, one had been identified in the Burkitt tumour cells, which would be shown to be an essential component of the cause of the lymphoma. Remarkably, it is a virus that also causes another illness, known to everyone.

Six years after Denis' first report of the clinical features of his lymphoma was published [5], two papers appeared in the same issue of *The Lancet* on February 1,

Fig. 16.1 Sir Alexander Haddow, CMG, FRS, FRSE. Director of the East African Virus Research Institute, 1953–1965. (By kind permission of the University of Glasgow Archives & Special Collections, University Photographic collection, GB248 UP1/565/1)

1964 [6, 7], reporting the successful culture in the laboratory of Burkitt's lymphoma (BL) cells. This was an important success because no member of the human lymphocyte cell series had ever been cultured in vitro. Cells grown in the laboratory would now provide a new opportunity to find the cause of the lymphoma and more especially whether it was due to a virus because viruses, unlike bacteria, require living cells in which to survive, reproduce and be studied. One article by Robert Pulvertaft, a retired pathologist, who had trained at the Westminster Medical School, described as an 'eccentric genius' [8], reported work done whilst he was visiting professor in the Department of Pathology at the University of Ibadan, Nigeria, where they had seen over 50 cases of Burkitt's lymphoma. During a 1-year visit, Pulvertaft collected specimens of tumour from the operating theatre and developed a method for culturing them. The object was to be able to study the cells further, including the possibility of virus infection. Pulvertaft was already retired when he went out to Ibadan in 1963 and whilst publishing further work on the histochemistry of the lymphoma cells it was left to others at the Medical School to follow up the virus question.

In the same issue of *The Lancet*, Anthony (Tony) Epstein reported, with his student Yvonne Barr,[1] the successful culture of BL cells, and acknowledges the Pulvertaft paper. Epstein had been working with BL cells since he had first heard Denis speak at the Middlesex Hospital in 1961. The tumour tissue was provided by Dennis Wright. He would go to the theatre where Denis was operating, collect the sample for diagnosis and send some to London. Epstein may have been stimulated into publishing his work earlier than he had planned when possibly *The Lancet* made him aware of the Pulvertaft paper. Epstein was a virologist, had been working on viruses causing cancer in chickens when he first met Denis and had been updated frequently by Denis about his observations of the altitude, temperature and rainfall dependence of the tumour distribution and the likelihood that it was a mosquito-borne infection that initiated the disease. Eight weeks after their February *Lancet* paper, Epstein, Achong[2] and Barr revealed to the world, in another short paper, that they had seen virus particles in the cultured BL cells [9] using electron microscopy. It became known as the Epstein-Barr virus (EBV) [10]. This was a sentinel observation and the first real indication that a virus might be involved in human cancer (Fig. 16.2).

Denis realised immediately the significance of the findings by Tony Epstein and his group. He had been in constant communication with them not least in ensuring they had a good supply of tumour material obtained at surgery. But it was not Denis' sort of research, being largely laboratory based, so he could not actively participate

---

[1] Yvonne Mollie Barr (1932–2016) had studied zoology at Trinity College Dublin and on arrival in Epstein's laboratory to do her PhD was given the most vital task of developing tissue culture of the lymphoma cells.

[2] Bert Achong (1928–1996) excelled at school in Trinidad and was given a scholarship to study medicine at University College Dublin. He joined Epstein's lab in 1963. He was given the task of developing electron microscopy of the cells in culture in search of a virus and would follow Epstein to Bristol when he moved there in 1968.

Fig. 16.2 M.A. Epstein, B.G. Achong and Y.S. Barr in 1964. In the summer of 1963, Tony Epstein (later Sir Anthony) of the Bland Sutton Institute of Pathology at Middlesex Hospital School of Medicine took on Yvonne Barr and Bert Achong as research assistants to carry out cell culture and electron microscopy studies, respectively. (With thanks to Professor Owen Smith, Dublin)

in it. He continued with his search for the most effective chemotherapy, and he together with Michael Hutt,[3] newly appointed Professor of Pathology at Makerere, and Ted Williams carried on plotting the worldwide distribution of the cancer.

Epstein, Achong and Barr reported in their *Lancet* paper describing the virus that they thought it belonged to the herpes group because it contained double-stranded DNA. In 1964, this was a well-known group of viruses causing a variety of infections such as blistering oral and genital disease (herpes), chickenpox and shingles. What surprised Epstein was that the virus existed in the cultured cells but did not seem to be causing any ill effects, as herpes viruses were known to do. Whilst trying to identify which type of herpes virus they were looking at, a fortunate series of events [3] put Epstein in touch with Werner and Gertrude Henle, two virologists working at the Children's Hospital in Philadelphia.[4] Cell cultures from the Epstein laboratory were sent to Philadelphia where the Henles were quickly able to show that the virus was previously unknown; in other words, it was a new virus within the herpes group. They were responsible for labelling it EB virus after Epstein and Barr, who had sent them the cell cultures.

---

[3] Professor Michael SR Hutt (1922–2000) was a pathologist who became a great friend of Denis and together they advanced the concept of the geography of diseases. He qualified from St Thomas Medical School and was in Makerere from 1962 to 1970, then returning to St Thomas' as Professor of Geographical Pathology.

[4] Werner Henle (1910–1987) and his wife Gertrude Henle (1912–2006) were from Germany, qualifying in medicine at Heidelberg, then emigrating to the USA in 1936 where they worked together at the Children's Hospital in the University of Pennsylvania on common viral infections in humans. Werner Henle's grandfather was Jacob Henle, one of the outstanding pathologists of the nineteenth century, who gave his name to the loop of Henle in the kidney, which he discovered.

The Henles then went on to investigate its relationship to the lymphoma. The first step was to detect antibodies to the virus in people's blood. When infected by a virus, our immune system, as part of our defence against infection, develops antibodies,[5] which are there to neutralise the virus, and these can be detected in blood. Viral and other antibodies may persist in blood long after an infection has been defeated. To determine if patients with BL had antibodies to EB virus in their blood, they were able to obtain serum from a Nigerian child with lymphoma who had been flown to the Clinical Center of the National Institutes of Health (NIH) in the USA for treatment. The Henles used a sensitive technique known as indirect immunofluorescence, and when testing the child's serum against BL cells, a reaction occurred confirming the presence of antibodies to the virus in the child. Next, they examined the sera of a large group of lymphoma patients, which they were able to obtain from the Lymphoma Treatment Centre in Mulago, the Kenyatta Hospital in Nairobi and from the NIH where Dr. Stevens had sera from patients in Nigeria. All of the patients who were judged typical BL cases showed antibody to EB virus [11], whilst those who had been treated and survived for over 3 years mostly had antibodies but at lower levels (titres) than the acute cases (Fig. 16.3).

Case proven? Was this a cancer caused by a virus? As work proceeded, it was clear that the story was not so simple but contained an element that would surprise virologists the world over. The Henles then tested children who did not have BL, including age- and sex-matched controls, siblings and neighbours of the affected cases, children from areas of both high and low BL incidence and other paediatric patients. Overall, more than 80% were positive, although the antibody titres (blood levels) were ten times higher in the BL children. Moreover, blood checked from children from many other parts of the world all contained antibodies to EBV, indicating that almost everyone is infected at some time – a remarkable finding and a major turning point in the story of the EB virus.

One notable feature of the age distribution was that African children had all acquired EBV antibodies by the age of 2 or 3, whilst in countries such as North America, acquisition of EBV infection and thus antibodies did not occur until adolescence or young adulthood and paralleled that of antibodies to other common viruses such as measles, mumps or polio seen in the pre-vaccination era. Working in parallel with the Henles in Philadelphia were Peter Clifford's group in Nairobi who had been collaborating with the Sloan Kettering Institute for some years [12] and with George Klein at the Karolinska in Stockholm. Denis thought very highly of George Klein[6] remarking that he was one of the most brilliant intellects that he had ever encountered. Klein worked initially with Clifford's group producing a series of

---

[5] Antibodies are small proteins made by our immune system in response to the antigens that are present on the outer envelope of viruses, bacteria, etc. They are there to neutralise the effect of the virus and once produced remain in the memory of our immune system for life, thus giving us some protection from future infection. Previous exposure to a virus can usually be detected by measuring specific antibody levels in blood.

[6] George Klein (1925–2016), a Hungarian tumour biologist who moved to the Karolinska Institute in Sweden where a chair in Tumour Immunology was created for him.

Fig. 16.3 Gertrude and
Werner Henle, Children's
Hospital, Philadelphia.
Circa 1964. The Henles
established that the virus
seen in the lymphoma, in
cultures sent to them by
Tony Epstein, was a unique
member of the herpes
family. (With thanks to
Professor Owen Smith,
Dublin)

over 40 papers on Burkitt's lymphoma, EBV and their immunology. Clifford and
Klein would work closely with the Henles becoming the key group on the immuno-
biology of Burkitt's tumour [13, 14]. The Nairobi group, given their more strongly
clinical and surgical interests then, in collaboration with Cornell Medical School in
New York, went on to test the serum of patients with a variety of other disorders
such as acute leukaemia, breast and lung cancer, using a different technique to the
Henles, and found no significant antibody levels with one striking exception,
patients with nasopharyngeal cancer.[7] This finding was later confirmed by the Henle
team. The virus story was developing rapidly but in an unexpected way. Now there
were two cancers very much associated with the EBV, which in turn appeared to
infect most of the world's population in childhood or adolescence.

The Henles thought that the almost universal distribution of antibodies to EBV
must mean that it was also the cause of a common but mild and self-limiting disease
and by a remarkable coincidence the clue to this came from an illness contracted by
one of the young technicians in their laboratory. Known as E.H. (Elaine Hutkin),
she had, like all laboratory staff, given a sample of her blood for work related to the

---

[7] Nasopharyngeal cancer arises in the area behind the nose and tongue and occurs mainly in parts
of Africa, China, the Far East and less commonly in the Eskimo.

EBV in January 1967. She had no antibodies to the virus, but in August of that year, she developed a sore throat, fever, enlarged lymph nodes in her neck and a rash and was diagnosed as having infectious mononucleosis (IM) or glandular fever. Serial blood tests showed the presence of EBV in her white blood cells and her antibody titres, which had been zero earlier in the year, rapidly rose confirming the presence of the virus. In order to confirm that EBV was the cause of IM, the Henles needed further blood samples from cases of IM. In this, they were fortunate because Drs. Robert McCollum and James Niederman of Yale University Medical School had been conducting a prospective study of IM in college students since 1958. They had collected blood from the students when they arrived at the college and then during and after they showed signs of glandular fever. These were made available to the Henles who showed that whilst antibodies to EBV were always absent before the disease once students had developed IM, antibodies to EBV appeared in blood as they had with E.H. This case was proven. It subsequently became known that the route of infection was through the mouth leading to the label 'kissing disease' [15], with the first clinical signs of the disease appearing 4–7 weeks after infection. Its peak incidence is at ages 15–25 years, the student years. In Africa, children aged less than 3 years were more frequently exposed to the virus, probably attendant on their living conditions, but did not develop clinical symptoms of IM. The virus infects B-lymphocytes, which are key to the development of the lymphoma and once infected we carry the virus for the rest of our lives [2].

Why it became known as the Epstein-Barr virus, and not Epstein-Achong-Barr, who were the authors of the first paper to describe the virus, was a matter that piqued Denis' sense of fairness. Bert Achong must have in reality been the first to see it because he was the electron microscopist in the laboratory. The electron microscope (EM), capable of much greater magnification than an ordinary light microscope, was not generally available in 1964, but Epstein, after visiting George Palade,[8] one of the early pioneers of electron microscopy at the Rockefeller Institute in New York, had managed to obtain an EM and taught Bert Achong, who had joined his lab in 1963, to use it. The likely explanation to authorship would seem to lie in Barr's role in culturing the BL cells [16]. She joined Tony Epstein at the Bland Sutton Institute a number of months before Achong in 1963 to do a PhD supervised by Epstein. Her task was to establish the culture of the BL cells, which with the help of Epstein she was able to do in a laboratory where cell culture was an important tool for studying viruses. 'It took 26 attempts before the Eureka moment on February 18th 1964'.[9] Many cell lines would be in use so hers became known as EB (Epstein-Barr) cells. When the Henles asked for the cell cultures to be sent to their laboratory in Philadelphia, they were labelled as EB1 because this was the first cell line successfully to be cultured by Barr and it was this cell line that was sent to the Henles. When it became clear that these cells carried a previously unknown virus, they

---

[8] George Emil Palade (1912–2008), a Romanian who spent most of his working life in the USA. Awarded the Nobel Prize in Physiology and Medicine in 1974 for work on the 'structural and functional organization of the cell'.

[9] Professor Owen Smith, Dublin, personal communication. 2020.

called it the EB virus (EBV) [17]. *The Lancet* paper describing the success in growing the cells, which was a major achievement accredited rightly to Mollie Barr, was thus written by Epstein and Barr only. Barr completed her PhD in 1966 and left the Epstein laboratory moving to Australia whilst Bert Achong remained and went on to co-author a number of papers and books with Epstein, moving with him to the Department of Pathology in Bristol Medical School in 1968. Achong was a highly capable scientist in his own right and in 1971 discovered a 'foamy virus' by EM examination of human cancer cells, which proved to be the first example of a retro-virus naturally infecting man. But Denis, not really understanding how research laboratories work and aware that Achong was from Trinidad, was uncomfortable that his name was not included.

Denis had followed the EBV story closely and thought that the case for a virus causing his lymphoma was strong. The age distribution of cases was similar to other viral diseases of childhood such as polio, and clustering of cases occurred in some areas of Africa. An insect vector such as the mosquito would fit with the climatic distribution of tumour cases. Moreover, viruses were known to cause cancer in a number of animal species, although these were RNA[10] containing viruses, even so he felt it was only a matter of time before the first virus causing cancer in humans would be found. Another crucial piece of evidence was provided by a very distin-guished group working together, which included George Klein, the Henles, Peter Clifford and Harald zur Hausen, who later would be awarded the Nobel Prize for his work showing that human papilloma viruses (HPV) were implicated in human cer-vical cancer. They were able to show that EBV DNA could be detected in all the biopsies of both Burkitt tumours and nasopharyngeal tumours but not in those from other cancers such as Hodgkin's lymphoma, myeloma, other lymphomas, thyroid cancer, etc. [18]. The evidence for the role of EBV in the cause of Burkitt's lym-phoma was becoming convincing. Subsequent research has implicated EBV in Hodgkin's lymphoma and a small proportion of cases of gastric cancer.

To explain why not everyone infected with EBV developed cancer, it was clear that there must be other biological influences contributing to the risk. A major study was undertaken by the IARC in 1971 in Uganda to determine whether children developing BL were infected with, or reacted to, EBV in a different way [19]. It was a heroic study for its day with blood samples taken from 42,000 children up to 8 years of age, stored in deep freezers and then recovered for examination as and when BL developed in one of them. The site chosen for the study was the West Nile region of Uganda because the incidence of BL was high and its occurrence well documented by Ted Williams who had accompanied Denis on the long safari in 1961. Ted and his brother Peter, both of whom spoke the local language, helped with the project. Between 1973 and 1977, 31 children developed BL but of these only 14 had given blood samples. Ten had high antibody levels, when compared with at least five control samples, long before tumour development, indicating that

---

[10] Ribonucleic and deoxyribonucleic acids are found in the nucleus of cells and carry the genetic code – genes. As such, they are essential for life.

children with a long and heavy exposure to EBV are at increased risk of developing BL. The four cases without raised antibody levels are typical of BL reported in countries with a temperate climate, now a well-recognised clinical picture [20]. The findings of the West Nile study strongly supported a causal relationship between the virus and BL but added to the need to search for additional determinants.

Denis realised quite early in the virus-cancer story that there was another factor, related to the specific conditions of temperature and rainfall, that determined population risk of developing BL [21]. There were places that met the climatic criteria for the lymphoma but did not report any cases. These included the island of Zanzibar off the coast of Tanganyika and Kinshasa, capital of the DRC. This added requirement for tumour initiation and growth was malaria. Who first suggested malaria is unclear, but it was Professor Haddow, who pointed out that malaria was endemic in areas of Africa where rainfall exceeded 20" and temperature never fell below 60° F. Greg O'Conor had suggested in 1961 that 'the possibility of a priming action of the reticuloendothelial system[11] by some parasite with subsequent malignant change must be seriously considered'. A key person in promoting this theory was Gilbert Dalldorf [22] from the Sloan Kettering Institute who worked mostly in Nairobi with Peter Clifford's group. Before the EBV findings had been reported in 1964, Dalldorf and colleagues set out in a long review of the evidence to date that they thought that BL might be a variation of childhood leukaemia, which was rare in the parts of Africa where BL occurred, and that the modifying factor was malaria. They concluded, 'May it be possible that an effect of chronic malarial infection of the reticuloendothelial system has been a different response to the etiological agent or agents of lymphoma in tropical climates?' In other words, some of the tissues involved in the body's defences might respond in a different way to a virus infection if those tissues had previously been exposed to an agent like malaria that had damaged them. Malaria is now known to promote the excessive growth (hyperplasia) of B-cells, which are also stimulated to grow by EBV, these two processes initiating the genetic and other changes leading to the development of BL [23] (Fig. 16.4).

The finding by Epstein's group of virus particles in cultured BL cells was published soon after Dalldorf's paper and diverted attention from his malaria theory, but Denis and his colleagues in Kampala, especially Professor Haddow, thought it a likely explanation for the particular distribution of the tumour. Haddow pointed out that the only places in the world where malaria was known to be holo-endemic, that is to affect almost everyone in a particular area, were tropical Africa and parts of New Guinea. Subsequently parts of Malaysia and the Amazon were found to be so affected. But Denis knew from his safari into East Africa in 1961 that there were places that did not fit into his rainfall/temperature theory, namely the islands of Zanzibar and Pemba off the coast of Tanganyika (Tanzania), and from his West African safari in 1962 the environs of Kinshasa in Zaire. Significantly in these areas, there had been successful eradication of malaria. They were to learn later that the incidence of BL had also been dramatically reduced following an intensive

---

[11] Reticuloendothelial system. Part of the immune system.

Fig. 16.4 Denis Burkitt
and Tony Epstein.
Circa 1970

anti-malaria campaign in Papua New Guinea. Thus, in areas of the world where malaria was endemic, then BL would be found. Denis thought that the reverse would be true in that in areas where there was no malaria, there would be no BL, but this did not prove to be so as it became clear that BL affected other populations.

Today, BL is accepted as a tumour that can occur all over the world and was the first human cancer to be related to viral infection. The endemic form is seen primarily in malarial areas of Africa in children aged between 0 and 15 years where it is he major childhood cancer. It has a reciprocal relationship with acute leukaemia, the commonest childhood cancer outside these African regions. Other forms, which are not clinically typical of BL but are still included under the BL classification, include sporadic forms, which occur in adults [24] usually over the age of 60 years. It is also a recognised problem in adults with compromised immunity such as is seen in HIV/AIDS [25] and in post-transplant patients where immunity is suppressed by anti-rejection drugs, but it remains a relatively rare tumour outside Africa.

Denis' role in all this, which he readily acknowledged was quickly taken over by more laboratory-based scientists, was his initial recognition of the lymphoma syndrome. Then his considerable work delineating the geographical distribution of the

Table 16.1 Viruses associated with cancer in humans (Oncogenic viruses) [26–28]

Virus	Virus family	Cancer(s)
Human papilloma virus (HPV)	Papillomaviridae	Genital cancer including cervical, penile, anal and oro-pharyngeal.
Hepatitis B (HBV) and Hepatitis C (HCV)	Flaviviridae	Liver cancer
Epstein-Barr Virus (EBV)	Herpesviridae	Burkitt's lymphoma, Naso-pharyngeal cancer, Hodgkin and non-Hodgkin lymphoma and other lymphomas.
Human T-cell lymphotrophic viruses (HTLV-1 and 11)	Retroviridae	Adult T-cell leukaemia
Herpes virus-8 Human immunodeficiency virus (HIV)	Herpesviridae	Kaposi sarcoma Increase risk of other cancer viruses
Merkel cell polyoma virus (MCPyV)	Polyomaviridae	Merkel cell carcinoma (Rare form of skin cancer)

tumour, which following discussions with Professor Haddow, pointed to a viral origin. From the anomalies in the EBV-lymphoma story, Denis realised that another factor must be involved, namely malaria (Table 16.1).

The table shows those cancers presently thought to be due to viral infections. Considering the ubiquity of viruses and the more than 50 years of research into the possibility that they might cause cancer following the Epstein discovery, the list of cancers known to have a virus involved is short. Probably around 15–20% of all cancers in humans can now be related to viral infection, whilst at present none of the really common ones such as lung, breast, prostate, bowel, etc. are included. If a virus is involved, as, for example, HPV and cervical cancer, then there is the possibility of developing a vaccine and protecting people from getting the cancer.

Denis' time in Africa was drawing to a close. With the facilities, and need, to continue his studies of the lymphoma now much reduced he knew he must take the opportunity to work on gathering information for his next major project.

References

1. Br J Haematol. 2012; 156:689–783.
2. Crawford DH, Rickson A. Cancer virus. The story of the Epstein-Barr virus. Oxford: Oxford University Press; 2014.
3. Henle W, Henle G, Lennette ET. The Epstein-Barr virus. Sci Amer. 1979:48–59.
4. O'Conor GT. Malignant lymphoma in African children: II. A pathological entity. Cancer. 1961;14:270–83.
5. Burkitt DP. A sarcoma involving the jaws in African children. Brit J Surg. 1958;46:218–23.
6. Pulvertaft RJV. Cytology of Burkitt's tumour (African lymphoma). Lancet. 1964;I(7327):238–40.
7. Epstein MA, Barr YM. Cultivation in vitro of human lymphoblasts from Burkitt's malignant lymphoma. Lancet. 1964;I(7327):252–3.

8. http://munksroll.rcplondon.ac.uk/Biography/Details/3665
9. Epstein MA, Achong BG, Barr YM. Virus particles in cultured lymphoblasts from Burkitt's lymphoma. Lancet. 1964;283(7335):702–3.
10. Epstein MA, Henle G, Achong BG, Barr YM. Morphological and biological studies on a virus in cultured lymphoblasts from Burkitt's lymphoma. J Exp Med. 1965;121:761–70.
11. Henle G, Henle W, Clifford P, Diehl V, Kafuko GW, Kirya BG, et al. Antibodies to Epstein-Barr Virus in Burkitt's lymphoma and control groups. J Nat Cancer Inst. 1969;43:1147–57.
12. Old LJ, Boyse EA, Oettgen HF, deHarven E, Geering G, Williamson B, Clifford P. Precipitating antibody in human serum to an antigen present in cultured Burkitt's lymphoma cells. Proc Natl Acad Sci. 1966;56:1699–704.
13. Klein G, Clifford P, Klein E, Stjernas J. Search for tumor-specific immune reactions in Burkitt Lymphoma patients by the membrane immunofluorescence reaction. Proc Natl Acad Sci. 1966;55:1628.
14. Gunven P, Klein G, Henle G, Henle W, Clifford P. Epstein-Barr virus in Burkitt's lymphoma and nasopharyngeal carcinoma. Nature. 1970;228:1053–6.
15. Hoagland RJ. Transmission of infectious mononucleosis. Am J Med Sci. 1955;267:262–72.
16. Epstein MA. Burkitt lymphoma and the discovery of the Epstein-Barr virus. Brit J Haematol. 2012;156:777–9.
17. Henle G, Henle W, Diehl V. Relation of Burkitt's tumor-associated herpes-type virus to infectious mononucleosis. Proc Nat Acad Sci. 1968;59:94–101.
18. zur Hausen H, Schulte-Holthausen H, Klein G, Henle W, Henle G, Clifford P, et al. EBV DNA in biopsies of Burkitt tumours and anaplastic carcinomas of the nasopharynx. Nature. 1970;228:1056–1.
19. De-The G, Geser A, Day NE, Tukei PM, Williams EH, Beri DP, et al. Epidemiological evidence for causal relationship between Epstein-Barr virus and Burkitt's lymphoma from Ugandan prospective study. Nature. 1978;274:756–61.
20. Ziegler JL, Andersson M, Klein G, Henle W. Detection of Epstein-Barr virus DNA in American Burkitt's lymphoma. Int J Cancer. 1976;17:701–6.
21. Burkitt DP. Etiology of Burkitt's lymphoma – an alternative hypothesis to a vectored virus. JNCI. 1969;42:19–28.
22. Dalldorf G, Linsell CA, Barnhart FE, Martyn R. An epidemiological approach to the lymphomas of African children and Burkitt's sarcoma of the jaws. Perspect Biol Med. 1964;7:435–49.
23. Magrath I. Epidemiology: clues to the pathogenesis of Burkitt lymphoma. Brit J Haematol. 2012;156:744–56.
24. Linch DC. Burkitt lymphoma in adults. Br J Haematol. 2012;156:693–703.
25. Zeigler JL, Miner RC, Rosenbaum E, Lennette ET, Shillitoe E, Casavant C, et al. Outbreak of Burkitt's-like lymphoma in homosexual men. Lancet. 1982;320(8299):631–3.
26. de-Thé G. Viruses and human cancers: challenges for preventive strategies. Environ Health Perspect. 1995;103(Suppl 8):269–73.
27. Harford JB. Viral infections and human cancers: the legacy of Denis Burkitt. Br J Haematol. 2012;156:709–18.
28. Esau D. Viral causes of lymphoma: the history of Epstein-Barr virus and human T-lymphotropic virus 1. Virol Res Treat (Virology (Auckl)). 2017;8:1–5.

Chapter 17
Out of Africa

Back in Mulago in December 1963, but without his family, Denis was moving rapidly towards leaving his job as senior surgical specialist with the Ugandan government and joining the MRC. Social life was interspersed with regular leaving parties for members of the ex-patriot community yet, whilst Denis' life was changing, he knew he still had important work to do. Earlier in the year, Professor McAdam, head of surgery in Kampala, had written to Sir Harold Himsworth, secretary of the MRC in London, setting out 'Proposed epidemiological studies of the geographical pattern of cancer in tropical Africa' to be carried out by Denis in conjunction with Michael Hutt in the Department of Pathology.[1] The argument Professor McAdam used was that, because of the success of the lymphoma work, Denis was finding it increasingly difficult to do justice to both his routine surgical work and his research interests. In reality, the ending of Colonial Rule in Uganda heralded the Africanisation of all the government services including the universities and health service. Denis was about to lose his job. For the present time, he needed to stay in Uganda to exploit his extensive network of contacts across equatorial Africa in his search for the key factors responsible for his lymphoma. In 1963, the search for the identity of a virus and evidence for the possible role of malaria were still in their early stages (Fig. 17.1).

Denis was now involved in a major programme of work with the ICRF (London). This included Professor Haddow, relating Denis' geographical patterns to viruses and their vectors, Dennis Wright on the histochemistry and ICRF scientists attempting to isolate the virus and culture it in various cell systems. Professor McAdam requested a salary of £2949 for Denis and the status of reader in surgery. In addition, monies were required for the overseas education of the Burkitt children, £540 per annum, travel costs for both work and family trips back to the UK, a full-time secretary, office equipment and an additional technician for Professor Hutt, all

---

[1] Wellcome Library, Papers of Denis Parsons Burkitt, WTI/DPB/B/9 'Continuation of cancer research' Letter from Professor Ian McAdam to Sir Harold Himsworth 21 January 1963.

© The Author(s), under exclusive license to Springer Nature Switzerland AG 2022
J. H. Cummings, *Denis Burkitt*, Springer Biographies,
https://doi.org/10.1007/978-3-030-88563-2_17

Fig. 17.1   Professor Michael Hutt and his wife Elizabeth

totalling £7834. At his showing of the lymphoma film at the ICRF Laboratories in London on September 26, Denis had met Brandon Lush from the MRC and HEO Hughes from the Department of Technical Co-operation, who were providing a grant towards Denis' work in Uganda. The outcome of the meeting had been that Denis would resign his government post immediately. Denis wrote to the Chief Medical Officer at the Uganda Ministry of Health, giving notice of his resignation and suggesting a date of March 31, 1964.

It was noted that Denis, in addition to giving up his clinical duties, would have to give up his private practice. For someone who had become one of Africa's best-known surgeons and who clearly enjoyed the immediate satisfaction that comes with curative operations and grateful patients, Denis made little of giving up this phase of his career. Writing later, in 1980, he would note, 'Moreover there was the possibility, subsequently realised, that political instability might seriously curtail the practicality of continued contact with rural medical centres. This became tragically true in the country that was my home for so long, Uganda'.[2] Denis saw that the political landscape was changing dramatically in Africa. Between 1960 and 1965, the principal colonial powers, the UK, Belgium and France, had relinquished their hold over Uganda, Kenya, Tanganyika, the Congo, Rwanda, Burundi, Northern

---

[2]Wellcome Library, Papers of Denis Parsons Burkitt, WTI/DPB/F/1 Burkitt DP, Autobiography Chap. 10 p. 2.

Rhodesia and Nyasaland, but what followed was a struggle for power between the main tribal and political groupings in some of these countries, which spilled over into violence affecting the whole community. There was also resentment at the religious, legal, cultural and political structures, beliefs and philosophy imposed by the occupying powers that the indigenous people saw as stifling their growth as independent sovereign nations with their own distinct national identities. Like it or not, Denis' days in Africa would be over by the end of 1965.

With the world changing around him, Denis knew he must focus on his strengths. These were the clinical aspects of chemotherapy for his lymphoma, which he now did in conjunction with his African colleagues at Mulago, and equally importantly the geographical pathology of other diseases in Africa that he would see as a surgeon and had been collecting data about for some time now. Work on the virus was in the hands of what Denis viewed as laboratory scientists. With his reputation securely established as the discoverer of the lymphoma, he was happy simply to be kept in touch with developments whilst reminding them of the emerging anomalies in the climate/virus/malaria theory highlighted by the worldwide occurrence of the tumour. He was confident about his future because his interests went beyond the lymphoma and always had done so. From the day he realised in Lira in 1947 that hydrocele of the testis had a very specific pattern of occurrence, later to be ascribed to infection with Filaria, every disease he came across in his clinical practice always made him ask himself about its geographical distribution. Whilst his brilliant clinical and epidemiological observations of the lymphoma and its likely viral origin had swept him into orbit around the world, he was now ready to give time to his other interests.

Writing to Sir Harold Himsworth from Johannesburg on November 15, 1961, with a preliminary note about his findings on the long safari, he had reported on the striking pattern of distribution of the lymphoma but ends the letter with, 'We are so grateful for the opportunity to do this investigation. We have also gleaned much information on many other surgical and medical problems as regards their distribution'.[3] Denis' curiosity about the distribution of common diseases in Africa had been re-enforced when Dr. George Oettlé had visited him in Kampala in 1957. Denis showed him cases of the lymphoma, which Oettlé was confident never occurred in South Africa, where he was head of the Cancer Research Unit of the National Cancer Association of South Africa. What he did tell Denis was that some cancers, such as liver and oesophageal, were much more common in some regions of Africa than others. This and the lymphoma quest had prompted Denis to start building up his network of government and mission hospitals and medical staff throughout Uganda who would send him annual reports of the incidence of various diseases. In preparation for the long safari in 1961, Denis had prepared a leaflet about the lymphoma, which was sent to all his contacts throughout East Africa asking them if they had seen the condition. Importantly and significantly before setting out on the safari, he had drawn up with his companions for the journey two

---

[3] Wellcome Library, Papers of Denis Parsons Burkitt WTI/DPB/D/1/3:Box 23.

proformas. One was designed to be completed from the actual records in operation registers, in which were listed operations known to be commonly carried out in these areas. The other proforma required information about a wide selection of conditions, where such accurate figures might not be available so, the local staff were asked to report these as 'unknown, rare or common'. These surgical conditions included were hernia repair, bladder and prostate problems, peptic ulcer, volvulus,[4] intussusception,[5] appendicectomy, ruptured uterus and ectopic pregnancy. Reporting the results in 1963 [1] with his colleagues from the long safari, Ted Williams and Cliff Nelson, they describe clear differences in the distribution of these conditions across Africa and add a note about two other cancers, oesophageal, which showed striking contrasts, and cancer of the penis: 'The incidence of this tumour varies enormously'.

Thus, when Denis was asked by Ian McAdam to draw up proposals for his future research, they were described as 'Proposals for extending cancer research – Kampala' and included what he would refer to as a 'low power view', namely a continuation of his safaris and collection of data from across Africa, and a 'high power view' concentrating on finding local factors that might be causing the condition. This latter dimension had been prompted by a 'phone call from Ted Williams saying that he had observed three children with jaw lymphomas all related and living within 500 yards of each other and a further seven cases all within the drainage area of one river. In his application to the MRC for future funding for Denis, Ian McAdam made the point that the methods used to investigate the cause of the lymphoma might usefully be applied to other cancers and suggests including bowel cancer, malignant melanoma, Kaposi sarcoma, gastric and oesophageal cancers, some other jaw tumours and rectal cancer, which latter cancer was not uncommon in people living around Kampala. Clearly Denis had the full support of the Head of Department, Professor McAdam, who strongly encouraged and may have contributed to the idea of broadening the search for the geography of other conditions.

Thus, 1964 became a year of urgent safaris. Eight in all starting on January 15, in the company of Bob Harris of ICRF London, to Northern Uganda, familiar territory for Denis who had been solely responsible for the health of this large area when he first arrived in Lira. All the safaris followed a similar pattern. Well planned in advance, each of the hospitals he visited were given a likely date, told the purpose of the safari and sent a detailed itinerary. The journeys lasted between 1 and 4 weeks depending on the distances to be travelled, often between 1000 and 2000 miles on poor roads. On arrival in Northern Uganda, Denis, with the help of Bob Harris, completed the proformas by asking the staff about the lymphoma and the other diseases he was now following up. They visited up to four hospitals in a day and stayed with people whom Denis had known in the past. At some hospitals, the staff would ask Denis' opinion about difficult clinical cases so he would do a ward round and

---

[4] Twisting of the bowel, seen more commonly in Africa than in the west.

[5] Condition usually in children in which a section of the small bowel tries to move down through the adjoining bowel, usually causing obstruction of the gut.

then an operating list, a practice that continued even after he had become full-time working for the MRC. Over the course of the year, Denis visited well over 100 hospitals and revelled in the whole concept of doing cancer research this way when conventionally he knew that much of it was laboratory and animal based. He was quietly pleased that through his epidemiological and clinical work, he was able to provide insights into the cause of major surgical diseases without recourse to a microscope. But with the news from surrounding countries disturbing, his visit to Lira carried a sense of finality. On January 21, 1964, he wrote, 'News of army revolt in Dar-es-Salam and later Tabora. Rather worrying. Drove on to Lira. Thought of the first time I did this trip and wondered if this were the last. Stayed in Lira Rest House where I had stayed on my first night in Uganda in 1946'.

Back at Mulago by February 1, Bob Harris left to return to the UK whilst Denis caught up on other duties. There was more unsettling news from Rwanda where the Tutsi minority were waging an unsuccessful insurgency against the ruling Hutu majority and suffering badly, a prelude to the genocide that would occur as Rwanda became a one-party state and all opposition supressed. He was able to watch events on Uganda TV for the first time, but it was all bad news, and the movies that filled the time between broadcasts were not much to Denis' taste. Then on February 8, he had what he called a very dark day in his life. It was precipitated by a 'phone call from Gerry Tewfik, the psychiatrist, who with his wife Tessa had repatriated to the UK and gone to visit Olive in Nettlebed. Olive had become lonely and to feel isolated after Denis had left and the girls gone back to school. When the Tewfiks arrived, it was clear to Gerry that Olive was depressed, so he prescribed an antidepressant for her, which had helped in the past. But she grew worse, and they all decided that she should go and stay with the Tewfiks in Bickley. Things did not improve, and the 'phone call from Gerry to Denis was to ask him to fly home immediately. He managed to get a cheap fare and arrived at the Tewfik's house 4 days later to find Olive in bed and very unwell. They all went to the local hospital where Olive was given ECT, but her recovery would take a further 6 weeks during which time Denis would stay with her, postponing a planned safari to Sudan. He used the time to draw up further proposals for his work and make a start on a book about the lymphoma which would be co-edited with Dennis Wright although not published until 1970 [2]. Olive longed for a home in England with her family, but before that could happen, there would be more partings. They agreed that she would return to Mulago and in future accompany Denis on all his safaris. She would join him 2 months later and enjoy some of the safaris.

On April 1, 1964, Denis started his new post working full-time for the MRC at Makerere. He saw it as the end of his clinical career. To add to his sense of loss, he had to move out of the house, No. 18 Mulago Hill, where he and Olive had enjoyed 16 years of family life. He was given No. 20 Nakasero Hill, which was not even his third choice of the places he had been offered. The faithful, and highly competent, Yusufu came to his rescue with moving and arranging all the contents of the house, and after a week, Denis was able to get back to enlarging his photographs and rearranging the Sudan safari for later in the year. His appetite for books to read about the Christian life led him to buy *Honest to God* [3] by Bishop John Robinson of

Woolwich, London. He was much troubled on reading it and was deeply upset that a senior leader of his adopted Church of England should suggest that the time had come for a radical recasting of traditional orthodoxy saying, 'in the process of which the most fundamental categories of our theology-of God, of the supernatural, and of religion itself-must go into the melting (pot)'. It raised serious questions for Denis about the meaning of faith, context of prayer, religious observance, sin and forgiveness and who would sustain him in the days ahead. At the end of the book, he might have found salvation in words quoted from Professor Herbert Butterfield's Christianity and History 'Hold to Christ, and for the rest be totally uncommitted'.

Denis managed to fit in a quick trip to Nairobi and then Dar es Salaam and Zanzibar but felt rather down and lonely and found meeting some of the staff at the hospitals he visited difficult. He resolved to try and persuade Olive on whom he depended, more than he had realised, to accompany him more of these journeys. It was their mutual interdependence that allowed them to share problems, manage the practicalities of family life between them and enjoy Africa with its landscapes and people. They read books together, mixed in a social circle where they were both comfortable and lead lives that were still patterned by a common faith. James, Denis' father, would have been pleased with the union he had urged them to make with haste as Denis went off to war. It was not perfect, but they worked at it, and each was quick to respond when they were in the wrong.

Aside from all this, Denis was in Mulago in time to greet the arrival of Tony Epstein who had come to check on the monkeys he had injected with lymphoma tumour cells from one of Denis' patients some 2 years earlier. Two of the first batch of three had developed bone tumours, and the plan was to biopsy the tumours and inject some of the material into other monkeys in an attempt to cultivate the virus. They had a memorable day visiting the primate colonies on the Ssese Islands, a long archipelago in the north-west of Lake Victoria, in order to capture young monkeys for their experiments. They left Kampala at 4:30 am with staff from the Virus Research Institute equipped with monkey boxes, travelled by road then motor launch, canoe and Land Rover meeting up with the local monkey-catchers on the main island. There were no suitable young monkeys to be seen, so Denis and Tony Epstein journeyed onwards but were hampered by the high level of water in the lake rendering many roads impassable resulting in them having to leave behind the Institute staff to work with the locals. They arrived back in Kampala at midnight. The next day, post-mortem examination of the monkeys with the lumps showed that the lumps were not true versions of the lymphoma, a big disappointment.[6]

Olive's arrival back at Mulago had been celebrated by everyone at a lunch where people brought presents as well as a contribution to the meal. Denis had put Olive to work immediately helping him with correspondence, typing papers and managing their social calendar. They left on safari to south-west Uganda a couple of weeks later on a familiar journey visiting Masaka, Mbarara, Kabale, Kisoro, Kisisi, Bushenyi, Fort Portal, Mubende, returning to Mulago after only 10 days. The new

[6] Wellcome Library, Papers of Denis Parsons Burkitt WTI/DPB/D/1/13.

team worked well despite Denis contracting a nausea and vomiting bug followed by a cold and having to spend 1 day in bed. Now he had his soulmate and personal nurse to keep him going. They both read Frank Morison's *Who Moved the Stone?*, a critical analysis of Christ's crucifixion and resurrection, which had become an influential book, and also Wilkerson's *The Cross and the Switchblade*. They were back into their usual routine.

Five more safaris followed in 1964, to Tanzania and Zanzibar, the coastal region of Kenya, the Rift Valley, Rwanda and finally the Sudan. The Epstein papers on the virus had now been published, and both he and Denis were in demand as speakers at scientific meetings. The family were helping him with his work and Judy especially with diagrams and maps for his lectures. She had been appointed Head Girl after only two terms at Westonbirt where she had moved from Highland in Eldoret in January 1964 and had been accepted at Homerton College, Cambridge, for training as a teacher. Their ties with Africa were slowly being cut. Denis was collecting large amounts of data on the occurrence of his selected diseases but did not have the statistical background to analyse it and would have to wait until he was joined by Paula Cook in 1966 on his return to the UK. Most of his writing was still about the lymphoma, its geography and treatment,[7] but his reports to the MRC[8] now contained detailed comments, although very much on a qualitative basis, of his observations of the occurrence of other cancers. Liver cancer seemed to be quite common throughout East Africa, but cancer of the oesophagus was seen five times more often in Kisumu, a large city in Western Kenya with a hospital of 200 beds than at Mulago where they now had 800 beds. Stomach cancer was the commonest cancer across the northern part of the Kivu-Burundi-Rwanda-West Tanganyika mountain ranges, mostly judged from surgical records with breast cancer rarely seen. Similarly, cancer of the penis, readily diagnosed and remembered for its radical treatment, varied greatly in occurrence, but Denis felt the distribution largely related to the practice of circumcision. It had a 'uniformly low incidence among all tribes which practice (male) circumcision' [4], but there were exceptions. It was unexpectedly low amongst the Karamojong who do not circumcise, but those who move from a low-risk area to one of high risk increased their risk. He felt that these observations would eventually provide clues as to the cause of these cancers and understanding them continually occupied his thoughts.

Later in the year, there was an incident between Olive and a local. She and Denis were at home when, according to Olive's account, 'Ranee[9] barked at an African & I told him it was because he was in our garden. He shouted back at me so fiercely & with such malice – I was quite taken aback. Felt strongly that Deny might have supported me – but he stood by my side without opening his mouth & allowed him to call me names. Felt thoroughly sad & upset & resentful. Deny said he thought he would have made a scene if he had spoken – but I feel this trait has followed me all

---

[7] Appendix 3 Papers for 1965, 1966.
[8] Wellcome Library, Papers of Denis Parsons Burkitt WTI/DPB/D/1/16.
[9] Family dog.

through our married life. His unwillingness to face the truth if it means a bit of a fuss. I hate injustice of any sort'.

Denis' diary note of these events reads:

> Olive got into an argument with an African who had crossed some of our shamba before. I felt argument futile & said nothing, hoping she would walk away. This gave the impression that I failed to support her, with much unhappiness & tears.

The solution. Together they both sought God's forgiveness and 'the Lord healed the break'. That evening, they went off to Bible class, which Olive was leading, but neither felt in a particularly Godly mood.

A few days later, on November 28, 1964, Denis set off on safari to Rwanda with John Church,[10] John Billinghurst[11] and Bryan Rogers[12] but not Olive because of reports of civil unrest in both Rwanda and the neighbouring Congo. Soon after they had set off from Mulago, a bouquet of flowers from Denis was delivered to Olive with a message from Denis citing Genesis 31:49. 'The Lord watch between me and thee, when we are absent one from another'. The travellers stayed in the Idelweiss (Edelweiss) Hotel in Kigali where Denis had been on his visit in 1948 but were saddened to note the empty houses of the former expatriate community. The safari was organised mainly by John Church, the eldest son of Dr. Joe Church who had worked as a missionary in Rwanda for 44 years and built a hospital at Gahini. It was very successful in that despite some bad weather making travel difficult, they were able to visit 12 towns but did not find a single case of the lymphoma in the mountainous country (Fig. 17.2).[13]

In January 1965, the East African MRC Conference was held at Makerere, where Denis had previously lectured on chemotherapy of the lymphoma. He was able to have a long chat with Sir Harold about his prospective move back to the UK. With his rising worldwide reputation, Denis then set off for a 25-day lecture tour starting with a visit to Astra in Germany, the pharmaceutical company which was still supplying him with cyclophosphamide, the most successful agent for treating lymphoma in Africa, then on to Geneva where he met Alan Lindsell at the WHO and so to London. Here, apart from catching up with the family, he gave a lecture at the Royal Society of Medicine section of Otolaryngology,[14] an area of surgery where he had very little expertise. But because the jaw tumours often invaded the nasal sinuses, he felt a tenuous connection with these surgeons. They were impressed, and to Denis' surprise and almost disbelief some months later, he was told that his lecture had been considered the best of the year, and as a result, he was to be awarded

---

[10] John C T Church MA, MD, FRCS (b 1931). Trained as an orthopaedic surgeon, was lecturer in anatomy at Makerere, then senior lecturer in orthopaedics and trauma in Nairobi. John, like his father, did medicine at Cambridge and in 1964 was working at Makerere.

[11] John Robert Billinghurst (1927–2012), physician colleague of Denis at Mulago.

[12] An English doctor training to be a surgeon.

[13] See Chap. 25 for John Church's vignette about events at the Rwanda-Uganda border on their return journey.

[14] Ear, nose and throat (ENT).

Fig. 17.2 John Billinghurst, on the right, in Butare, December 1964

the Harrison Prize, the first of many he would accept over the next 28 years. From London, he journeyed to Dublin where he was now greeted as a returning hero, given a special lunch by what he called 'my professors', lectured at Trinity, at the Royal College of Surgeons and to the Pathology Society before heading to Edinburgh again to lecture to the surgeons then returning to London after 5 days away. There he caught up with Robin Lush and Henry Bunjé from the MRC to finalise arrangements for his return to the UK, next to Tony Epstein at the Middlesex before leaving for Frankfurt en route to Athens, Cairo and then Eritrea. In the capital, Asmara, of this mountainous country by the Red Sea, he spent a very valuable day talking to the medical staff at the hospital and giving a talk before heading for Addis Ababa the following day. There he had lunch with Mrs. Barry, tea with the Prices, dinner with Dr. Ende and Mrs. Barry and then saw the Kennedy film. On Sunday, he was taken to church and had a walk in the hills, and on the Monday, he visited various hospitals, lectured to the Medical Association, his 13th talk of the trip, and met the staff from the Ministry of Health. But they were not as overwhelmed as some by his charisma and enthusiasm. The Medical School at Makerere proposed Denis on three occasions for the Haile Selassie Reward (Prize), but it was never given to him.

Denis' reluctance to refuse an invitation to speak to even the smallest and least informed of audiences was not primarily motivated by financial gain, although in places such as North America, he would usually receive an honorarium. Nor was it just an ego trip, despite the feedback and in Denis' case often adulation, that speakers may receive from their audience. Denis' faith, which he fed daily with Bible reading and prayer, kept him genuinely humble. His motivation was the story he had to tell about the cause and treatment of a horribly disfiguring cancer in children, pictures and slides from his hobby photography to show them, epic journeys to recount and a message which said 'You can do it' because you do not need to be in a laboratory nor an expert in statistics to change the world. And he was an engaging and captivating speaker.

Aside from this, he concentrated his efforts on writing papers, mainly about the lymphoma, with nine published in 1965 and seven in 1966. He was the sole author on the majority but set down his new ideas about other cancers in East Africa in papers in the *British Medical Journal* and the *International Journal of Pathology* with Michael Hutt [5, 6] and about response to therapy with Hutt and Dennis Wright [7]. These were solid papers that would have gone through the normal refereeing process in respectable journals where he would have been able to rely on his co-authors to handle the statistics and graphs whilst he always had a ready supply of clinical photographs and hand-drawn maps. There was also some urgency to his work with the move back to the UK scheduled for the end of the year necessitating much reorganisation on his and Olive's part. They needed to find a house to buy, although in the short term they would return to Cedarwood in Nettlebed. An important criterion for them was that 'A keen church should weigh heavily on our choice of house'. Money was now more of an issue with the loss of his private practice. Outside of Denis' control, in his Spring Budget in April 1965, the UK Chancellor of the Exchequer, James Callaghan, facing a widening balance of payments deficit had decided to make considerable changes to the tax system, which Denis calculated would lose him an extra £100/year from his MRC income of £2949. Whilst he and Olive were careful with money, a habit acquired during their upbringing, they had both inherited money and been able to make some investments whilst in Africa. But in June, his financial advisor reported that the value of these investments had fallen by £1200, around £10,000 today, taking changes in inflation into account, not helpful when you are looking to buy your first house.

Nor was time on his side. Denis had two more lecture tours booked and three safaris arranged during the rest of the year, all of which took considerable planning, which he insisted on doing himself. There was quick trip with Dennis Wright in early May to visit the Sloan Kettering Institute in New York, then on to Washington to the NIH and the Armed Forces Institute of Pathology conference where he and Dennis Wright spoke. There they renewed acquaintance with Greg O'Conor before heading back to Africa, arriving early morning of May 17. They were met by Olive only to leave the next morning with her on an 18-day safari to the central region of Kenya. The trip followed the usual pattern of travel and hospital visits, but a highlight for them, in anticipation of their wedding anniversary, was a stay at Treetops Lodge north of Nairobi in the Aberdare National Park, made famous by the visit of Princess Elizabeth and her husband Philip the Duke of Edinburgh on February 5, 1952. During that night, King George VI had died and Princess Elizabeth awoke to the news that she was now Queen and Head of the Commonwealth. The Burkitt's enjoyed the visit greatly, both the luxurious accommodation and the variety and numbers of game they saw on drives through the park.

In Kampala early in June, Denis caught up with his lymphoma patients on the ward. Although having relinquished his post as government surgeon, a case had been made, in view of its importance, that Denis should be allowed to admit patients to the wards in the new hospital for treatment and have access to operating theatre time when necessary. On his return from Kenya, there were seven lymphoma patients, only two of whom were doing well, to his disappointment. Moreover, 2 weeks later, after his return from a quick lecture visit to London, one of the first patients he had treated in 1960, Namusisi, returned with a recurrence. There was more to do before he could hand over this work to the new cancer institute run by Dr. Zeigler. Then a 2-week safari to Tanzania with Bob Harris in July allowed Denis to update, in a report to the MRC, his latest view on the lymphoma and more extensively 'Observations and impressions of disease distribution'.[15] He was casting his net more widely to diseases other than cancer, which he was fairly sure would show distinct patterns of distribution in East Africa and was building up a network of mission hospitals whose doctors reported their experience regularly to him. They were mostly from Uganda, Kenya and Tanzania, but already he had one hospital in Ethiopia similarly engaged. Although he had not made it clear to the MRC at this stage, his plan on returning to the UK was to use this information to contrast with cancer incidence in the west. After all he had solved the problem of one cancer already, and with exciting implications for other tumours, why should he not now go on and find the cause of others?

For the lymphoma, his travels in Kenya had confirmed altitude dependence, and he was now able to plot his final graph of the 302 lymphoma cases he had gathered since 1958. This took the form of a histogram, which showed the largest number of 45 cases occurring at the age of 6 years but with a long tail to the right including sporadic cases up to the age of 55 years. This age range was not what he had observed originally and for years he had referred to the lymphoma as a childhood cancer. But the cases in adults he had seen were consistent with the experience of cancer specialists in the USA and the UK, and he could happily leave following up the implications for the cause of the lymphoma to those who would come after him. Other diseases now had captured his attention. The existence of localised areas of unusually high incidence of oesophageal cancer in parts of Kenya contrasted with the experience of hospitals he had visited in Southern Tanzania where staff could

---

[15] Welcome Library Papers of Denis Parsons Burkitt WTI/DPB/D/1/19.

remember seeing only one case a year or not seeing the disease at all in some centres such as Iringa. The pattern of distribution of cancer of the penis fitted into his previous observations on the relation to circumcision but with notable tribal exceptions such as the Lugbara and Acholi. Breast cancer was rarely seen at any of his reporting hospitals contrasting with northern Sudan where it was the most common cancer recorded. Now included in his report are observations on inguinal hernia, volvulus, uterine fibroids, appendicitis and other more tropical conditions, all of which allowed Denis to conclude that, particularly for oesophageal cancer, the search should begin for significant factors in the environment.

After a sundowner in Kampala one evening in July, Denis was reading his Bible, as he usually did, and came across Hebrews 12:1–2.

1. Therefore let us also, seeing we are compassed about with so great a cloud of witnesses, lay aside every weight, and the sin which doth so easily beset us, and let us run with patience the race that is set before us,
2. looking unto Jesus the author and perfector of our faith,

Upon which he noted 'Encompassed by witnesses shows heaven is not a special conception, but another, a spiritual dimension'.[16]

Perhaps he had always just thought that heaven was a place without thinking through the concept. Now at the age of 54 years, he was able to grow in faith and in separating the myths from reality to have strengthened considerably the cohesion between the teachings of Christ and his everyday life. Something of *Honest to God* had remained in his subconscious.

A short break in August 1965 with the family and Guy and Dawn Timmis to a game park near Ishasha on the east shores of Lake Edward provided a reminder of the beauties of Africa which they were about to leave behind. More importantly, the members of the ex-patriot community, many of them they would never see again, were now leaving almost every week, which both Olive and he found somewhat unsettling. Denis busied himself with writing helped by Judy and Cas in the office whilst Olive packed crates for sending to the UK by sea. The Mercedes was serviced and handed over to the AA for shipment, but they kept their small Fiat for everyday use. As they departed they sold it for a small sum to the Plumtrees, who were medical staff at Mulago. A brief safari in October accompanied by Olive, a nurse and two friends took them to Arua where they were present at the Independence Day Celebrations at which the Prime Minister, Milton Obote, spoke at length. Denis had a profitable time with a Russian surgeon at the hospital looking at his data on lymphoma cases. Onwards across the Northern Province, they visited towns they knew well, ending in Lira where the house in which he and Olive had lived 19 years earlier had become the Prime Minister's lodge. Finally Denis journeyed to Sudan in November for the last safari where in Khartoum he attended the Sudan Association of Surgeons meeting, gave the keynote lecture and was made an Honorary Member of their society.

---

[16] Diary July 3, 1965.

The biggest and most heartfelt send-off was the departure of Leslie Brown, the last archbishop appointed from outside Uganda. A large convoy of some 20 cars left All Saints Church at 08:30 on Tuesday, November 23, bound for Entebbe. There the airport was crowded with people 'who love L & W.[17] some weeping, including Mrs. Obote. How greatly God has used them & how we thank Him for their friendship'. It marked the end of an era.

Returning to Kampala, there was one last major conference to prepare for under the banner of the UICC on the topic of 'The Treatment of Burkitt's Lymphoma' [8]. This was scheduled to take place from January 5 to 7, 1966, and was attended by many of the big names in the lymphoma story to mark Denis' time in Uganda. Some of the stories about the meeting are apocryphal, but there are a number of documented highlights. With Denis' imminent departure from Africa, the meeting provided a forum to review his contribution to the lymphoma story and cancer research in general. For those attending it was the chance to acknowledge his unique achievements at Mulago Hospital, Makerere Medical School, to the social fabric of a rapidly evolving country and to many who were there just to listen, as their personal friend. Denis gave four talks, an unusual complement at any international meeting, on the clinical features of the lymphoma, its epidemiology, responses to chemotherapy and long-term survival of the children. He demonstrated a case where a short remission had been seen when a patient had been injected with plasma from one recovered from the lymphoma. On the evening of the first day, the Minister of Health gave a reception for conference participants, although he did not in the end attend himself, followed by the Conference Dinner. The next day, Denis, ever a showman on these occasions, paraded 10[18] healthy-looking children who had been successfully treated for the lymphoma, together with photographs of their initial presentation. The children were brought from their homes, sometimes over long distances, and were provided with overnight accommodation. A significant logistical exercise but a demonstration that impressed the delegates. An evening reception at Entebbe was sponsored by the IARC, followed the next day by a lunch at Denis' house, prepared by Olive, for key friends including Professor Haddow who had returned from Glasgow where he had been made Professor of Administrative Medicine.

The final afternoon of the conference was chaired by the Minister of Health and concluded with a speech by Professor Haddow [8] in which he said,

'I have, myself, been in the field of cancer research for very nearly 40 years: never in this time has the situation been more encouraging or exciting. During this time, I have attended hundreds of scientific meetings, yet with the utmost sincerity can I say that never have I attended a conference more fruitful or more inspiring than this.' 'My four years as President (of UICC) have brought many unexpected chores, but many satisfactions and rewards, among these awards outstanding in my recollection will always be this conference on the chemotherapy of Burkitt's

---

[17] Leslie and Winifred.
[18] Diary records 'Over 10', Autobiography Chap. 8 p. 5 says '23'.

Lymphoma. On the first day he (Burkitt) welcomed us and thanked us for coming, but this is wrong; it is we who should thank him for the privilege to take part in a meeting of such inspiration. No-one who witnessed it can ever forget the demonstration yesterday of his patients – we hope cured. African children – an experience expressingly moving and of the utmost practical significance, not only for these children and for their parents and families, but also for us of intense theoretical importance, proving that gross pathology can be reversed by purely chemical means.'

> I would add a further word – spare his blushes! – a tribute to Burkitt; a tribute which we all must share - to him an outstanding surgeon and, I will say, physician, and outstanding observer and researcher, but perhaps, beyond all else, what has most impressed us in his kindness and his humanity towards his patients, and we are all very proud of him.

Denis' place in the history of Ugandan, and world, medicine was assured and further recognised by a letter written on December 20 on behalf of the Permanent Secretary and Chief Medical Officer of the Ministry of Health at Entebbe:

> Dear Mr. Burkitt,
>   We learn with regret that you are leaving Uganda, a country for which you have served so well. I take this opportunity to extend the deep appreciation of the Ministry for the very good services rendered by you during your long stay in Uganda and your esteemed contribution to research, especially in connection with cancer.
>   The Minister has requested me to inform you how much he appreciated the part you played as a member of the Butabika Official Visitors' Panel. We have noted with pleasure and appreciation all your very sound recommendations.
>   The Minister and indeed the whole Ministry wish success in your future undertakings and hope that you will not forget Uganda for which you have done so much.
>   Yours sincerely.
>   P.S.B. Muganwa.
>   For Permanent Secretary/Chief Medical Officer.

Olive and Denis were booked every evening for the remaining 2 weeks of their time at Mulago with farewell events. At a party at Mengo, they were given a cheque for 280 shillings,[19] which in a paternal way he found to be so kind, to be followed the next evening by a buffet supper organised by Dennis and Gill Wright for the whole Pathology Department. Raeburn Murray, an artist and also doctor at Mulago, had been commissioned by Denis and Olive to paint Yusufu, a fine portrait which is today in the house of one of Denis' grandchildren, Lucy. On Sunday 16th January 1966, Denis took his last service at Mulago, and in the evening, prayers were offered for the Burkitts, which they found very moving. Visitors from the MRC arrived the next day, but Denis' time to talk to them was limited because they were preparing to move out of their house and were selling furniture and other household items. In the evening, Dr. Abdul Adatia and Sheila, an Indian dentist, invited them out for a very good dinner at the Grand Hotel together with the Hutts and Edna Powell, followed the next day by a special CMF meeting where prayers were again offered for them. On Wednesday 19th, they moved out of their house to stay with the Hutts and

---

[19] Would have been East African Shillings just before Uganda issued its own coinage. 20 EAS = £1 Sterling.

advertised their remaining household effects in the local paper. They were 'Flooded with requests. Telephone going all day'. Olive was by now becoming rather tired, so that evening, Denis went alone to a dinner for them at the College of General Practitioners, but the next day, she was able to attend a farewell lunch given for her at the High Commissioner's Residence. In the evening, the Surgical Division of the Medical School said farewell at a party at Professor McAdams,[20] at which they were given a silver candlestick. Further lunches and tea parties followed, Denis did his final enlarging in the Medical School darkroom, cleared his office and on Tuesday, January 25, more than adequately dined, feted and prayed for Olive and Denis left Kampala for Nairobi on their journey out of Africa.

In Nairobi, they were met by William (Burkitt), stayed with Arthur and Sheila Scotchmer and the next day flew to Cairo for the start of 19 days visiting places they had always had in mind to see. Starting with the Treasures from Tutankhamun's tomb in the Egyptian Museum, then on to Jerusalem where they were met by Arthur and Alice Boase, who gave Denis some very useful carpentry tips. They walked through the Old City, up to the Mount of Olives, then taxi to Caiaphas' House, the Church of the Holy Sepulchre and on to the tomb of Lazarus. For someone who knew his Bible so well, this was a magical time, but Denis did not miss the opportunity to visit the impressive Eye Hospital and a leprosarium and ask the local doctors about lymphoma cases they may have seen. They then journeyed through Jordan to Bethlehem, Jericho and to Petra, which Denis described as 'the most impressive human made thing I have seen'. Here they noticed a big change in temperature. Now it was cold for those used to the equatorial climate of Uganda. They would have to get used to it. Back in Jerusalem, they saw the Dead Sea Scrolls, and Denis lectured at Arthur's Hospital before entering Israel and visiting the Weizmann Institute where he spent a day talking to staff and gave a further lecture. On to Tel Aviv where they joined a tour taking them to Nazareth and then to Tiberius on the Sea of Galilee where they stayed a night. They were invited into the house of a local women who discovered on talking to Denis that she had read about him in the local paper that day. Capernaum was followed by Athens and finally back to London on Monday, February 14. A train took them to Victoria Station and bus to Nettlebed where it was bitterly cold and too late to see Denis on a television programme. For Olive, there was some sadness at leaving Uganda, but the whole exercise was a positive one for her because at last, after 19 years, she would be able to build a home for them all in England.

This was a faint hope because Denis would be back visiting Uganda in 5 months and within 5 years would be travelling the world with a new story to tell, arguably of greater importance than that of the African childhood lymphoma. For him, the search for the cause of this awful disease had been compelling but had been a diversion from Denis' real intention, which was to find the causes of the major diseases that varied in their occurrence across the world. Lymphoma was not really a surgeon's disease. There was no operative treatment yet Denis had ended up spending

---

[20] Who had been knighted in the New Year's Honours.

his every waking hour for several years pursuing its cause, developing treatments and as a result having the disease named after him. It had led to him becoming an international star in the world of science and medicine. As his days in Africa came to an end, he turned his attention to the diseases that were killing larger numbers of people, not just in Africa but worldwide.

References

1. Burkitt DP, Nelson CL, Williams EH. Some geographical variations in disease pattern in east and Central Africa. East Afr Med J. 1963;40:1–6.
2. Burkitt DP, Wright DH. Burkitt's Lymphoma. Edinburgh/London: E and S Livingstone; 1970.
3. Robinson JAT. Honest to God. London: SCM Press Ltd; 1963.
4. Cook P, Burkitt DP. An epidemiological study of seven malignant tumours in East Africa. MRC Report January 1970.
5. Hutt MSR, Burkitt DP. Geographical distribution of cancer in East Africa – a new approach. BMJ. 1965;2(5464):719–22.
6. Burkitt DP, Hutt MSR. An approach to geographic pathology in developing countries. Int J Path. 1966;7:1–6.
7. Burkitt DP, Hutt MSR, Wright DH. The African lymphoma: preliminary observations on response to therapy. Cancer. 1965;18:399–410.
8. Burchenal JH, Burkitt DP, editors. Treatment of Burkitt's tumour, UICC monograph series, vol. 8. Berlin Heidelberg/New York: Springer-Verlag; 1967.

Chapter 18
'Time and Chance Happen to All Men'

Was there a sense of anti-climax as Olive and Denis arrived back in England? Although Denis had made his reputation in Uganda, his return to England was something both he and the family were well prepared for and to which they were looking forward. For Denis, it was the natural progression of his work towards establishing the reasons for the widely varying incidence of the major chronic diseases throughout the world. He did not waste a moment. The day after their arrival, February 15, 1966, he headed to London and his new office at 172 Tottenham Court Road, then on to the MRC Head Office to meet Sir Harold and others who had facilitated his research over the past 5 years. After signing customs documents for his car, he collected it but found that some damage had occurred in transit. But his irritation with the delay at customs was nothing when compared with his realisation when driving back to Nettlebed in the cold, fog and rain that this was a big contrast with the warm, dry and relatively uncongested conditions he had been used to in Uganda. Life was going to be different working in the UK, very different, but with Olive, the family and his faith to support him he had little doubt that he could progress beyond the lymphoma to other cancers and conditions affecting people everywhere. There was no looking back now (Fig. 18.1).

He was highly successful in this. Within 5 years, Denis would be writing a paper about a cancer he had rarely seen in Uganda, large bowel cancer, and propose a cause, namely lack of dietary fibre, which, as he moved into his new office in February 1966, was something he knew nothing about. This paper, entitled 'Epidemiology of Cancer of the Colon and Rectum' [1], would become his most cited. The concept of diets that are low in fibre reducing life expectancy by predisposing people to major killer diseases like coronary heart disease, stroke, diabetes, bowel cancer and increasing blood pressure and cholesterol was ground-breaking although would invite controversy. But when Mr. Burkitt spoke, he had to be listened to. He already had one cancer named after him and was a persuasive speaker.

How did he progress from the lymphoma to working on these conditions and developing an overarching theory for their cause? It was Himsworth at the MRC

© The Author(s), under exclusive license to Springer Nature Switzerland AG 2022
J. H. Cummings, *Denis Burkitt*, Springer Biographies,
https://doi.org/10.1007/978-3-030-88563-2_18

Fig. 18.1 Denis, with a map of Africa, in his office at 172 Tottenham Court Road, London

and McAdam, the Professor of Surgery at Makerere, who realised that what Denis was telling them was not just that his lymphoma had a distinct geographical distribution around the world but that variations in the occurrence of other diseases could hold important clues as to their causation. Denis had started with his observations of hydrocele of the testis whilst in Lira in 1946; then in 1951 at Mulago, he realised that the emergency cases he was operating on were very different from the cases he had seen whilst a surgical registrar in England. The lack of appendicitis and perforated peptic ulcer in Uganda was confirmed by the study he did with his brother Robin, then a surgeon in London. The childhood lymphoma, which he first described in 1958, had then taken over his life because it was cancer, a condition feared the world over, and its distribution in East Africa suggested a viral cause. Although Denis' name would forever be associated with the lymphoma and he would travel the world for the rest of his life responding to invitations to talk about it, for him it was only one dimension of a bigger story. His safaris ostensibly to delineate more carefully the pattern of distribution of the lymphoma had always included question-naires about other diseases and discussions with doctors, wherever he met them, of their experience of other cancers particularly.

Professor Himsworth had realised, from as early as January 1961 when he first met Denis, that the geography of cancer prevalence and possibly other diseases could provide a new and complementary approach to finding their cause alongside the emerging cellular and molecular biological sciences. He had no hesitation in appointing Denis to be an External Member of the MRC's staff when he returned to England and to provide him with the resources he needed, although these were mod-est. In what Denis described as a small office at 172 Tottenham Court Road in London, his team comprised a PA, clerical assistant and Paula Cook, a geographer and statistician. Importantly nearby in the MRC Statistics Unit was Richard Doll. They were already acquainted having met at cancer meetings around the world. Doll, with Austin Bradford Hill, had done pioneering work on the link between smoking and lung cancer [2] and had gone on to show that radiation caused leukae-mia. He was one of the world's leading epidemiologists and statisticians, and Himsworth realised that Denis needed his input. Doll had just, in 1966, been elected to Fellowship of the Royal Society (Fig. 18.2).

All of the data that Denis had gathered about cancer and other diseases on his safaris and from the monthly reports that came to him from his contacts in the mis-sion hospitals in East Africa were qualitative, not really precise enough for detailed statistical evaluation.

For example, when in Baghdad in 1966, Denis records cancer occurrence as:

Stomach:- <u>Much</u> more stomach than oesophagus. Reverse in Karachi.
Oesophagus:-? 45 a year. Mostly <u>from south</u>.
Mouth:- Common.
Breast:- <u>Very</u> common c.f. Khartoum.
Penis:- Almost unknown. Some in uncircumcised Christians.
Liver:- <u>Rare</u>. c.f. Khartoum.
Kaposi:-? 1 a year.
B.C.C. (Basal cell carcinoma):- <u>Very</u> common.? 10 a week. Probably 4 times as
    common as skin epithelioma. <u>Reverse to Karachi</u> situations.
Melanoma:- <u>Rare.</u>
Lung:- on increase.
Bladder Papilloma:- <u>Rare</u>.[1]

Whilst these observations show quite striking patterns of cancer occurrence, the more scientific acquisition and analysis of data concerning cancer incidence was

---

[1] Notebook labelled 'Far East 1966' family archives.

Fig. 18.2 Professor Sir
Richard Doll (1912–2005),
CH, FRS. (With the
permission of the *British
Journal of Cancer* [3])

developing rapidly. The recent establishment by the WHO of their International
Agency for Research on Cancer (IARC) in Lyon under the Directorship of John
Higginson was a powerful endorsement of the value of such information. Denis,
never a statistician, found in Doll someone with whom he had a common interest,
and in turn he merited Doll's respect for his lymphoma work. Denis' aim was to set
up small cancer registries largely in the poorer parts of the world and on April 13,
1966, met with Doll and Himsworth who agreed to his proposals. He and Doll
would meet frequently in the coming years during which time Doll would keep
Denis aware of data on cancer incidence in the 'developed' world. Doll had been
collecting the numbers for some years from 32 cancer registries in 24 countries and
would publish later, in 1966 as a Technical Report 'Cancer Incidence in Five
Continents', the first volume of what would become the Bible of cancer statistics
[4]. What Doll thought of Denis and his approach is not recorded, but Denis' enthusiasm for collecting data on cancer and other diseases was not diminished. Over
time, he would establish a network of over 140 locations, mainly rural hospitals in
Africa and some in India, from which he would receive monthly reports on the
occurrence of various diseases. Denis believed that great ideas came from small
beginnings.

Spending time settling into a new job was not something Denis did. He immediately started reorganising his life in London whilst carrying on writing, giving talks,
travelling and meeting people. House hunting did take up more time than he
expected. A house in Oxfordshire that both he and Olive felt was right for them was
trashed by the surveyor, but eventually they found The Knoll in Shiplake on the
River Thames, close to Henley of rowing regatta fame and with good rail links to

London. Their offer of £14,500 was speedily accepted, and they were able to sign the contract before Denis left for Uganda where he, not unexpectedly, received a great welcome.

> Coming back into Mulago I felt as if I had only left it yesterday and everywhere I went I met old friends and seemed to shake more hands per hour than I have done in weeks in England. I saw patients on the ward and had long discussions with many different people in connection with my research work.

He left with John Billinghurst on safari through Kenya into Tanzania and then to Blantyre in Malawi where he was to speak at a conference. Formerly Nyasaland it had very recently become a republic with Dr. Hastings Banda as President. The lecture was one of many he would give during the 3 weeks he was in Africa, but they were no longer just about the lymphoma. Everywhere he visited, he was collecting information on other cancers, setting up cancer registries in the smaller hospitals and looking at geographical and lifestyle factors that might help him to understand their distribution. So, his talk was about 'Pathology along the Nile'. Such was his reputation that during the conference he was summoned to see Dr. Banda at his private residence where, along with the Chief Medical Officer, Professor Michael Gelfand and one other speaker, they spent 40 min in discussion about the history and peoples of the country. Denis was impressed by the President who had studied medicine both in the USA and Scotland, worked as a general practitioner in the north of England and was a committed Presbyterian. Banda was full of praise for the early Portuguese pioneers and explorers, which Denis felt was unusual in the leader of an African country.

The day after the conference finished, they flew to Salisbury, Rhodesia (Zimbabwe), where after a short meeting they travelled on to Johannesburg and the South African Institute for Medical Research for talks with George Oettlé. Back at Mulago a few days later, an operating list had been arranged for Denis, which pleased him as he had few opportunities now to keep up his surgical skills. Leaving for London the next day, he was asked if he would look after a young Tanzanian boy called Raspice, who was travelling to London for heart surgery. Raspice had never been away from home, but for the trip to England, he had been taught to eat with a knife and fork, dressed in a smart shirt and suit and presumably told what to expect. Denis was amazed by the boy who seemed completely unfazed by the flight, spoke very little, although Denis could speak Swahili, and reacted to being met at Gatwick by an ambulance and its crew with equanimity. Leaving Raspice at the Hammersmith Hospital, Denis went straight to his office for a few hours, collected the keys for The Knoll and took the train to Oxford where he received a great welcome from Olive, Cas and Rachel. Moving house would occupy some of his time over the next few weeks since he no longer had the assistance of Yusufu and his team. The Knoll proved to be a good house with a cellar, enough space for Denis to set up a workshop where he could do his carpentry and a room for his photography. He was busy with a series of distinguished visitors over the next few weeks including Rosemary Wiggins from the Sloan Kettering Institute, Bergstein from Holland to discuss Cancer Registry, John Higginson from IARC Lyon, Greg O'Conor with

Alan Lindsell from the WHO, who took him out to dinner, Tony Epstein and there were frequent meetings with Richard Doll. All this with a major lecture tour only a few weeks away.

The journey he undertook between October 16 and December 5, 1966, was a highly organised lecture tour during which he usually asked his hosts what were the common cancers they were seeing. Doll however reminded him that he needed well founded quantitative data corrected for age and sex on which to base any theory as to causation, a message that was probably endorsed by John Higginson from the IARC and George Oettlé in Johannesburg. Denis was never going to be able to obtain such information himself and as with the lymphoma when the search for a cause was taken over by the virologists and pathologists, he was content to leave the detail to the professional epidemiologists, whilst he saw himself as pointing to the direction that should be followed.

The lecture tour started in Singapore and was followed by meetings Hong Kong, Tokyo, Dunedin and Christchurch New Zealand, Sydney, Port Moresby and Goroka in New Guinea, Brisbane, Townsville, Melbourne, Adelaide, Bangkok, Karachi, Baghdad and finally back to London. This daunting schedule included 23 formal lectures, participation in panel discussions and the usual chairing of sessions, meeting the press, Embassy invitations, dinners and compulsory excursions, with which those who regularly undertake such busy but less onerous, programmes will be familiar. Denis was unfailingly courteous, even enthusiastic, throughout delighted to be meeting so many people and likely buoyed up by being the centre of attention most of the time. Only once, in the 80 pages of his handwritten account of the journey sent back to Olive, nor in his diary, the detailed notes of his itinerary, the people he met and observations on disease occurrence that he made, does he complain. Towards the end of the trip when he is in Brisbane, he finds he is due to give a lecture in Townsville some distance further up the east coast of Australia, which detour evinced the following remark whilst he was there: 'In a way the 900 mile journey each way and two nights here was hardly worth it to address a small group but I have enjoyed it and found it interesting and the time to relax, read and write has been useful'.

There were between 15 and 20 doctors in the audience. Denis' had the ability to get the best out of such situations, which many people would find frustrating. No matter how small the audience, he enjoyed having their attention and his story was now beginning to change.

The Tokyo meeting, at the start of his tour, was by contrast a glittering affair with the Crown Prince and Princess to open the conference, which was the Ninth International Cancer Conference of the UICC. There were over 600 delegates including many of Denis' colleagues from around the world – Frank Horsefall, Joe Burchenal, Dennis Wright and George Oettlé, Clifford from Nairobi, Thomson and Burn from ICRF, drug company chiefs Armstrong and Johnson from Lilly, Professor Schlesinger from Jerusalem, Richard Doll and Professor Haddow who was President. Denis kept to his brief and talked about long-term remission and cure of the lymphoma. He had time to go to several sessions where there were papers he could not understand, but those on viral carcinogenesis intrigued him and made him realise that the lymphoma story was proving to be a major stimulus to cancer research. He felt he needed to keep abreast of developments and to continue to contribute where he could. Thus, with George Oettlé, he scheduled a visit to New Guinea as part of his tour of the Far East and Australia/New Zealand. With regard to the lymphoma, this was a special part of the world because jaw tumours in children similar to those observed by Denis in Africa had been observed since 1960 [5]. Whilst sporadic cases of Burkitt's lymphoma were now being reported from many countries, Papua New Guinea was unusual in that they had now documented 37 cases and as in Africa it was the commonest childhood cancer. The condition was more prevalent in the coastal regions, whilst the area around the capital was almost tumour-free. Matching these observations to the local climatic conditions confirmed what Denis and his colleagues had reported from Africa that the distribution was depended on temperature and rainfall. They spent 4 days there. In Port Moresby and Goroka, Denis gave four lectures before George left for Sydney whilst Denis headed for Brisbane. As they said goodbye, little did Denis realise that he would never see him again. George would die a few months later, aged 49 years, after surgery for complications of long-standing rheumatic heart disease. They had first met in Mulago in 1958 before Denis published his early paper on the lymphoma and had forged a strong bond based on their common interests in cancer epidemiology. He was a person to whom Denis had always felt close. Oettlé was a pioneer in the field of cancer epidemiology using the contrasting populations in South Africa to show differences in disease risk. Denis had learnt a lot from him.

Towards the end of the tour, Denis was frustrated by a strike at Qantas Airlines and had to cancel a visit to Perth, added to which he received a letter from Olive to say their house in Nettlebed had been burgled. Their engagement ring along with Denis' gold cufflinks, grandpa's silver watch and some money had been stolen. The news troubled him all day not so much because of the loss of these items of sentimental value, but he knew that Olive would take it badly and he was not there to support her. Alitalia Airlines came to the rescue, and Denis was able to make it to Karachi where he gave a delayed lecture at the Pakistan Medical Association conference. Attended by a distinguished audience, he was presented with the Stuart Medal of the British Medical Association by Sir Arthur Porritt. Home via Baghdad his first appointment was with Richard Doll, to catch up on all the news.

Leaving Africa and relocating his work and family back to the UK would have been life-changing for most people and to some extent it was for Denis. He no longer had a regular surgical practice and could not use the skills he had spent many years acquiring. He had left many good friends behind in Kampala, at the mission hospitals around Uganda and in the rest of East Africa, whom he had helped and who had contributed information for his lymphoma studies. The lifestyle of the white colonial ex-patriot community with its multiple servants to help in the house and care for children, carefully guarded social circle, club, religion and even the climate were now gone. It had been a very successful 20 years for him. He returned to the UK with a worldwide reputation in the burgeoning field of cancer research. At the age of 55 years, many people would have been happy to put on cruise control and let events carry them along. Such a life was not for Denis. He had work to do, finding the reason why other cancers had such distinctive geography, for which he had the full backing of the MRC in London and regular contact with and the support of one of the world's leading epidemiologists, Richard Doll. He would soon be given an extra member of staff, Miss Shah, to help with typing, would move into new, bigger offices at 172 Tottenham Court Road and even a pay rise together with £100 per annum London allowance, backdated a year. Progress with the lymphoma was now the province of others better qualified than he to follow the virus-cancer story. He was able to keep in regular contact with Tony Epstein, also in London, who was invited regularly to meetings where the lymphoma was discussed. Denis was also working on the definitive book on the subject with Dennis Wright.

What gave Denis confidence was the belief that having made the major break-through in finding the cause of one cancer he could do the same for others. He was also a man with enormous energy, selfless commitment to help people whoever they were and a firm belief that he was doing what God wanted for him. He knew that there would always be problems ahead but those outwith his control were dealt with by prayer and Bible reading. The 6-day Arab-Israeli war in June 1967 caused him a lot of concern, and he was happy when a cease-fire was declared so quickly. When Cas brought home a boyfriend called Andrew and spent an hour on the phone to him one evening, he wrote, 'Although he is a nice boy, I cannot help wishing Cas wasn't so keen on him as he is not a Christian. Not even nominally'. Also out of his control was the continuing unrest in several of the countries of East Africa, with which he had become very familiar and which had provided vital evidence for the cause of his lymphoma. Doors were now closing to him, and his safaris would become less fea-sible and less frequent.

July and August 1967 saw him on one of these last safaris to East Africa. Driving through Nairobi, then to Lusaka, he arrived finally in Zanzibar where he felt uncom-fortable owing to the civil unrest but was treated like a VIP. He was able to meet the local CMO, who was ex Makerere, who warned him about the difficulties he would now face. In his letter to Olive, Denis wrote,

> The medical set-up is almost impossible as far as my work is concerned. The hospital has been virtually taken over by Red Guard type of Chinese, doctors, sisters, technicians. They treat all out-patients. They are without their wives, and never go anywhere except in groups, presumably for security. They live in a communal mess, appear to have little money and all wear the same simple uniform. They have no contact with the non-Chinese staff and cannot even drink coffee with them but go to their own exclusive room.

On to the nearby island of Pemba things were no better. The hospital at Chake Chake was very run down, primitive and run by Bulgarian and Hungarian doctors. Picking up a copy of the latest issue of the *Chinese Medical Journal*, he noted that the first four articles were entitled:

1. 'Hold High the Great Red Banner of Mau Tse Tung's Thought and Actively Participate in the Great Socialist Revolution'
2. 'Never Forget the Class Struggle'
3. 'Sweep Away All Monsters'
4. 'A Great Revolution That Touches People to Their Very Souls'.

Following which the articles of medical interest appeared.

Fig. 18.3 Dr. John Zeigler at the Lymphoma Treatment Centre in Uganda with some of his young patients. (With thanks to Professor Owen Smith, Dublin)

There was no possibility of setting up a cancer registry here, so Denis returned to Dar es Salaam where he gave a lecture, met the Minister of Health, obtained a visa for Rwanda, then met up with Guy and Dawn Timmis. They headed north-west across Tanzania to Hombolo before returning to Kampala for the opening ceremony of the new Lymphoma Treatment Centre under the Directorship of John Ziegler [6]. Ian McAdam made a speech about Denis' pioneering work and unveiled a plaque dedicating the centre to him (Fig. 18.3).

Denis' intention was then to meet up with John Billinghurst again and revisit Rwanda and Burundi but was told that in Burundi all Europeans were to stay where they were for their own safety and could move only on urgent business, requiring a pass from the governor of the district. They could expect roadblocks manned by the Jeuniere youth movement who had been called up to defend the country. Denis set off the day after the opening ceremony and reached Ishaka in south-west Uganda by early evening. Crossing into Tanzania the next day, he met up with Billinghurst en route to Bukoba on the south-western shore of Lake Victoria, where the hospital was not in a good state. After doing a ward round and having the car serviced, they drove to Ndolage where they stayed the night. Denis awoke the next morning feeling unwell with nausea, but they pressed on to the hospital in Murgwanza close to the border with Burundi. Denis was beginning to feel anxious about crossing into

Burundi but was well enough the next day to operate on a urethral stricture. From there they crossed into Burundi without any difficulty, apart from the endless slowness and inefficiency of the border officials. Crossing Burundi and into Rwanda, they headed for Kibuye where Denis was once again in action.

> After lunch we drove the 20 odd miles to Kibuye. The sole doctor was just about to operate on a baby with? intestinal obstruction, and was all scrubbed up. He asked me to do it. It turned out to have all sorts of congenital defects, and Paget spotted it was a Mongol. After opening the abdomen and finding the defects and learning that there were also cardiac defects we did the minimum not wanting the child to live as a mental cripple.

The safari ended without incident, apart from at the borders, and Denis returned to Kampala 10 days later and from there to The Knoll by the end of August. After a day at home, doing useful jobs, he was soon back in his office enjoying his work and was in what might be called his comfort zone. Within a few weeks, that would start to change.

How did food and diet come into Denis' thinking? He makes regular comments about what he eats on the safaris and his enjoyment of the cuisine of the many cultures he encounters. But in his search for the possible causes of the cancers, about which he was now gathering data, there is never any mention of diet. This is perhaps not surprising given that the food consumed by the rural dwelling African at that time was quite monotonous. He would have been aware of diet in the context of deficiency diseases in childhood from his former colleague and friend, Dr. Hugh Trowel, one of the world's leading authorities on Kwashiorkor. That diet might be important in determining cancer risk does not seem to have entered his thinking.

However, when sitting in his office on Thursday, September 14, 1967, a life-changing event occurred. The 'phone rang. It was Richard Doll who said that he had a man in his office whom he felt Denis might be interested to meet, a retired naval physician Captain T.L. (known as Peter) Cleave. Doll remarked on the 'phone that 'Statistically I could pull much of his work to pieces, but nonetheless I have a hunch that he may be right'. That afternoon, Cleave met Denis in his office and put to him his ideas about diet and how he felt that the diet as eaten in most 'Western' cultures was responsible for many common chronic diseases (Fig. 18.4).

Cleave (1906–1983) had been born into a naval family and after qualifying in medicine was commissioned into the navy in 1927 where he remained all his working life, retiring in 1962 as Surgeon Captain and Director of Medical Research. He became known initially for his use of bran to treat constipation in sailors, which he started when medical officer on the battleship King George V [7]. This approach to the management of constipation was not new in that the laxative properties of bran and wholemeal bread had been known for centuries. But Cleave was an original thinker, a great admirer of Charles Darwin, and from his observations of the health of contrasting populations around the world developed a theory of the cause of 'modern' diseases. In a key paper by him in 1956 in the *Journal of the Royal Naval Medical Service*, he sets out his belief that 'our neglect of natural principals is the root cause of the ills of Western society' [8]. In this long and discursive article, Cleave says that our diet has moved from its natural state because of 'cooking and …the concentration of the food by machinery'. He singles out carbohydrates

Fig. 18.4 Surgeon Captain
T L (Peter) Cleave, RN
(Healthscams.org.uk)

because of the milling of wheat to produce white flour and the refining of sugar cane
to produce raw sugar, which processes he suggests lead to overconsumption and
'dental decay, peptic ulcer, diabetes, obesity, constipation and intestinal toxaemia'.
On retiring from the navy, Cleave began work on a book that was to encapsulate all
his ideas and bring him to the attention of the world more than any of his previous
writings. *Diabetes, Coronary Thrombosis and the Saccharine Disease* [9] was published in 1966. In it, Cleave suggests that overconsumption of white flour and sugar,
a rapid change in diet on an evolutionary scale, has led to a group of modern diseases he groups under the title of 'Saccharine Disease'. Fibre gets barely a mention
in the book, is not in the index nor is it mentioned in his main table listing the various diseases of his saccharine theory.

There is no record of the conversation that took place between Denis and Cleave
on that Thursday afternoon in 1967. His diary note is just 'Visited by Surgeon Capt.
Cleave'. But years later, when he is writing his autobiography and going through his
diaries to jog his memory of events for his various biographers, he adds, 'Time &
chance happen to all men' clearly signalling that, in retrospect, it was a momentous
meeting. It was the first of many discussions he would have with Cleave, of which
Denis notes that there were inevitably arguments about 'minor aspects of his thesis
over which we disagreed'. In fact, to his credit, Denis was never convinced by
Cleave's sugar hypothesis, which subsequent research has not shown to be quite so

holistic [10].[2] However, Cleave was an uncompromising and at times irascible character whom Denis handled with kid gloves at all times in order to minimise quite heated arguments and the correspondence which would ensue if Cleave thought his theory was being challenged. But Denis pays tribute to Cleave's enquiring mind, courage to challenge current orthodoxy and wrote that 'No-one has had a more profound influence on my professional career'. More importantly, beyond sugar, white bread and the refining of foods, there was an important concept he learnt from Cleave. If a number of diseases occur together either geographically or in the same individual, they are likely to have a similar cause. This made sense to Denis because he had originally had to fight to get people to accept that his lymphoma syndrome, affecting many parts of his young patients' bodies, was a single disease with a single cause and incorporated the jaw tumours with which the story had begun.

As the table shows, Cleave had identified in his first book dental caries, peptic ulcer, obesity, diabetes, coronary thrombosis and 'colonic stasis' as 'saccharine diseases', some of which were, and still are, major killers in countries that had developed a Western style of living. Fibre was added to the second edition [11] in 1969, due to the influence of Neil Painter, a London surgeon, and bowel cancer would be included only in the third edition and then somewhat grudgingly [12], perhaps as a result of his conversations with Denis. The initial meeting with Cleave was followed by many more, although it is probable that Denis did not read *Diabetes, Coronary Thrombosis and the Saccharine Disease* until somewhile afterwards. His first move was to appraise Doll of the content of his conversation with Cleave and the idea that for all these diseases of modern society, there was a single cause. Doll was not convinced by the all-embracing nature of Cleave's ideas, and at this point, fibre was not part of the story. It is likely that he encouraged Denis, knowing that his interests were mainly surgical, to focus on large bowel cancer and other bowel diseases such as constipation, appendicitis and diverticular disease, which for Denis was home territory and was a much more manageable concept (Table 18.1).

Throughout the rest of 1967 and into 1968, Denis continued to travel, mainly to the USA, lecturing about his lymphoma but was now starting to gather as much information as he could about large bowel diseases. He was frequently taken on ward rounds where he found the surgical beds full of people who had had appendicectomies, surgery for bowel cancer and for diverticular disease, gall bladder removal and various operations for peptic ulcer. This was in marked contrast to what he had been accustomed to see in Uganda, but no one could explain why this should be. Unconvinced by the Cleave hypothesis, Denis was occupied with his book on the lymphoma, co-edited with Dennis Wright. Importantly, in 1968, he was pursuing other ideas. He was the author of 14 papers in respectable journals, half of which were now about cancer epidemiology more broadly to include those that been attracting his interest for some years such as cancer of the oesophagus, stomach,

---

[2] These WHO Guidelines implicate sugar principally as a cause only of dental caries and as a contributor to obesity in children through its calorie content.

Table 18.1 The development of Cleave's concept of the saccharine diseases as given in the three versions of his book in 1966, 1969 and 1974 [13]

1966 (p. 11) [8]
Dental caries and parodontal* disease ('pyorrhoea')
Peptic ulcer
Obesity
Diabetes
Colonic stasis, with its complications of varicose veins and haemorrhoids
Coronary thrombosis
Probably many primary *B. coli* infections and certain other conditions
1969 (p. 12) [10]
From overconsumption:
Diabetes
Obesity
Coronary thrombosis
Primary *B. coli* conditions
From the removal of fibre:
Dental caries and parodontal* disease
Colonic stasis, with its complications of varicose veins, haemorrhoids and diverticular disease
From the removal of protein:
Peptic ulceration
1974 (p. 17) [11]
1. *By the removal of fibre*:
a. Simple constipation (intestinal stasis), with its complications of venous ailments (varicose veins, deep vein thrombosis, haemorrhoids and varicocele), diverticular disease and, in part, cancer of the colon
b. Dental caries (in conjunction with the taking of sugar) and periodontal disease
2. *From overconsumption*:
Diabetes
Obesity
Coronary thrombosis
Primary *E. coli* infections and gallstones
3. *From the removal of protein*:
Peptic ulceration

The table shows clearly the absence of fibre initially and subsequently its progressive inclusion in Cleave's thinking (*parodontal = periodontal)

liver, penis, skin and Kaposi's sarcoma [14]. Large bowel cancer does not yet get a mention nor does diet. How did Denis come to regard fibre, rather than the refined carbohydrate of Cleave, as key to the prevention of bowel cancer and to be critical for other common diseases? There were a number of other people he would meet who would prove crucial in developing his understanding of these conditions.

References

1. Burkitt DP. Epidemiology of cancer of the colon and rectum. Cancer. 1971;28:3–13.
2. Doll R, Hill AB. Mortality in relation to smoking: ten years observations of British doctors. BMJ. 1964;1:1460–7.
3. Kinlen L. Sir Richard Doll – a personal reminiscence with a selected bibliography. Br J Cancer. 2005;93:963–6.
4. Doll R, Payne P, Waterhouse J, editors. Cancer incidence in five continents: a technical report. New York: Springer-Verlag; 1966.
5. Booth K, Burkitt DP, Bassett DJ, Cooke RA, Biddulph J. Burkitt lymphoma in Papua New Guinea. Br J Cancer. 1967;21:657–64.
6. Now the Uganda Cancer Institute. https://www.uci.or.ug/
7. Cleave TL. Natural bran in the treatment of constipation. BMJ. 1941;1:461.
8. Cleave TL. The neglect of natural principals in current medical practice. J R Nav Med Serv. 1956;42:55–83.
9. Cleave TL, Campbell GD. Diabetes, coronary thrombosis and the saccharine disease. Bristol: John Wright and Sons; 1966.
10. World Health Organization. Guideline: sugars intake for adults and children. Geneva: World Health Organisation; 2015.
11. Cleave TL, Campbell GD, Painter NS. Diabetes, coronary thrombosis and the saccharine disease. 2nd ed. Bristol: John Wright and Sons; 1969.
12. Cleave TL. The saccharine disease. Bristol: John Wright and Sons; 1974.
13. Cummings JH, Engineer A. Denis Burkitt and the origins of the dietary fibre hypothesis. Nutr Res Rev. 2018;31:1–15. https://doi.org/10.1017/S0954422417000117.
14. Burkitt DP, Hutt MS, Slavin G. Clinico-pathological studies of cancer distribution in Africa. Br J Cancer. 1968;22:1–6.

Chapter 19
In Transition to a New Theory

Whatever else that was on Denis' mind early in 1968, he was still very much involved with work on his lymphoma. A safari to Ghana in West Africa early in the year started in Accra where he was treated like royalty. Met at the airport by Cofie George and members of the Burkitt Tumour Project, he was taken straight to the VIP lounge where he was surrounded by the assembled press and television cameras. He gave numerous interviews, then exited the airport, without going through customs or immigration, into a waiting Mercedes and to the house of his host, Dr. Johnson, a barrister and Methodist Local Preacher. He noticed en route that all signs of former President Nkrumah had gone following his overthrow in 1966 and the realignment of the country with the West. At the hospital, he was introduced to the leaders of the lymphoma project and told they had now documented over 60 cases. Twelve patients had been brought to the hospital for him, most of whom were well or 'cured', and all were photographed by him. What was now more interesting to Denis was that there were areas of Africa having similar climatic conditions to Ghana, but where the lymphoma was rare. He was now more than ever convinced that a second factor was needed, namely malaria, which in its severe form may prime the reticuloendothelial system promoting excessive growth (hyperplasia) of B-cells. In the presence of the virus this could lead to cancer. Ghana was one of the world's countries most affected by malaria. Denis' case for malaria being involved was argued by him in a paper entitled 'Etiology of Burkitt's Lymphoma – An Alternative Hypothesis to a Vectored Virus'. In this, he presents the evidence in a way that belies his protestations about being left behind in these matters by all the clever scientists who followed him into the virus and lymphoma story [1]. The paper, given the information available in 1969, is a classic dissection of the hypothesis, although the cause of the lymphoma proved not to be either virus or malaria. Both are needed for the African form of the disease.

Cleave and his ideas were not foremost in his mind as he was now engaged with the MRC in setting up a centre to co-ordinate the work of the cancer registries that he had helped to get started across Central Africa. In talking to Doll, he made it clear

© The Author(s), under exclusive license to Springer Nature Switzerland AG 2022
J. H. Cummings, *Denis Burkitt*, Springer Biographies,
https://doi.org/10.1007/978-3-030-88563-2_19

that he was now collecting information on several cancers, but Doll reminded him that he would need more precise data on incidence than simply whether the condition was present or not. A 2-week lecture tour to the USA in June commenced the day after the assassination of Bobby Kennedy, meaning that attendance at his talks was not as numerous as usual because of interest in the funeral. But he found visits to the USA, whilst tiring, quite reassuring in other ways. At Wheaton College in Chicago, the Principal spoke at the graduation ceremony of the importance of Christian belief, following which there were choruses by the college football team with a gospel message. 'It was such a totally normal experience to be in a university setting where all aspects of life are orientated firmly Godward'. Denis' faith was now well integrated into his life. Perhaps because of this, or the passing of the years, he was becoming quite conservative in his views on patterns of behaviour especially amongst the young. At M D Anderson Hospital in Houston, he was pleased to note that skirts came down below the knee, there were no bikinis in the swimming pool and boys' hair was cut short. 'Like England 5 or 10 years ago'.

Back at home, he was preaching regularly at the local parish church, had frequent meetings in connection with his position as Vice President of the CMS and was still trying to persuade his daughter Cas to break off from her increasingly amorous relationship with the non-Christian Andy.[1] Denis was also busy at home with practical tasks in their new house, at which he was very good. In London, he continued to write papers and prepare for another safari to East Africa in August and a conference on the lymphoma in September. There was no suggestion that diet was in his mind at this time. Then came a major change in his thinking.

Denis had agreed to give a lecture to the British Society of Gastroenterology (BSG) at their conference in London on November 7, 1968 and thought that members would be more interested in his new work on other cancers rather than the lymphoma, and so chose to talk about oesophageal cancer. Although there are no archives of these early BSG meetings, it is highly likely that Mr. Neil Painter (1923–1989), a London surgeon, was speaking in a symposium at the conference about his work on diverticular disease[2] of the large bowel. It had been suggested to Denis by Dr. Harold Dodds, from Reckitt and Coleman, that he get in touch with Mr. Painter because of their common interests in the large bowel and so Denis had written to him on November 5 about colorectal cancer and his theories as to its cause. He must have heard Painter lecture at the BSG because the next day he wrote again to him saying, 'I was very interested in your comments suggesting a relationship between diet and diverticular disease...particularly ...the work on rats fed with

---

[1] The pressure was resisted by the young couple, and Denis agreed to their engagement on December 31, 1970.

[2] Diverticular disease of the large bowel is a common condition increasing with age and consists of thickening of the muscle layers of the bowel, leading to increased luminal pressure and the protrusion through the lining of the bowel of small pouches, which may become inflamed. The main symptoms are abdominal pain and a change in bowel habit in uncomplicated disease.

low and high residue[3] diets'.[4] These studies carried out in the USA in 1949 [2] were on groups of rats fed diets with differing amounts of what the authors called roughage, which can be equated with fibre. They were lifetime studies and showed that diverticula developed with age, very much as they do in humans, but that this process could be almost entirely eliminated by feeding the animals fibre. The authors also suggested that these healthier rats would have lower intracolonic pressures and would thus be less likely to develop diverticula.

Painter had done postgraduate research in Oxford with Dr. Sydney Truelove between 1959 and 1963 and had read the American paper describing the rat studies. He had gone to Oxford to measure pressures in the bowel of patients with diverticular disease using a new technique of intraluminal pressure recording. Little was known about the cause of the condition, which was usually treated with morphine when patients were admitted to hospital in pain. Painter showed that whilst there were no differences in resting pressures in the sigmoid colon, the region most commonly affected by the condition, when morphine was given there was a greater increase in pressures in those areas with diverticula. Morphine was thus not the correct approach to treating the condition. The study in rats made Painter realise that where a diet was low in fibre/residue, the amount of residual material in the bowel would be correspondingly low. This would then require thickening of the muscle in the wall of the bowel to propel the inspissated (dried, thickened) contents, with attendant higher pressures developing and then diverticula. In the rat study, adding fibre to the diet had ameliorated this problem. His paper on the dangers of using morphine was published in the *British Medical Journal* in 1963 [3] and was the first of a series he would write introducing a completely revolutionary idea that the key to good treatment of diverticular disease was adding fibre to the diet (Fig. 19.1).

Painter, in his MS thesis [4] in 1962, concludes by saying, 'It may be surmised that a low residue diet leads to narrowing of the colon and that allows segmentation and subsequent pressure production to occur more frequently…From observation of the African colon, which are usually more distended and do not get diverticular…It is therefore possible that a bulky diet lessens the probability of segmentation occurring and hence prevents the onset of diverticulosis'. A few months before the November meeting of the British Society of Gastroenterology in 1968, a review by Painter of his work was published in the *British Medical Journal* [5] in which he recommends treating diverticular disease with bran. This paper was almost certainly read by both Cleave and Denis, who as members of the British Medical Association would have had the journal delivered to them weekly. Since Cleave had used bran to treat constipation in sailors in the Royal Navy, he shared a common interest with Painter, the outcome of which was that Painter would contribute a chapter for the second edition of '…the Saccharine Disease' and with Cleave's encouragement Painter would start a clinical trial of bran in diverticular disease. The results of this

---

[3] The term residue in this study is an early synonym for fibre and a low-residue diet would be low in fibre.

[4] Wellcome Library Papers of Neil Painter PP/NSP/Accession no 624 Box 2 File 'Burkitt and Fibre'

Fig. 19.1 Neil Painter
outside the Manor House
Hospital in London. Circa
1980 (Wellcome images)

study were published in 1972, and it became one of the iconic papers of the fibre story [6].

Painter was not just another enthusiast in the long history of people who believed in the beneficial effects of bran. His work with Truelove on the cause of diverticular disease was original. The recommendation to treat the condition with a high-residue diet was a turning point in the management of the diverticular disease, treated for the previous 100 years with low-residue diets and morphine for the pain. Painter is an important figure in the development of the fibre story and clearly was working on it 10 years before Denis or Cleave wrote anything on the subject [7]. It is likely that he put the idea of fibre into the mind of Denis, who was already primed to the importance of diet by Alec Walker. Neil Painter has probably not received the recognition for his work that it deserves.

Denis could not meet Painter until after he had been to West Africa in December, so they met in January 1969 and discussed diet and bowel cancer. Now 3 years after his return to the UK, there were a number of reasons for the emergence of this cancer as the primary target for Denis' next major endeavour. Doll after meeting Cleave had told Denis to focus on bowel diseases because he knew that was where many surgeons' interests lay. After hearing Painter talk and meeting him, Denis probably felt that diverticular disease of the colon had been fully explored leaving the problem of bowel cancer, a major killer in the West, to be solved. With his background, Denis felt no qualms about tackling another cancer. Thus, from the time he had met Cleave in September 1967 to his discussions with Painter in early January 1969, Denis' thinking had taken a major step forward. He had now heard about fibre and something of its importance for the large bowel. He would meet Painter on many occasions in the future, they would write five papers together, but they were never good friends because Painter did not subscribe to that brand of evangelical Christianity that merited Denis' respect. Painter was a successful London surgeon somewhat to the right of centre in his politics, conventionally dressed, portly, who enjoyed his whisky and liked to tell stories. As the fibre idea took hold in the medical and nutritional community, he would be invited to talk about his studies. His lectures were given with many asides and humour. His favourite 35 mm colour transparency was of the stool from a patient with diverticular disease from the Manor Hospital, which was run for the benefit of patients from the Trades Union Movement. His caption for the slide was 'a Trades Union Movement'.

Immediately after his meeting with Painter, Denis decided to make a precis of '…the Saccharine Disease' to help him with the next step in his quest for the cause of bowel cancer. This was to be the gathering of information about bowel function in order to test Cleave's refined sugar and starch hypothesis. His plan, in testing Cleave's ideas, was to send a brief summary of the book together with a questionnaire to all his contacts in Africa and around the world. There were at least 140 government and mission hospitals, largely in rural areas, from which he gathered views on Cleave and information about disease incidence in their location.[5] The conditions listed by Cleave in the first edition of his book in 1966 as 'consequences of the refining of carbohydrates' were diabetes, coronary disease, varicose veins and haemorrhoids, peptic ulcer and urinary tract infections. Denis' questionnaire also asked about surgical conditions of the large bowel such as colon cancer, diverticular disease, appendicitis in addition to peptic ulcer and gall bladder problems.

Excluded from Denis' questionnaire was dental caries, which Cleave had thought an important saccharine disease. Why Denis was apparently not interested in caries, despite his having been a frequent visitor to the dentist over the years and having had one tooth removed with a hammer and chisel without local anaesthetic, is not explained. Nonetheless, Dr. Cubey from Mkomaindo Hospital in Tanzania, in responding to the questionnaire, added a note at the end of his answers saying that 'Dental caries is common in the local community. Even amongst the majority peasant farmers'. It would never feature in Denis' thinking, despite caries being one of the commonest diseases in the world and correctly associated with sugar intakes, as Cleave believed. But Denis did now include, for the first time as a result of his discussions with Cleave and Painter, requests for information about the local diet. There were also questions about the number of beds in the hospital, annual

---

[5] A copy of the questionnaire was also included in communications between the Medical Assistance Programme of Chicago (MAP) and the mission hospitals to whom it sent surplus supplies of drugs and equipment donated by US companies.

admissions to those beds and outpatient attendances, which in some hospitals reached huge numbers. The hospital in Ilesha, Nigeria, which Denis had visited in 1962, with only 158 beds was recorded as having 280,000 outpatient attendances annually and doctors in several hospitals indicated over 200,000 all seen with courtesy and good humour but with the barest of resources, as Denis had often noted.

The replies to questions asking 'How often do you see...' make fascinating reading varying from yes or no to 'very rare', 'quite often', 'never in the villages', 'only in higher social levels' and 'Not as often as I used to see during my training in the U.S. and U.K.' The absence of coronary heart disease is very striking, for example, Joe T writing from Mvumi Hospital, Dodoma, in Tanzania notes, 'In the last ten years which covers almost 30,000 admissions to the hospital we have not seen a single case of proved coronary thrombosis and nothing even vaguely suggesting this'. There is a treasure trove of information in the answers to the questions, and reading them gives some insight into why the patterns of disease in different parts of the world must have impressed Burkitt, Cleave, Oettlé and others who followed them. The number of questionnaires sent out is not known, but 58 were returned from 17 countries, mostly in Africa but also including Hong Kong, India, Iran, Ceylon (Sri Lanka) and New Guinea, and 15 were accompanied by letters expanding on the answers to the questions, mostly expressing agreement with the Cleave story. Yet, despite their compelling nature, the data are entirely qualitative and do not give sufficient information to draw conclusions about exposure to lifestyle and environmental factors that might predispose to disease risk. But for Denis, the contrasts between Africa, together with the other countries he had visited, and the occurrence of the major killer diseases in Europe and North America, of which he now had first-hand knowledge, were so great that their qualitative nature was never a cause for concern to him.

There was a sting in the tail for those receiving the questionnaire. On the last page, Denis asks: 'In the light of the enclosed memorandum it is obviously important to estimate bowel transit time[6] and bacterial content of stools in different communities living on different diets and subject to different disease patterns'. He asks them to carry out the studies, for which he will provide the necessary materials and then outlines how to do the test. Thirty responded affirmatively. The suggestion to do transit studies owed little to Cleave who was neither a fibre nor large bowel aficionado and had probably never thought about transit time through the gut.

Where did the idea of measuring transit and bowel function come from, and why is it important? The time it takes for material to pass through the bowel is an important measure of bowel function and key determinant of bacteria growth and metabolism in the large bowel, the principal site for the gut biomass or microbiome. The idea to measure it came most probably from someone he had met in South Africa on the long safari in 1961, whilst in Johannesburg, Dr. Alec Walker (1913–2007) of the

---

[6] Transit through the gut is the time it takes for material to pass from mouth to anus. It is measured by giving the volunteer something inert with a meal and recovering it in faeces. Most commonly, the markers used in this test were plastic pellets made radiopaque by the addition of barium, allowing excretion of the pellets to be measured by x-ray, thus minimising the handling of faeces.

South African Institute for Medical Research. Walker had been working on fibre and human health for more than 15 years and was a colleague of George Oettlé.[7] He was a small sparely built man with a ruddy complexion and in later life a mane of white hair, with boundless energy and enthusiasm for his work. He never received the credit he deserved for his original studies because of a natural shyness and unwillingness to tell people that very often he had already done the experiments they were proposing. He was of Scottish Presbyterian origin with degrees in physics and in chemistry who emigrated to South Africa in 1938 and, stimulated by the major differences in health he observed in the contrasting populations, began a lifetime of work into diet, nutrition and health. His early work was on calcium metabolism, but he is best known for his series of papers from 1948 onwards on cardiovascular disease (coronary heart disease and stroke) and the role of fat, cholesterol and essential fatty acids in contributing to, or ameliorating, the risk of these diseases. He combined epidemiological studies with laboratory work and human feeding studies.

Fibre first came to Walker's attention when he learnt about the widespread use of laxatives in many countries to correct an 'inadequate bowel movement'. He was aware of Cleave's letter to the *British Medical Journal* in 1941 about using bran to prevent constipation in sailors and the important pioneering work by Williams, Olmsted, Cowgill and Anderson in the USA during the 1930s on the mechanism by which various sources of fibre could benefit bowel habit [8].[8] He started to work on the effects of fibre by measuring the change in stool output in his volunteers and transit time through the bowel. In 1956, he mentions studies he had done of transit in Black and White groups using the red dye carmine[9] which showed that the time for the dye to pass through the gut of the adult 'Blacks' was only half the time seen in the 'Whites' and stools were characteristically bulkier in association with the faster transit. Walker assumed quite correctly that this was due to the higher intake of cereal foods and pointed out 'that (they) rarely suffer from a large number of diseases which exact a high toll of mortality and morbidity among western peoples. Such diseases include appendicitis, gall stones, peptic ulcer, certain types of cancer, and atherosclerosis and coronary heart disease' [9]. Here, in 1956, was the fibre story in embryo, a more credible hypothesis than that of Cleave and backed by preliminary experimental evidence.

Walker, an avid reader of the literature, had seen the work of Wynder and Shigematsu from the Sloan Kettering Institute about environmental factors in bowel cancer. They suggested that the composition of the contents in different parts of the

---

[7] The exact date is uncertain but must have been before 1969 because by then Denis realised the importance of bowel function and was ready to use Walker's techniques to study it. He had probably first met him in 1961 during the long safari but was not at that time receptive to ideas about nutrition and health, being focussed on the lymphoma, viruses and chemotherapy.

[8] See [8] for more information about this.

[9] Carmine is a red dye, a capsule of which was given by mouth to subjects participating in Walker's studies, following which they were asked to observe when their stools became tinged with red. It was one of the early methods for measuring transit, quite crude but in the populations that Walker was studying differences in transit were great.

large bowel, and the variation in the time these contents spent in contact with the lining of the bowel where cancer arises, was probably determined by diet [10]. They did not mention fibre because there were no fibre data in food tables at the time. Oettlé's similar observations and the work of Professor REO Williams' microbiology group in London must have put the idea of gut transit and the role of the contents of the bowel clearly into Alec Walker's mind and then into that of Denis. Denis had got to know Alec Walker much better when visiting South Africa and he would become 'a loyal and valuable colleague. I know of no-one who has better knowledge of the literature on fibre'.[10] 'I have often in medical literature been credited with originating the hypothesis that dietary fibre is protective against large-bowel cancer, but pride of place should go to Alec Walker rather than to me'. Denis clearly learnt a lot from Walker, who may have been the first person to mention the word fibre to Denis. However, Neil Painter planted the idea of fibre and diet into Denis' thinking about large bowel diseases (Fig. 19.2).

After meeting Painter early in 1969, Denis wanted to inform himself as to how a healthy or unhealthy bowel worked and how to measure it. He could not undertake the complicated studies of bowel muscular activity that Painter had done. Measuring transit time appealed to him. Denis' change in outlook and approach to his research at this time should not be underestimated. For over 20 years, since the days of the hydrocele, he had been plotting the distribution of diseases geographically and from such data finding environmental or other factors that would provide an explanation for their varying prevalence and from that suggest a suitable preventive strategy. Now he wanted to do some physiological measurements to help him understand more about the bowel, and soon he would be suggesting mechanisms that might be at work in the bowel to cause cancer. He would then propose a hypothesis based on his physiological observations, particularly bowel habit, stool weight, transit time and the gut bacteria. Epidemiology, especially of the 'none, little or lots' variety, was not really a basis for public health policy, but with Richard Doll's statistics and Denis' new physiology and resultant hypotheses to explain all the facts, the arguments for prevention became much stronger.

The key people whom he met in the early weeks of 1969 before sending out the questionnaire on March 10 were Neil Painter, who understood fibre and gut motility, and Professor Basil Morson at St Mark's Hospital, probably the UK's leading gut pathologist at the time. At the Wright-Fleming Institute, St Mary's Hospital were Professor REO Williams and his pioneering group of gut microbiologists including Vivienne Aries, JS Crowther, Bo Drasar and Mike Hill, and there was Dr. George Misiewicz at the MRC Gastroenterology Unit, a medical scientist with unrivalled knowledge of gut motility. All of these people were in London, giving Denis easy access to their expertise. The St Mary's Group were working on the cause of bowel cancer. They believed that gut bacteria, through their metabolism of bile salts, were important and had obtained stool samples from Ugandans, who have a low risk of bowel cancer, and compared them with those in England, where risks are much

---

[10] Autobiography Chap. 13, p. 22

with Alex Walker in S. Africa
about 1983

Fig. 19.2 Denis Burkitt and Alec Walker in South Africa 'about 1983' (Wellcome images)

higher [11]. They showed significant differences in the composition of the bacteria in the large bowel, which make more than half of its contents [12]. These studies alerted Denis to the importance of the gut contents as key to understanding bowel cancer and the need to find some way of studying bowel function in Africa and other areas of the world. He visited George Misiewicz at the Central Middlesex Hospital to learn about measuring transit time and was shown a method that was relatively simple to perform in the field. Participants in these studies would be asked to swallow 25 small pellets impregnated with barium and to collect their stools for 5 days thereafter. The specimens could then be taken to an x-ray department where the numbers of pellets excreted each day counted. From that transit could be calculated either as the average time each pellet spent in the gut or when a certain number, usually 80%, had been passed out in their stools (Fig. 19.3).

So, the scene was set for Denis to start his studies of bowel behaviour in relation to cancer incidence. To learn more about the transit technique, he decided to test it on his family. Olive remembered taking part in this pilot study. She helped Denis with the design of plastic bags that could be suspended over the rim of the toilet so positioned as to catch bowel content only. Samples were secured, labelled with the date and time of collection and kept in the family refrigerator until the study was

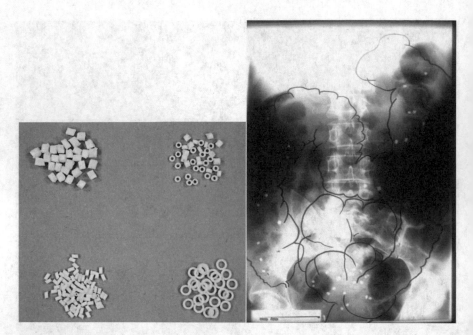

Fig. 19.3 (a) Various radio-opaque plastic pellets given to volunteers to take by mouth for transit measurement. (b) Abdominal x-ray showing transit markers in the bowel (JHC)

completed after which Denis took them to London to get them x-rayed. He was then able to calculate a gut transit time and was pleased with the results. Learning the intricacies of stool collections, which skill he would go on to use in Uganda and in Henley with local schoolboys, was a remarkable departure for someone whose development of ideas about both the lymphoma and now Western diseases had been based entirely on observational studies, backed by pathology. The motivation to test the Cleave hypothesis was driven probably by lack of belief in the sugar story, an instinct that served him well. The move to testing bowel function, to strengthen the bowel cancer and diet hypothesis, came from his discussions with Walker, Painter, Misiewicz and others, all of whom used experiments in humans to try and establish, or reject, their theories about the cause or nature of various diseases. This gave him the confidence to do the transit studies himself. He found he enjoyed doing them, calculating the results and getting Judy to draw the results in diagrammatic form for his papers and talks, which were now equally about the lymphoma and the causes of other cancers. His talks about bowel function would become legendary, but he revelled in the somewhat incongruous nature of the topic for scientific meetings, which on occasions led people to forget that he was a pioneer in the field of cancer research.

Thus, in 1969, the two strands that made up his career overlapped. He finished work on the book about the lymphoma, which he was editing with Dennis Wright, and it was sent to E&S Livingstone for publication in 1970. It comprised 22 chapters by the key authorities in the lymphoma story and was prefaced with a Foreword by Harold Himsworth, in which he said that 'people are seldom moved out of the ruts of their conceptions about the cause of disease, or anything else, until brought into the realisation that there was evidence to the contrary'. He cited the cause of Burkitt's lymphoma as just such an experience and concluded by saying that no explanation for the cause of any cancer would ever be adequate unless it took account of the geographical and other factors that contributed to its understanding. 'In a sense, this tumour may well prove to be something of a Rosetta stone' [13].

Whether it was due to the continuing extreme pace at which Denis lived and the demands he made on himself or the huge cultural changes, including in his diet, that moving from Uganda to England had brought, he developed quite severe migraine. He had experienced only one attack whilst in Africa but was now getting them every 2 months or so, requiring him to rest. He described one attack as 'Nasty attack of cerebral migraine with hallucinations & patchy loss of memory'. This caused him to book an appointment with the family doctor. Dr. Foster examined him, did some blood tests and said there was nothing to find except that his blood pressure was above average. No medication was prescribed. Denis was reassured and made no changes to his hectic lifestyle. With the lymphoma book now out of the way, Denis was free to give more time to his developing ideas about the diseases of the West. Earlier in the year, he had met Professor John Yudkin at Queen Elizabeth College in London. Yudkin was in the Cleave mould in that had been warning of the dangers of excessive sugar consumption for more than 10 years. He was best known for his book about sugar entitled *Pure, White and Deadly: The Problem of Sugar* [14], but Denis remained unconvinced by the incrimination of sugar as a major cause of ill-health. More to his liking was the work of Dr. A Rendle Short, a surgeon who had been one of Cleave's teachers at the Bristol Royal Infirmary, and who had written a scholarly article for the *British Journal of Surgery* in 1920 on 'The causation of appendicitis' in which he concluded 'that the cause is the relatively less quantity of cellulose eaten' [15]. Cellulose is a key component of fibre. Appendicitis was definitely on Denis' agenda following the observation he made with the help of his brother in 1952 during his early years in Africa. These showed that operations for acute appendicitis were almost unheard of at Mulago, whereas for brother Robin in London, it was the commonest surgical emergency. Denis had now talked to many people about his emerging ideas for the cause of cancers, other than the lymphoma, and their association with common diseases like diabetes, obesity, coronary heart disease, peptic ulcer and venous disorders. He liked to work out his ideas by writing about them so sent off a paper to the Pakistan Medical Forum, partly to fulfil an obligation he had to write for them following an earler visit to Karachi. In this paper, for the first time, he puts together all the conditions he felt had widely differing incidences across the world and suggested that these disease patterns could be related to diet [16]. This was 1969, less than 4 years since he had left Africa.

Back on safari in August, Denis' plan was to visit as many as possible of the mission hospitals that had replied to his questionnaire and expressed willingness to take part in the transit studies. In the first 18 days, he visited around 20 hospitals in Kenya, Uganda and Tanzania, then headed for Durban where he met another key player in the fibre story, George Campbell, who had a different perspective on the cause of chronic diseases, particularly diabetes. The first edition of Cleave's book *The Saccharine Disease* included as co-author Dr. George D. Campbell (1925–1998) who was a believer in sugar as a problem nutrient and later came to claim that he was one of the originators of the fibre hypothesis [17]. He was born into a wealthy family of sugar plantation owners in Natal, South Africa, studied medicine and after postgraduate work in Philadelphia and Edinburgh settled in Durban as a physician with a special interest in diabetes. Contrary to opinion at the time, he encountered diabetes regularly in both the Indian and Zulu peoples of Natal and in 1958 set up the first diabetic clinic in Durban which attracted 650 patients in the first 18 months. He was very successful in devising treatment strategies for people who were largely very poor, ate high carbohydrate diets and in the case of the Indians had a high prevalence of obesity. Campbell was a man of wide interests which extended beyond diabetes. He was one of the first people to warn that a drug that eventually became known as thalidomide might not be safe for human use, was a reputable marine biologist writing a series of papers about shark attacks, spoke fluent Zulu and Xhosa and was a Fellow of the Royal Society of South Africa (Fig. 19.4).[11]

Campbell did original research on the blood sugar responses to test meals and by the end of 1961 was the author of eight letters or papers about diabetes in Natal Indians and the Zulu and its treatment, by virtue of which he came to the attention of Cleave. In December 1962, Cleave wrote to Campbell sending him his latest

Fig. 19.4 GD Campbell, Magda Campbell, his wife, with Peter Cleave. Taken by Helen Cleave in the garden of Cleave's house in Fareham, Hampshire, circa 1970 (Wellcome images)

---

[11] See [8] for sources about GD Campbell.

book, which was about peptic ulcer [18], asking him if he would co-operate in writing a book about sugar and refined carbohydrate foods. Campbell replied that it would be both a pleasure and an honour to work with Cleave although they did not meet until 1966, Cleave declining repeated invitations to visit South Africa. Campbell believed 'passionately that diabetes was somehow related to sugar'[12] and like Cleave, fibre was not on his agenda despite his later assertion to the contrary. It was Painter who brought the idea of fibre into Cleave and Campbell's thinking and in the chapter on diverticular disease, in the second edition of the book, it is accepted that removal of fibre leads to the formation of diverticula although Cleave, almost certainly, qualifies this by saying, 'We consider ....that both the aetiology and pathogenesis of diverticular disease have their origin in our modern, refined diet, and that the condition is a manifestation of the saccharine disease' [19]. Cleave would defend his theory to the end and beware anyone who suggested it be modified in any way.

Denis met Campbell in Durban in 1969, probably for the first time, and after lecturing at the University spent a couple of days visiting surrounding hospitals, then was invited to the beach cottage of Campbell and his wife Magda for 3 days. There was plenty of time for discussion about diabetes and its cause. The prevalence of diabetes was ten times higher in the local Indian population when compared with those in India and sugar consumption, freely available in Natal, much higher. When asked about fibre intake in the sugar cane cutters, Campbell pointed out that the workers simply chewed the cane and discarded the structural components. There was another important observation that Campbell made to Denis, which was new to him, 'The rule of twenty years'. He pointed out that there seemed to be a remarkably uniform incubation period of exposure of a population to a diabetogenic factor before the disease appears [20]. Denis thought that this might be true for other conditions as well. He returned to England via Johannesburg, where he met one of the local surgeons, Dr. Cedric Bremner, and also Alec Walker whom he was now getting to know. Diabetes was something that was not really on his agenda as he pursued the cause of bowel cancer, but it would become part of the fibre hypothesis once he and Hugh Trowell became reunited on his next visit to Uganda.

Meeting people from around the world in 1969 and studying the replies he had received to his questionnaire made Denis realise that now he was away from Africa his broader approach to the geography of Western diseases was opening his eyes to far greater opportunities to improve health than he had been able to do with his lymphoma work. Whilst the major resources of hospital medicine were concentrated on treating disease, a highly lucrative occupation in some countries for those in specialist medicine, little effort was put into prevention. Guided by Cleave's holistic hypothesis but with regard to the experimental work, to which he had been introduced as he visited many of the leading scientists in their fields, Denis developed a new hypothesis. He proposed that 'the nature of intestinal content and the

---

[12] Professor Jim Mann, University of Otago, New Zealand, Personal communication. Professor Mann, one of the world's leading authorities on diet and diabetes, lived in Cape Town until 1975 and knew Campbell for many years.

resultant behaviour of the bowels, were directly related to a person's risk of developing many of our commonest diseases. Bowel content is in turn more dependent on the fibre in our diet than to any other component of food'. This was perhaps something of an over-simplification of the causes of diseases such as diabetes, coronary heart disease, stroke, obesity and less spectacular conditions like varicose veins that had attracted his interest. Diet was something he had never thought about in this or any other health-related context. Fibre was a new concept for him, but he was already correct in assigning a principal role to it in determining bowel function.

To learn more about it Denis looked up the word fibre/fiber in the Cumulative Medical Index, an annual list of biomedical journal articles grouped by authors and topics, but could find nothing about such a component of the diet although a great number of papers about fibre optics. This omission added to Denis' growing belief that fibre and diet were largely neglected as primary risk factors for many diseases. The world of science and medicine in later years looked back at these ideas, which Denis started to promulgate, and labelled him as a rather naïve although enthusiastic believer in a particular dietary fad. That is to underestimate him considerably. By the end of 1969, he had already been awarded six medals or prizes for his lymphoma work,[13] the first of many, he would be elected to Fellowship of the Royal Society in 1972, and his name was a bye-word in the world of cancer research. Importantly, his motivation was to save the world, not so much now from its sins but from the burgeoning incidence of the killer diseases of Western society, also starting to appear in the developing world. He was sure he knew the primary cause of these diseases. This was a task worthy of his status and the belief that prevention was the key to the future, not treatment, a significant change in his approach.

A few days after his return from the African safari where he met George Campbell, Denis was in the air once more this time heading for Rio de Janeiro in Brazil. There he was to speak at a Cancer Congress held under the auspices of the American Cancer Society. His lectures to the Congress were about the lymphoma, but listening to him talk was a member of the organising committee, Dr. Ronald Grant, who recorded an interview with him. During this, Denis started to talk about the geography of cancers other than the lymphoma and discussion came around to bowel cancer. Dr. Grant was wise enough to realise that what Denis was saying, about the importance of the contents of the bowel, transit times and diet in the cause of bowel cancer, made sense and he was aware that colon cancer was now one of the commonest cancers in North America increasing every year. He invited Denis to speak about it at a major cancer conference to be held in January 1971 in San Diego. This invitation made Denis start to learn in a more serious way about bowel cancer and its epidemiology particularly in Africa.

> Taking my cue from Alec Walker…I developed the hypothesis that perhaps deficiency of fibre in western diets was the major factor responsible for the high prevalence of this disease in more affluent societies.

---

[13] See Appendix 2, Curriculum Vitae

He would go on to write a paper, published in 1971 [21], based on the content of this lecture and his studies of bowel function that would become the first to set out the case for fibre being involved in any condition other than constipation. But before that, in December 1969, Denis sent to *The Lancet* an article, which he said was based on his 'thinking aloud' ideas, under the title 'Related Disease – Related Cause?' [22]. In it, he credits Cleave with the idea that where diseases occur geographically and chronologically together, they are likely to have a common cause. Whilst Cleave had suggested that this might be excessive consumption of over-refined carbohydrates, Denis now adds with 'removal of the unabsorbable cellulose content', cellulose being a significant component of all naturally occurring fibre in the cell wall of plant foods. He writes in *The Lancet* about large bowel diseases, namely appendicitis, diverticular disease, polyps and cancer and postulates that there is a common risk factor. This he believes is the amount of stool, the gut bacteria, transit time and intralumenal pressures, all of which he says can be altered by changes in diet and that the key to their control is 'unabsorbable fibre'.

This paper and the one on bowel cancer, which would follow 18 months later, were his first serious step down the dietary fibre road, a cause that would occupy him for the rest of his life and lead to him becoming known as 'The fibre man' [23]. In the year of the first moon landing this was no small step for Denis but a complete reorientation of his thinking and way of approaching problems by combining epidemiology with experimental work in the field and the role of diet in prevention. Whether it would be 'One giant leap for mankind' would require the work of others and the passing of many years before its true importance for health would gradually become accepted internationally [24].

References

1. Burkitt DP. Etiology of Burkitt's lymphoma – an alternative hypothesis to a vectored virus. JNCI. 1969;42:19–28.
2. Carlson AJ, Hoelzel F. Relation of diet to diverticulosis of the colon in rats. Gastroenterology. 1949;12:108–15.
3. Painter NS, Truelove SC. Potential dangers of morphine in acute diverticulitis of the colon. BMJ. 1963;2:33–4.
4. Painter NS. Diverticulosis of the colon. MS Thesis. University of London: 1962.
5. Painter NS. Diverticular disease of the colon. BMJ. 1968;3(5616):475–9.
6. Painter NS, Almeida AZ, Colebourne KW. Unprocessed bran in the treatment of diverticular disease of the colon. BMJ. 1972;2:137–40.
7. Burkitt DP. Related disease-related cause. Lancet. 1969;2:1229–31.
8. Cummings JH, Engineer A. Denis Burkitt and the origins of the dietary fibre hypothesis. Nutr Res Rev. 2018;31:1–15. https://doi.org/10.1017/S0954422417000117.
9. Walker ARP. Some aspects of nutritional research in South Africa. Nutr Rev. 1956;14:321–4.
10. Wynder EL, Shigematsu T. Environmental factors of cancer of the colon and rectum. Cancer. 1967;20:1520–61.
11. Aries V, Crowther JS, Drasar BS. Bacteria and the aetiology of cancer of the large bowel. Gut. 1969;10:334–5.

12. Banwell JG, Branch W, Cummings JH. The microbial mass in the human large intestine. Gastroenterology. 1981;80(5 Part 2):1104.
13. Burkitt DP, Wright DH. Burkitt's lymphoma. Edinburgh/London: E & S Livingstone; 1970.
14. Yudkin J. Pure, white and deadly: the problem of sugar. London: Davis-Poynter Ltd; 1972.
15. Short AR. The causation of appendicitis. Br J Surg. 1920;8:171–88.
16. Burkitt DP. The challenge of geographical pathology. Pak Med Forum. 1969;4:13–8.
17. Campbell GD. Cleave the colossus and the history of the "saccharine disease" concept. Nutr Health. 1996;11:323–9.
18. Cleave TL. Peptic ulcer. Bristol: John Wright and Sons; 1962.
19. Cleave TL, Campbell GD, Painter NS. Diabetes, coronary thrombosis and the saccharine disease. Bristol: John Wright and Sons Ltd; 1969.
20. Campbell GD. On the causation of diabetes. In: Cleave TL, Campbell GD, Painter NS, editors. Diabetes, coronary thrombosis and the saccharine disease. Bristol: John Wright and Sons Ltd; 1966.
21. Burkitt DP. Epidemiology of cancer of the colon and rectum. Cancer. 1971;28:3–13.
22. Burkitt DP. Related disease-related cause? Lancet. 1969;2(7632):1229–31.
23. Kellock B. The fibre man. Tring, Herts: Lion Publishing; 1985.
24. Reynolds A, Mann JI, Cummings JH, Winter N, Mete E, Te Morenga L. Carbohydrate quality and human health: a series of systematic reviews and meta-analyses. Lancet. 2019;393:434–45.

Chapter 20
A 'Flash of Understanding'

Now fired with enthusiasm by his new belief that fibre in the diet could prevent the major large bowel diseases, Denis was back in Africa in March 1970 to start doing his bowel transit studies and attend an important meeting in Kampala. There he would renew his acquaintance with a former colleague who would bring much needed expertise to the diet concept. The Uganda Medical Association had planned a special meeting to celebrate the centenary of the birth of its most famous doctor, the former medical missionary and founder of Mengo Hospital, Sir Albert Cook. Denis had been invited to speak on what everyone was expecting to be about his lymphoma, which had brought him to worldwide acclaim and put the medical school at Makerere into the forefront of cancer research and treatment. Denis had other ideas. Also invited to the meeting was Hugh Trowell who had been the senior physician at Mulago, a colleague of Denis, but had left hastily in 1958 following the attack on his wife. Trowell had gone on to become an ordained vicar and country parson in the Church of England. He had been both Cook's friend and his personal physician and had felt honoured to be asked to deliver an appreciation of Sir Albert's remarkable achievements [1].[1] A second lecture was to be about his own work on kwashiorkor. Denis and Hugh met before the meeting had started, and Denis, who always carried a camera, asked Hugh if he would go and stand with him in the old Ward 1 where Hugh had shown Denis his first case of jaw tumours. They posed side by side for a photograph and exchanged news. Hugh noticed that Denis was giving a second lecture the next day to students and staff not attending the meeting and asked if it was also about the lymphoma. Denis replied that it was not but instead about the epidemiology of large bowel diseases (Fig. 20.1).

Hugh was intrigued. On his arrival before the meeting, he had toured the wards where he had once been senior physician and paediatrician and to his surprise now found patients who had been operated on for appendicitis or admitted with stroke,

[1] Wellcome Library, Papers of Hugh Trowell, PP/HCT/A.5.

© The Author(s), under exclusive license to Springer Nature Switzerland AG 2022
J. H. Cummings, *Denis Burkitt*, Springer Biographies,
https://doi.org/10.1007/978-3-030-88563-2_20

Fig. 20.1 With Hugh Trowell in Denis' garden, circa 1980

diabetes and coronary heart disease and noted that many of the Africans were obese, unheard of in the 1950s. He listened to Denis' second lecture, which was based on Denis' recent paper in *The Lancet* 'Related Disease-Related Cause' [2] and was notably excited. After the lecture, Denis talked to him about his new work, of meeting Doll and Cleave and the theory he had built up following his discussions with Walker, Painter and many others. Hugh immediately appreciated the significance of what he was doing because the list of diseases that Denis showed in his talk was almost exactly the same as that in Hugh's book *Non-infective Disease in Africa* [3], which he had written on his return to England in 1958. It is a carefully researched and referenced description of the occurrence, nature, pathogenesis and treatment of diseases such as coronary heart disease, diabetes, peptic ulcer, hypertension and bowel and urinary tract disorders. There is no significant discussion of diet or fibre in the book. Hugh was intrigued by the rarity of these conditions in Africa whilst they were, and are, major problems in most industrialised countries, but he had no ready explanation for this. Meeting Denis and hearing him speak at the Cook Centenary Meeting gave Hugh the first real insight into the possible cause of these related diseases, namely a refined diet depleted of fibre. For Denis here was a former colleague and friend with a physicians' knowledge who would now add an important new dimension to the developing dietary fibre hypothesis. Trowell's book had been published 6 years before Cleave's first edition of '…the Saccharine Disease', which emphasised the importance of excess consumption of sugar and refined starchy foods. If Cleave had read Trowell's book, he makes no mention of it in any of his writing.

On returning to England, Denis and Hugh would meet again and together develop the fibre story. The Burkitt-Trowell partnership became well known and was very productive. Neither was convinced about the saccharine hypothesis of Cleave, and they quickly moved away from it. Their first book together, published in 1975, was entitled *Refined Carbohydrate Foods and Disease: Some Implications of Dietary Fibre* [4][2]; the latter phrase in a lower sized font so as not to antagonise Cleave, although they gave full credit to him in the preface. The conditions discussed, chapter by chapter, are a combination of Denis' bowel diseases and Hugh's non-infective diseases. Now the compass of the fibre story (Table 20.1).

Before all this could happen, Denis had to get his bowel transit studies done. On arrival in Uganda, in advance of the Centenary Meeting, he had gone straight to stay with Dick Drown, a long-standing family friend and Chaplain at King's College Budo located a short distance south-west of Kampala. Denis had warned Dick of the purpose of his visit to Uganda and the morning after his arrival gave a talk to the assembled school before breakfast about his proposed transit studies, managing to get 26 students and 5 staff to volunteer. Leaving Budo in a hired Volkswagen, he drove north-east to Amudat on the border with Kenya where he was met by Dr. David Webster who ran the mission hospital. Full of enthusiasm they recruited 30 Pokot[3] patients from around the hospital and to their complete amazement asked them to take the transit study pellets and collect the next three stools they passed. Denis was somewhat concerned that these hospital patients were eating a rather atypical diet of maize meal and asked if he could visit a village well away from

Table 20.1   The 'fibre deficiency diseases'

	Disease	Postulated causative factors associated with a low fibre diet
Large bowel	Cancer	Reduced intestinal content, drier consistency, low stool bulk, delayed transit time, high intraluminal pressures, starved bacterial flora, degraded bile salts
	Diverticular disease	
	Polyps	
	Appendicitis	
	Ulcerative colitis	
Metabolic	Diabetes	High sugar and refined carbohydrate diet. Less likely – Deficiency of polyunsaturated fats
	Obesity	
	Atherosclerosis, coronary disease	
Other	Varicose veins	Economic development
	Dental caries	Sugar and refined carbohydrate (flour)

Diseases likely to have a common cause based on his paper to the Albert Cook Centenary Meeting (Burkitt [5])

---

[2] See Appendix 3 where the Burkitt-Trowell books are listed.

[3] Pokot – the local people who inhabited this region.

Amudat. There would not be time to do proper transit studies, but at least Denis hoped he could get a look at some of the stools, often passed by the roadside in these rural areas, and confirm the nature of their bowel function. Such is the way science progresses, at times. Denis, David Webster and one of the hospital staff drove north as far as they could into the Moroto region, then followed a cattle track on foot through thorn scrub until they came to a manyatta[4] typical of the Karamojong people who inhabit the region. They are a pastoral tribe living mostly on meat, milk and blood from their cattle. Denis and the party from Amudat arrived whilst the men of the village were meeting to discuss topical issues at a baraza and were astonished at such a visit being unannounced. Even more so when Denis produced his camera only to start taking pictures of the local faeces (stools). On return to Amudat, Denis spent a further night with his hosts before returning to Mulago with 90 bags of collected stools from Budo and Amudat, in the back of the car. When x-rayed and transit time calculated, it was shown that the Pokot had rapid transit times to go with their bulky stools but the schoolchildren's times were much slower.

In Nairobi, a few days later to receive his Honorary DSc from the University of East Africa, the first of six he would be given, he met Olive who had just flown in. She and Denis had agreed to spend a week at the coast near Mombasa to celebrate her 50th birthday and enjoy some time together. Such recreation did not come naturally to Denis, but the Shelly Beach Hotel[5] provided superb food, and the two of them enjoyed walking on the beach. Denis read *The Reluctant Missionary* during the first couple of days and started work on a paper about diseases of Western civilisation which would be published later in the year under the title 'Are Our Commonest Killing Diseases Preventable?', one of 11 substantive papers in 1970, only two of which were about the lymphoma. He also took advantage of the time away from the energetic pursuit of his new theory, from travel, which was still mainly in connection with the lymphoma story, and from his duties in the Church and CMF, to do some thinking. The result was a eureka[6] moment. Many years later, in 1983, when he and Hugh Trowell were working on their third book [6] Trowell wrote to Denis and asked him when he had first thought that fibre was the key to the story of Western diseases rather than Cleave's refined carbohydrates. Denis had replied:

Dear Hugh
    I well remember after I had met you in Kampala in 1970 Olive & I went for a few days holiday to Mombasa. I vividly remember that when there it came almost as a flash of under-

---

[4]A tribal village, comprising temporary huts surrounded by a thorn stockade usually with an enclosure for cattle.

[5]Shelly Beach Hotel closed in 1997 because of local violence and has remained closed although is now scheduled for an expensive refurbishment.

[6]Archimedes (287–212 BC) on stepping into his bath suddenly realised that the amount of water displaced from the bath was exactly equal to the volume of his body, meaning that the volume of even the most irregular shapes could be determined by this simple process. He is said to have climbed out of the bath and run down the street shouting 'eureka'.

standing that you cannot have refined carbohydrate without removing the fibre, & that the fibre was the crucial thing not the sugar (Fig. 20.2).

Trowell had an ulterior motive with his question. He had suggested to Denis that he, Trowell, write an introductory chapter to their next book[7] outlining the history of fibre and 'unravelling the tangled story'. By 1973, it become clear to Trowell that Denis was getting the credit for the fibre story and travelling the world to preach the gospel of fibre. So, in the opening chapter by Trowell in their second book (1981), the flash of understanding does not get a mention, but he writes a very good summary of the evolution of ideas about fibre, giving due prominence to his early book on *Non-infective Disease* and his solution to the problem of defining fibre, which Denis had asked him to do.

During his stay with Olive at the Shelly Beach Hotel after the Albert Cook meeting, Denis made a special trip to Kaloleni Hospital about 20 miles north of Mombasa. There was a mission hospital in one of the poorest parts of Kenya run by Dr. David Milton-Thompson, who after service in the navy had become a medical missionary. He had worked initially in China then, after the communist takeover, in Kenya where he would remain for 30 years. Denis had never signed Kaloleni Hospital up

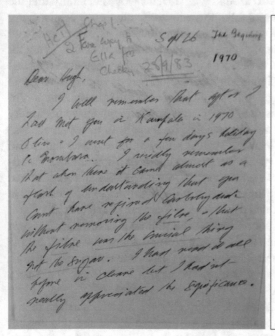

Sept 26

Dear Hugh,

I well remember that after I had met you in Kampala in 1970 Olive & I went for a few days holiday in Mombasa. I vividly remember that when there it came almost as a flash of understanding that you cant have refined carbohydrate without removing the fibre, & that the fibre was the crucial thing not the sugar. I had read it all before in Cleave but hadn't really appreciated the significance.

Note by Hugh Trowell at top right;

'The Beginning

1970'

Other notes at the top in pencil are by Ethel Nelson.

Fig. 20.2 'Flash of understanding' letter from Denis to Hugh Trowell. Dated 'Sept 26' by Denis and some years later, assigned by Hugh Trowell as '1970' when he was starting to write about 'Western diseases' in 1983. Denis relates the story of the inspiration he received whilst on holiday after the Albert Cook Memorial meeting, in 1970 (Wellcome Library PP/HCT/C/2/4/1)

---

[7] They wrote and edited three books together: 1975 *Refined Carbohydrate Foods*; 1981 *Western Diseases*; 1985 *Dietary Fibre*.

to help him with his studies but wanted to see the hospital and its doctor. He made this pilgrimage after meeting David's younger brother, Surgeon Lieutenant Godfrey Milton-Thompson, RN,[8] at St Mark's Hospital in London when he had visited Professor Basil Morson and met Dr. George Misiewicz, who had told him how to do bowel transit studies. Godfrey Milton-Thompson was at the time Honorary Research Fellow at St Mark's doing work on the digestion of cellulose and may have surprised Denis by telling him that cellulose, a key component of fibre, was readily digested by bacteria in the large bowel. Although this had been demonstrated by a number of scientists earlier in the twentieth century, this was likely the first time that Denis became aware that fibre was in no way inert roughage but might be playing an active role in the gut particularly in conjunction with the resident bacteria in the colon – a major concept that would later come to be recognised as a new dimension to normal digestion, bowel function and healthy eating.

Back in England after the 5-week safari cum holiday in Uganda and Kenya, he was pleased to learn that the MRC had confirmed his appointment up to the age of 65, another 6 years, and that, after a prolonged gestation, his book with Dennis Wright about the lymphoma was about to be published. There would be four more overseas trips during the rest of 1970 around which Denis wove his family life, the fulfilment of requests for lectures and talks to various church groups. A preaching appointment in York Minster in October was a high point of the latter, where he spoke about preventive medicine at a service of 'Rededication and Thanksgiving for all persons taking part in the Health Services in the York area'. Equally important for Denis was writing papers for significant medical journals of which 11 would be published in 1971. How he managed to fit so much into his life is never a matter for comment by him, except the occasional 'busy' note in his diary, nor by his family and seemingly was never a point for discussion with those who met him. He was also immensely practical maybe learning such skills from his engineer father and the year he spent at Trinity in this speciality. He had inherited his father's very well-equipped workshop and would tackle almost any job around the house as he had done in Lira in 1945, making artificial legs for polio and accident victims. He avoided electrical work although fitted a new Xpelair fan to the kitchen at The Knoll along with a new door and re-laid floor. During the year he also managed his own car servicing, fitted a new fireplace to the living room, made a new bench for his electric saw, repaired a leaking central heating tank, painted the outside woodwork of the house and enjoyed some heavy work in the garden, (Fig. 20.3). But a very different life from that in Mulago.

In the garden he found a new assistant, Philip Howard, who had arrived on the scene as Judy's boyfriend. In a similar vein to remarks about the relationship between Cas and Andrew, Denis wrote, 'Judy out with Philip Howard. A nice boy, very fond of Judy, but not a Christian. May she be given wisdom and strength'.

---

[8] Surgeon Vice Admiral Sir Godfrey Milton-Thompson (1930–2012) who was, as part of a distinguished medical and naval career, Professor of Naval Medicine and between 1982 and 1984 Deputy Medical Director General of the Armed Forces with special responsibility for the Navy during the Falklands conflict.

Fig. 20.3 Denis in his workshop at The Old House, Bussage, into which he and Olive moved on October 13, 1978

Despite this, to Denis, serious weakness in Philip's character, the relationship between him and Judy flourished, and they would marry in 1971. Perhaps their union was destined to happen related as they were through the ornithological prowess of their forbears. Judy was the granddaughter of James Parsons Burkitt who, on account of his studies of the robin in his large garden in the early 1900s, had been named by David Lack as one of the seven most distinguished pioneers into the life of birds. Included amongst the other six is Eliot Howard, a noted amateur ornithologist and author of *Territory in Bird Life*.[9] Eliot was of a similar age to James and, whilst director of the family steelworks in Worcester, spent much of his time visiting the coast of Donegal in the north-west of Ireland to study the natural history of the area and had married a local girl, Anne Stewart, daughter of a well-established Donegal family. It seems highly likely, given their overlapping interests and reputations, and the proximity of Donegal and Enniskillen that James would have met Eliot Howard. He would have had no inkling that his granddaughter would marry a distant cousin of Eliot.

---

[9] Howard HE. Territory in Bird Life. London: John Murray; 1920. Henry Eliot Howard (1873–1940) was born into a wealthy family who owned the Lloyd and Lloyd steelworks in Worcester. He was educated at Eton, then went into the family business but had a lifelong interest in birds and became a well-known amateur ornithologist who was first to describe the territorial behaviour of birds. He wrote four other books about birds including a two-volume treatise on *The British Warblers* and, in addition, *An Introduction to the Study of Bird Behaviour*, *The Nature of a Bird's World* and *A Waterhen's World*.

The high-level interest in birds and their behaviour was not passed down the generations, but the Howards were distinguished in other ways.[10,11] After Marlborough and Oxford, where he studied geography, Philip decided to teach first at Windsor Grammar School, then at Bloxham School near Banbury where he met Judy. After their marriage, they both taught in Nairobi, but on return to England, Philip added to the family's honour by becoming one of the country's leading landscape gardeners.[12] In Philip Denis found a companion with whom he could share practical things and with whom he could entrust the future of his eldest and much-loved daughter Judy.

A trip to the USA in May 1970 took Denis to Washington, Chicago, Grand Rapids, staying once more with the Blocksmas, this time in their lakeside cottage, then back to Chicago where he met, for the first time since his marriage in 1943, Walter Thompson who had been his best man. In Houston at a cancer congress he was given a copy of Glemser's book [7] about the long safari, published in the USA as *Mr Burkitt in Africa*, which title he described as awful. It was the first of three books that would be written about Denis and his work. Then to New Orleans and Tulane Medical School to deliver the 13th lecture of his tour. Back in England for 2 weeks before departing once again for Africa, there were meetings with Cleave and Alan Lindsell of the MRC and a gathering for 70 people at The Knoll organised by Cas. Still not a party animal, he felt he was a great failure during the evening, which finished at 01.30 h and wrote to her the next day apologising for his behaviour.

Despite feeling unwell, Denis departed for Durban on June 17 to talk to GD Campbell about diabetes and his wife, Magda, who told him that she was having trouble with George. Denis tried unsuccessfully to build bridges between them and blamed George for being rather selfish. On to a meeting in Johannesburg where he gave two talks and then had meetings with Alec Walker, which he described as very valuable. He refers to Walker rather formally as ARP Walker yet he had met him on several previous visits to the South African Institute for Medical Research and had learnt about bowel behaviour and transit time measurement from him. They would go on to publish seven papers together, and Alec with his wife, Betty, would visit the Burkitts at home and stay with them. Hearing Christiaan Barnard lecture about his success with heart transplantation, he was able to have a discussion with him about the epidemiology of coronary heart disease, which apparently turned more into an argument about prevention. The two men, both distinguished in their own speciality, were at opposite ends of the broad spectrum of healthcare services across the world. Home via Bloemfontein, Nairobi and Mulago, Denis then enjoyed 3 months

---

[10] Luke Howard (1772–1864) was a pharmacist by profession who set up a successful chemical manufacturing business, initially Allen and Howard, then Howard and Sons.

[11] Luke Howard's interest in clouds started when he was still at school in 1783. At this time, there were volcanic eruptions in Iceland and Japan that created dramatic skies and cloud formations. At the age of 11, this fascinated him and was the start of a lifelong study of the clouds earning him a place in history. Sotheby's Auction Catalogue, Treasures, London July 6, 2016, p. 154.

[12] His company Graduate Gardeners Ltd. won six gold medals at the Chelsea Flower Show between 1974 and the early 1990s.

without any major travel and was able to catch up with many people including Michael Hutt who was now back from Uganda and had taken up the position of Chair of Geographical Pathology at St Thomas's Hospital Medical School. A trip to Slad near Stroud to Philip's home to meet his parents went well with Denis describing them as nice people in a lovely home, thus paving the way for Philip to ask for Judy's hand in marriage a few weeks later. 'Of course I agreed'.

As the weeks without travel went by, Denis decided to write another paper for *The Lancet* [5] in which he would take the fibre story beyond bowel disorders to incorporate the whole gamut of Western diseases. He was becoming increasingly sure they were all diet related. The paper entitled 'Relationship as a Clue to Causation' is very Burkitt. He uses evidence from the cause of acute infections, Charles Darwin's writings, his own experience with the lymphoma and his discussions with people around the world to make the point that if two or more disorders occur together in the same person and have a distinct geographical distribution, then they are likely to have a common cause – a concept he had learnt from Cleave, but now with a somewhat different group of conditions. He includes in his theory not only those diseases of the bowel but those of the cardiovascular system, diabetes and obesity. He lays the cause of bowel diseases as being directly due to diet and particularly those characterised by low fibre, although his lack of understanding of exactly what is dietary fibre is reflected in his use of several different terms to describe it, i.e. high- or low-residue diet, unabsorbable cellulose, unabsorbable fibre, indigestible fibre, dietary cellulose and just, fibre. In his discussion of the cause of bowel diseases, he credits Walker, Morson, Wynder, Dimock and Oettlé, all of whom he had met during the year whilst Campbell, who was committed to the theory that diabetes was somehow related to sugar, was a big influence in Denis' thinking about the metabolic diseases.

October 21 saw Denis heading for Kabul at the start of a journey of astonishing intensity given that he was alone and largely unaided in its organisation. How the itinerary was put together is not clear but must have been initiated by him being invited to one or two conferences following which the word would have been passed around that 'Burkitt is coming', so let us ask him to come and speak to us. Invitations he always found hard to refuse. Around all this he programmed visits to some of his mission hospital contacts. The whole safari was done on a limited budget with Denis travelling alone by the most economical method and cheapest class and staying mostly with his hosts to avoid hotel costs. The journey went from Kabul to Jalalabad in East Afghanistan, then into Pakistan and India stopping at Peshawar, Rawalpindi, Islamabad, Karachi, Bombay, Pune (Poona), then back to Bombay and on to Vellore in South India for a CMA conference and visit to a leprosy centre. Then, as detailed in his diary for the week commencing Monday, November 2, 1970:

Monday. Left Vellore 7. Taxi broke down. Got bus then another taxi. In time by 15 mins Madras airport. To Nagpur. Night in hotel.

Tuesday. Drove Nagpur to Achalpur, Howard's[13] Hospital. Then to Basim. Hot flat
  country.
Wednesday. Conference at Basim. Then drove 170 miles to Nagpur. Some rest in
  rest room. 2.30 am plane. Delhi.
Thursday. Morning at All India Institute for Medical Research. Early to bed.
Friday. Up 4.45. Flew 6.15 to Benares (Varanasi) via Lucknow and Allahabad.
  Drove to Kachua. Stayed with John and Evelyn Hobson.
Saturday. Spoke three times at medical conference. Then drove to Benares and
  stayed with John and Erika Bentley.

He spent time to see the Ganges on the Sunday before heading north-west close
to the border with Pakistan and the Himalayas for a brief stop at Amritsar. Driving
to nearby Ludhiana, he met physicians and pathologists in the morning and lectured
to surgeons in the afternoon. Next, 60 miles south-east, came Chandigarh with two
more lectures and a formal dinner. Arising at 5 am the next day, he was driven to
Delhi in time to give two more lectures and, finally, only 19 days after leaving
England was able to return home.

On this safari, Denis had fulfilled around 24 pre-arranged engagements, been
introduced to countless people, given at least 17 lectures and followed a punishing
schedule of travel on his own. Approaching his 60th birthday and with the African
childhood lymphoma acknowledged worldwide to be his discovery, he would have
been thoroughly justified in allowing himself more time with his family, church
activities, reading, photography and doing jobs in the house and garden. His health
was good, although always something that worried him. He recorded 13 attacks of
migraine during 1970. But he was now in pursuit of the solution to a problem,
Western diseases, that had far greater implications for health than had the childhood
cancer, and he was increasingly sure that he knew what the answer was, namely diet
and specifically fibre. He still had much to learn about this, but over the next few
months, he would become convinced that he had the evidence to show that lack of
fibre was the cause of bowel cancer and would report this to the world.

Pleased as always to be back home, he was met at the airport by Olive and Cas
and started to catch up with the paperwork in his office. With speaking engagements
to fulfil and meetings with people arranged, he was busy but was telephoned a few
days later to say he had been awarded a prize in the USA and would he come to the
award ceremony? This was the Robert de Villiers Award of The American Leukaemia
Society, and the ceremony was to be in San Juan, Puerto Rico, 10 days later. He
went, grand lecture in hand but felt lonely at a meeting of haematologists. He was,
in his mind and with his work, now moving away from the lymphoma and in noting
in his diary at the end of the year what had been the highlights for him it was clear
that the fibre story was central to his thinking. He was a man with a new message,
this time for all the world.

---

[13] Possibly refers to Howard Somervell (1890–1975), a multitalented surgeon, who was part of the
1922 and 1924 expeditions to Everest. He was a talented sketcher and painter and worked mainly
at the Christian Medical College in Vellore, India.

References

1. Elizabeth Bray, Hugh Trowell: Pioneer Nutritionist 1904–1989 Chapters 31 and 32.
2. Burkitt DP. Related disease-related cause? Lancet. 1969;2(7632):1229–31.
3. Trowell HC. Non-infective disease in Africa. London: Edward Arnold; 1960.
4. Burkitt DP, Trowell HC. Refined carbohydrate foods and disease: some implications of dietary fibre. London: Academic Press; 1975.
5. Burkitt DP. Relationship as a clue to causation. Lancet. 1970;2(7685):1237–40.
6. Trowell HC, Burkitt DP, Heaton KW, editors. Dietary fibre, fibre-depleted foods and disease. London: Academic; 1985.
7. Glemser B. The long safari. London: The Bodley Head; 1970.

Chapter 21
The Gospel According to Burkitt

Arriving in San Diego for the Cancer Conference early in 1971, to which he had been invited by Dr. Grant a year earlier, Denis' brief was to talk, for the first time, about bowel cancer. It would mark another turning point in his life. The lecture was well received with many scientists attending the conference approaching him afterwards not just to meet him but also to tell him about their own work. These conversations continued into the next day causing Denis to remind himself of the need to be humble when in receipt of such adulation. It was an important meeting for Denis giving him immediate confirmation that people were going to take his new work seriously. He had taken the first step to becoming known as the Fibre Man. After another week in which he visited Jackson, Louisville, Philadelphia, Chicago, Boston, Washington and finally Bermuda, now giving his bowel cancer talk to audiences who were apparently enthralled, he arrived home reassured that he was heading in the right direction and that this was a big story.

His first priority back in his office was to write the paper based on his lecture in San Diego, at the invitation of the journal *Cancer*, the official publication of the American Cancer Society and one of the highest-ranking journals in the field of oncology (cancer). It was the same journal that had published his key paper on the lymphoma with Greg O'Conor 10 years earlier. He spent some time in the first weeks of the year writing and revising the paper, asking the opinion of his colleagues and telling himself that if it was rejected, he would still be able to try another journal. He was clearly nervous and apprehensive about how the scientific world would view his theory for the cause of another cancer and felt much more isolated in this than he had been when working on the lymphoma in Africa where, as a hands-on surgeon, he had personally looked after the young children. Richard Doll must have been a steadying influence and unrivalled source about cancer epidemiology, but to suggest that our diet might be implicated in the cause, and more importantly the prevention, of one of the commonest cancers in the Western world and how that might come about, was brave and challenging. The world of cancer research was focussed mainly on the cellular and molecular mechanisms involved,

© The Author(s), under exclusive license to Springer Nature Switzerland AG 2022
J. H. Cummings, *Denis Burkitt*, Springer Biographies,
https://doi.org/10.1007/978-3-030-88563-2_21

on environmental hazards such as radiation and toxins and of course viruses, about which Denis was happy to debate.

Denis sent the paper off to *Cancer* in mid-March, and it was published in the July 1971 issue of the journal under the title 'Epidemiology of Cancer of the Colon and Rectum' [1]. It details the incidence of colorectal (large bowel and rectum) cancer worldwide, which data he had obtained from Richard Doll not his Mission Hospitals, showing major variations ranging from 3 to 6 cases/100,000 per year in various African countries to 40–50 in the USA, Scotland and parts of South America.[1] A more than tenfold variation in risk. Denis then summarises the evidence for an environmental agent being important. He writes that when people move from a low-risk area to one of high risk, such as the Japanese moving to the west coast of the USA or Africans similarly to the USA, then as the dietary customs of their country of adoption are accepted, so, by the second generation of immigrants, does the risk of bowel cancer increase.

Denis makes the point that what is most likely to determine the health of the lining of the large bowel, where this cancer arises, is the composition and activity of the contents of the bowel and especially the interaction of the resident bacteria with dietary residues. For someone with no training in either microbiology, nutrition or gut physiology, this shows remarkable insight. He follows this with the results of his studies of stool bulk and bowel transit time that he had done in Uganda and Shiplake which revealed marked contrasts between the rapid transit of the African villagers and their greater daily stool weight when compared with that of the children from the English boarding school. He discusses various possible mechanisms and presents a diagram to represent the relation between diet, a possible series of events and bowel cancer. He cites the abundant evidence in the literature for the effect of fibre, mainly bran at this time, on bowel habit[2] and concludes that the close relationship between bowel cancer and the refined diet characteristic of economic development 'suggests that the removal of dietary fibre may be a causative factor'.

The paper was a great success and is another of the iconic papers in the fibre story. It became Denis' most quoted work, surprisingly being referred to in medical and scientific publications four times more often than his most cited paper on the lymphoma and was a Citation Classic 1981 [2].[3] In his response to the Citation award, published in *Current Contents* the summary of all knowledge for those wanting to keep abreast of the burgeoning scientific literature of the time, Denis made the point that the geography of bowel cancer, then the commonest cancer in North America, pointed to an environmental cause. This was likely to be lack of dietary fibre, which he correctly surmised had a fundamental influence on large bowel function as measured by transit time. He pays tribute to Cleave for introducing him to

---

[1] Age-standardised incidence/100,000 for cancer of the colon and rectum in men aged 35–64 years.

[2] Bowel habit. Usually refers to the frequency of defaecation, consistency of stool and ease of passage.

[3] A paper that became a classic in its field and selected from amongst the most highly cited papers from previous years. The Epstein and Barr Lancet paper on the isolation of a virus from Burkitt lymphoma cells had already been selected as a Citation Classic in 1978.

the important role of refined carbohydrate foods but credits Alec Walker for calling his attention to fibre. 'The credit which has often been attributed to me for originating the fibre hypothesis for this disease, should by right go to him' [3]. Humble as ever but in this case, he was paying tribute to a man who never received the recognition he deserved for his early work on diet and disease, nor did he ever seek it.

Is it a good paper? Today, most journal editors would have turned it down without even sending it out for review. The questionnaire that Denis had posted to his mission hospital colleagues was not validated, and response rates are not given. Answers are entirely qualitative. There is no proper statistical analysis of the data. The most advanced calculation is a mean (average) not even accompanied by a standard deviation even less a test of significance and P value. The bowel transit studies were done in a small number of adults and children and then extrapolated to the world, whilst there are no measurements of dietary intake. But it is well referenced and cogently argued. Importantly, it combines epidemiological observations, however anecdotal, with the development of a hypothesis about the possible cause of bowel cancer, which he then puts to the test with his transit studies.

This was the era when studies of the environmental causes of disease were dominated by the views of Austin Bradford Hill who laid down criteria for the likelihood of an association between two variables in a study pointing to causation. These were, and still are, the strength of the association, consistency across all studies, temporal relationships, etc. [4].[4] Sixth and eighth in his list, and often ignored because of today's compartmentalisation of medical science, are two very important criteria, biological plausibility and supporting experimental studies. In other words, does the idea make sense in the light of what is known about how the body works, and has the proposal been tested by experiment, preferably in human studies? Denis addressed both of these in his paper in that he proposed a working theory, based on his discussions during the previous year with many leading scientists in the areas of gut microbiology, cancer epidemiology and diet, which he then tested in his transit studies. He does not recount ever hearing Bradford Hill speak although may have read his paper in the 1965 *Journal of the Royal Society of Medicine*.

Fifty years later, is the Burkitt hypothesis for the prevention of bowel cancer still credible? Most certainly, yes, although there are other dietary and genetic factors that are important. But Denis should be given credit for his suggestion that diet could be involved in both the cause and prevention of this cancer. For bowel cancer, the role of the gut bacteria and the effect on transit time are crucial. If your diet is high in fibre, comprising whole grain cereals, lightly processed fruits, vegetables, pulses and legumes, then you are less likely to get bowel cancer [5]. It is an original and cogent piece of work showing a remarkable understanding of large bowel physiology and microbiology and fulfils criteria one, two, six and eight of Bradford Hill.[5] All this only 5 years after he had left Africa.

---

[4] Bradford Hill's nine aspects of the association between two variables that are important to consider when interpreting likely causation are strength, consistency, specificity, temporality, biological gradient, plausibility, coherence, experimental evidence and analogy.

[5] Strength, Consistency, Plausibility, Experiment.

The fibre story, about which Denis was now very enthusiastic, was much bigger than just bowel cancer. During 1971, he wrote papers for the medical and scientific press not only on this cancer and about the lymphoma but cancer in general, diverticular disease and appendicitis. These would soon be followed by similar articles on varicose veins, hiatus hernia, gallstones, deep vein thrombosis and haemorrhoids. From where did the inspiration to follow these leads come? From the people he met or already knew. Principal amongst them and the person most responsible for widening the fibre story beyond the gut and the surgical conditions that Denis understood well was Hugh Trowell. Having retired from his duties as Vicar of Stratford-Cum-Castle in Wiltshire, Trowell was a frequent visitor to Denis' office in London and with Peggy his wife, now showing the early signs of dementia, to Burkitt's house in Shiplake. They had renewed their friendship, based on their shared faith and professional life as doctors in Africa, when they met in Kampala at the centenary celebration of the birth of Sir Alfred Cook. It was at this meeting that Trowell had opened Denis' eyes to the likelihood that physicianly disorders such as cardiovascular disease, diabetes and obesity were also part of the fibre story given their rarity in Africa and similar geographical distribution to Denis' surgical conditions of the bowel. Trowell was energetic in his pursuit of the evidence for his ideas, had already written several books and was a born editor. They had complimentary knowledge and skills, got on well and both were beginning to realise that the fibre story could overturn established ideas on the prevention and treatment of many common diseases, if they could convince the medical world of its veracity. They agreed to write a book together under the title *Refined Carbohydrate Foods and Disease*, which was published in 1975, and whilst paying homage to Cleave moved fibre to centre stage in the story. On reading it, Cleave would be furious.

Another regular correspondent with Denis was Neil Painter who, encouraged by Cleave, had run a clinical trial of bran in the management of diverticular disease [6]. The study started in December 1967 and ended in March 1971. He recruited 70 patients from his clinic at Manor Hospital in London who were told to eat high-fibre breakfast cereals such as Weetabix, wholemeal bread and fruits and vegetables, to reduce their sugar intake and take two teaspoons of bran three times a day. At the end of the trial, a questionnaire was sent to the patients, and some were also interviewed. The diet and bran intervention improved bowel habit and relieved abdominal pain and distension. A success. Sixty-two of the patients had continued with the trial for an average of 22 months. As was the case with Denis' paper on bowel cancer, there was no statistical evaluation of the data by Painter and colleagues, and it is easy to criticise the design of the study. It lacked a control group. The patients were asked to make several dietary changes, and there was no attempt to measure what they ate, thus making it difficult to ascribe any improvement solely to fibre. The study was not blinded in any way in that the authors both conducted it and assessed the response of the patients themselves. When submitted to the *British Medical Journal* for publication, it was sent to Richard Doll for his opinion. He acknowledged that the study had many faults but recommended it be published because he realised its novelty and that the outcome, improving the patients' symptoms, was likely to be due to the intervention.[6] This study and Painter's earlier work on pressures in the bowel led to a significant change in the management of symptoms in diverticular disease, which continues today, although good clinical trials are few. The work described by Painter in the paper is another of the iconic events in the fibre story and for Denis added weight, through experiment, to his general theory about fibre and the prevention of bowel diseases.

Also influential in Denis' thinking at this time was Dr. Ted Dimmock whom he had met at Mengo Hospital in Kampala where he was helping out as he had done at various mission hospitals. In 1936, Dimmock had worked on the benefits of bran in treating constipation in his MD thesis for the University of Cambridge. Whilst the subject was somewhat mundane, the background was one of the common uses of purgatives, which practice by the medical profession he found to be deplorable. Dimmock's understanding of the role of fibre in the gut was very much state of the art. He summarised his work in a paper in the *British Medical Journal* in 1937 [7] in which he described his treatment trial of 121 patients with constipation in 90% of whom he was able to restore normal bowel habit with bran. Although the work was thoroughly carried out this was not an original observation, but Dimmock writes about the causes of constipation observing that it is more frequent in women than men, especially for women at the menopause, can be due to lack of physical activity and unsuitable diets. He emphasises that prevention is the best strategy and, correctly notes, that fibre from vegetable foodstuffs has a lesser effect on bowel habit than bran because 'the fibre of green vegetables and ordinary foodstuffs is more readily broken down in the alimentary tract than that of wheat bran'. The emphasis on prevention spoke volumes to Denis as did the importance of a good bowel habit as a sign of a healthy diet and lifestyle. That fibre from different dietary sources might have contrasting effects in the gut and on other aspects of metabolism did not register with him because he still did not really understand anything about the nature of fibre. He would eventually ask Hugh Trowell to look into this for the book they were writing, but it would be 40 years after Dimmock before work would recommence on the chemistry and physiology of different fibre sources [8] stimulated then by the emergence of the fibre hypothesis of Denis and colleagues.

Calling in to see Denis at his London office early in 1971 was Dr. Kenneth Heaton [9][7] who brought gallstones into the story. Denis had met Ken Heaton earlier on a visit to Bristol at the invitation of Tony Epstein, by now Professor of Pathology at the Medical School, to talk about his new interest, fibre. Heaton was a gastroenterologist who had become interested in the physiology of bile when early in his research career he had spent time in the USA at Duke University, North Carolina, with Leon Lack and MP Tyor. He was amongst the first to show that the ileum (final part of the small bowel) was crucial to the circulation of bile and to understanding

---

[6] Professor Sir Richard Doll, University of Oxford, personal communication, circa 1972.

[7] Dr. Kenneth W Heaton, DSc, FRCP (1936–2013), graduated with first-class honours from Cambridge and went on to become a gastroenterologist in Bristol. He had a distinguished career working initially on the cause of gallstones and then on large bowel disorders. He was amongst the first to study the effect of fibre on bile and its components.

Fig. 21.1 Dr. Ken Heaton
(1936–2013), Reader in
Medicine at Bristol. (By
kind permission of Sue
Heaton)

our susceptibility to gallstone formation. He was a disciple of Cleave and soon
became interested in the role of diet in determining the composition of bile. He
showed that slow transit predisposes to gallstones and that the mechanism was
through the formation of deoxycholic acid, now known as the Bristol hypothesis.
When the Royal College of Physicians convened a Working Party to report on the
'Medical Aspects of Dietary Fibre' in 1980, Ken Heaton was appointed its Honorary
Secretary [10]. For Denis, it was enough that fibre was involved, never mind how,
and that the diet he was promoting would prevent not only bowel diseases but also
gallstones. It fitted with his African experience. He had rarely operated on a patient
with gallstones and when he did it would have been a member of the ex-patriot community in Kampala. There was no enquiry about gallstones in any of the question-
naires Denis drew up for his geographical pathology studies. This was a disease of
what he was now calling 'Refined Carbohydrate Foods' in obeisance to Cleave and
would merit a chapter by Heaton on the subject in the forthcoming book with
Trowell (Fig. 21.1).

Later in the year, when he was in Edinburgh attending a British Cancer Council
Conference, Denis met Martin Eastwood who, like Ken Heaton, was a gastroenterologist with an interest in diet and would bring a further dimension to the fibre
hypothesis. As a young doctor, Eastwood had gone to work in the Department of
Biochemistry of the Edinburgh Medical School in the laboratory of Professor GS
Boyd whose interest was in bile acids and in cholesterol metabolism. Eastwood
started work with Boyd by showing that bile salts[8] in the lower gut were bound to a

---

[8] Bile salts/acids are made in the liver and secreted into the bile, which then passes into the upper
gut during meals. They are important for the digestion of fat.

non-absorbable material he thought might be bacteria. Eastwood had recently visited a local brewery and been impressed that the residual material in the mash tun[9] looked like large bowel contents, an opinion he presumably kept to himself at the time. He used this residue in a laboratory experiment to show that binding was in fact to the vegetable fibrous material that remained after the malting of the maize and barley [11], essentially to fibre. He went on to demonstrate that this binding was to lignin which is the component of fibre that knits the constituents together to give structure to the plant cell wall. Eastwood followed up his early work with clinical studies of the effects of fibre on blood cholesterol, on bowel function and on diverticular disease but realised that the importance of lignin to the fibre story was much more fundamental. As had been clear to scientists working on the digestibility of animal feeds for over 150 years, the role of fibre was not just to add cellulose, pectin and other carbohydrate polymers to the diet to provide energy for the bowel bacteria. These molecules, which are long and at times highly branched chains of sugars, are, with lignin, forming fibres that give the plant cell wall its structure. The cell wall encloses nutrients such as starch, protein, essential fats, vitamins and minerals so the nature and amount of fibre has an important effect on the digestion and nutritional value of the feed. To the animal nutritionist, the fibre content of feeds, especially highly lignified cereal fibre, would impair digestion, but for human nutrition, Eastwood proposed that fibre present in fruits, vegetables, cereals, legumes and pulses gave the food its physical properties. This meant we have to chew it, and that in the gut water would form gels with it, attract ions such as calcium to bind to it and affect the rate and efficiency with which nutrients from the diet were absorbed. Lower down the bowel, bacteria could live on the surface of these cell walls and break them down, thereby moderating bowel function [12].

For Denis, this was fascinating. Whilst not fully understanding the work that Eastwood was doing, he realised that fibre could, through its physical nature, affect cholesterol metabolism, which he had delegated to Hugh Trowell to work on for their book. Fibre also promoted good bowel function, essential to the bowel health story he was now preaching to all the world. The wonders of fibre were growing by the day. The physical properties proved difficult to investigate further, but the importance of gels forming in the gut in moderating the absorption of glucose and insulin responses was exploited for the benefit of diabetics [13, 14]. The role of the cell wall was taken up by Ken Heaton in a classic study of apples and the blood glucose responses [15]. With his group in Bristol, he gathered ten healthy subjects and fed them three different test meals all based on apples. One was whole normal apples, the second the same amount but pureed and third the juice, each test meal having the same content of carbohydrate. The removal of the fibre to produce juice led to the subjects feeling much hungrier after the meal, with higher initial insulin responses leading to lower blood glucose at 90 min, known as rebound hypoglycaemia. A clear demonstration of the potential of fibre, when present naturally as the plant cell

---

[9] Mash tun. A large vessel used in the brewing of beer in which the malted and ground barley is mixed with water.

Fig. 21.2 One of Denis'
lecture slides outlining the
way in which fibre
('roughage')
**might affect bowel
function and so prevent a
number of large bowel
diseases. (Wellcome
Library WTI/DPB/C/8)**

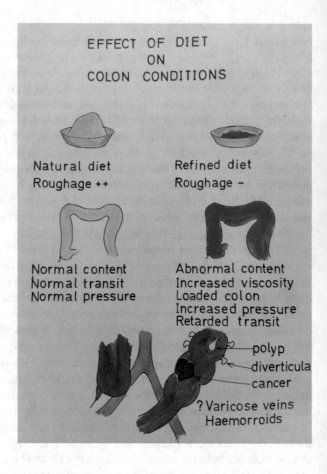

walls of whole foods, to moderate digestion and metabolism in what would be considered a healthy way (Fig. 21.2).

Denis was delighted with this. Everything he learnt and heard about fibre was good. He was no longer doing experimental work himself now, apart from some final bowel transit studies at the local St Christopher's School, but was building up his fibre theory through his meetings with people who were much more accessible from his base in London than from Africa. Notable amongst these acquaintances was Sir Ernst Chain[10] who with Howard Florey at Oxford had worked out the chemistry of penicillin and its antibacterial action, for which they shared the 1945 Nobel Prize in Medicine with Alexander Fleming. Chain had become interested in fibre after hearing Denis speak and through his chairmanship of the Scientific Committee of the British Nutrition Foundation contacted him to discuss the possible use of bagasse, the residue left after the refining of sugar cane, which Beecham Research Laboratories were interested in as a possible cholesterol-lowering agent. Denis was

---

[10] Sir Ernest B Chain (1906–1979), Founding Professor of Biochemistry at Imperial College London.

clearly flattered by the interest of so distinguished a scientist in his work. They met four times, on the final occasion Denis bringing along Hugh Trowell because much of the discussion was too detailed in relation to what work might be done and the mechanisms involved in cholesterol lowering by bagasse. The action was thought to be due to bile acid binding to components of the fibre, initially lignin, as this is present in very high amounts in sugar cane fibre. For Denis, it was the big picture that mattered and here was another fibre-containing dietary ingredient, perhaps rather obliquely so, that might benefit health. Chain pursued the bagasse idea actively with experiments in rats [16]. The work was taken up by Dr. Ian McLean Baird[11] and a distinguished team of microbiologists and nutritional scientists at the West Middlesex Hospital in London who did a series of controlled feeding studies in human volunteers [17, 18]. Whilst the bagasse caused the expected increases in stool output due to its fibre content and increased faecal bile acid excretion, there were no changes in blood cholesterol and so bagasse slipped quietly from the diet pages of the Sunday magazines.

The year 1971 included three safaris, to East and South Africa, Ethiopia, Iran and India, which kept Denis away from home for 11 weeks. He continued to gather data on the geography of diseases and wrote a compelling paper on the cause of appendicitis for the *British Journal of Surgery* in which he suggested that the change from a high- to low-residue (fibre) diet was largely responsible for its high prevalence in Western countries [19]. At the end of the paper, he suggests that the undue refining of dietary carbohydrate and increased sugar consumption might be important. This was clearly a nod to Cleave who was increasingly unhappy about Denis' modifying his refined diet/sugar hypothesis and moving to fibre being the key. They talked regularly on the telephone, engaging in long arguments about the evidence. Denis did not like this at all having spent his life avoiding conflict with people wherever possible. As Denis moved on, looking at the causes of venous disorders such as thrombosis and varicose veins, his ideas diverged even more from those of Cleave. Trying to keep Cleave on board proved increasingly difficult. 'Upset Peter Cleave totally sending him references on thrombosis. He put down the telephone in the middle of the conversation. I hate upsetting people'.[12] Denis' working colleagues were now no longer Wright, Burchenal, Oettgen, Harris, Epstein, but Richard Doll, Hugh Trowell, Neil Painter, Alec Walker and Peter Cleave together with Paula Cook and Michael Hutt at St Thomas's Hospital Medical School, London. Unlike with the lymphoma where Denis had ownership of the story, he never claimed this for fibre. He became much more collegiate, and his papers from now on no longer had him as the sole author but were co-authored with his inner circle of colleagues, except for Cleave, with whom he never wrote anything.

---

[11] Dr. Ian McLean Baird (1923–2014), gastroenterologist who trained at St Andrews and Manchester Medical Schools, then settled at the West Middlesex Hospital in London. He had a strong interest in nutrition and worked on the role of low-calorie diets for obesity, on iron metabolism and on fibre and diverticular disease.

[12] DPB diary, December 8, 1971.

There can be little doubt that were it not for Denis' success with the lymphoma the impact of his work on fibre would have been much less. He acknowledged this readily saying,[13] 'I had one enormous advantage that had been denied Cleave and others. Because of my previous work in the field of cancer I was allowed to speak when others were not'. Had there been no Burkitt's lymphoma, then when Burkitt was inspired by Cleave, he would have been considered just another food crank. This truth was illustrated well when he was invited to address a meeting of the Edinburgh Medical Society, the oldest student medical body in the world. This is an occasion when the speaker and office bearers wear dinner jackets. Olive accompanied him, and after the meeting, a student spoke to her, not knowing who she was. He remarked, 'That was a good talk but no-one except Burkitt would have been allowed to say that'. Denis got the message.

His reputation continued to grow and with it his status as a pioneer in the causes of cancer, always a big story. Now with fibre, he made the world start to think that prevention was as important as treatment, which in the case of bowel cancer was through simple modifications to diet. In 1971, he was awarded the Walker Prize of the Royal College of Surgeons of England, and Richard Doll told him that he and Professor Michael Stoker, President of the ICRF from 1968 to 1979, were intending to propose him for Fellowship of the Royal Society, regarded as the highest honour that can be given to a living scientist in the UK. The Fellowship was awarded the following year with the citation [20]:

> As a practising surgeon he recognized the characteristics of a new syndrome in African children which is now known as Burkitt's lymphoma. By personal investigations throughout the world he demonstrated that the tumour occurred commonly only under specified climatic conditions in Africa and New Guinea and has subsequently been responsible for investigations into its possible association with virus infections and holo-endemic malaria. He demonstrated the responsiveness of the tumour to chemotherapy and that cure could be achieved by small doses. He has shown that many diseases vary unexpectedly in incidence over short distances and that a wide variety of diseases vary in incidence in the same way between underdeveloped and developed countries. Some of these may be associated with a low roughage diet. His studies have been responsible for renewed world interest in the geographical distribution of cancer and other apparently non-infectious diseases.

Thus, lymphoma led the way, but the Royal Society acknowledges his pioneering work in geographical pathology and even at this very early stage in its development, fibre. With this accolade, his billing as a speaker almost invariably attracted expectant audiences. What they now heard about was the gospel of dietary fibre according to Burkitt.

---

[13] Wellcome Library, Papers of Denis Burkitt, WTI/DPB/F/1 Autobiography, Chap. 13, p. 23.

References

1. Burkitt DP. Epidemiology of cancer of the colon and rectum. Cancer. 1971;28:3–13.
2. Garfield E. Introducing citation classics; the human side of scientific reports. Curr Contents. 1977;1:5–7.
3. Burkitt DP. Citation classic – a sarcoma involving the jaws in African children. Curr Contents/ Life Sci. 1981;21:20.
4. Hill AB. The environment and disease: association or causation? Proc R Soc Med. 1965;58:295–300.
5. World Cancer Research Fund and American Institute for Cancer Research. Diet, nutrition, physical activity and colorectal cancer. London: World Cancer Research Fund International; 2017. www.wcrf.org
6. Painter NS, Almeida AZ, Colebourne KW. Unprocessed bran in the treatment of diverticular disease. BMJ. 1972;2:137–40.
7. Dimmock EM. The prevention of constipation. BMJ. 1937;1(3982):906–9.
8. Cummings JH, Southgate DAT, Branch W, Houston WH, Jenkins DJA, James WPT. The colonic response to dietary fibre from carrot, cabbage, apple, bran and guar gum. Lancet. 1978;1:5–9.
9. Pomare EW, Heaton KW. Alteration of bile-salt metabolism by dietary fibre (bran). BMJ. 1973;4(5887):262–4.
10. Royal College of Physicians of London. Medical aspects of dietary fibre. A report of the Royal College of Physicians. London: Pitman Medical; 1980.
11. Eastwood MA, Hamilton D. Studies on the adsorption of bile salts to non-absorbed components of diet. BBA. 1968;152:165–73.
12. Eastwood MA. Vegetable fibre: its physical properties. Proc Nutr Soc. 1973;32:137–43.
13. Jenkins DJA, Hockaday DT, Howarth R, Apling EC, Wolever TMS, Leeds AR, et al. Treatment of diabetes with guar gum – reduction of urinary glucose loss in diabetics. Lancet. 1977;2(8042):779–80.
14. Ellis PR, Dawoud FM, Morris ER. Blood glucose, plasma insulin and sensory responses to guar-containing wheat breads: effects of molecular weight and particle size of guar gum. Br J Nutr. 1991;66:363–79.
15. Haber GB, Heaton KW, Murphy D, Burroughs LF. Depletion and disruption of dietary fibre. Effects on satiety, plasma-glucose and serum-insulin. Lancet. 1977;310(8040):679–82.
16. Morgan B, Heald M, Atkin SD, Green J, Chain EB. Dietary fibre and sterol metabolism in the rat. Br J Nutr. 1974;32:447–55.
17. Walters RL, Baird IM, Davies PS, Hill MJ, Drasar BS, Southgate DAT, et al. Effects of two types of dietary fibre on faecal steroid and lipid excretion. BMJ. 1975;2:536–8.
18. Baird IM, Walters RL, Davies PS, Hill MJ, Drasar BS, Southgate DAT. The effects of two dietary fiber supplements on gastrointestinal transit, stool weight and frequency, and bacterial flora, and fecal bile acids in normal subjects. Metabolism. 1977;26:117–28.
19. Burkitt DP. The aetiology of appendicitis. Br J Surg. 1971;58:695–9.
20. https://collections.royalsociety.org/DServe.exe?dsqIni=Dserve.ini&dsqApp=Archive&dsqDb =Catalog&dsqSearch=RefNo==%27EC%2F1972%2F09%27&dsqCmd=Show.tcl

Chapter 22
Fibre Launched – But Is Controversial

It was varicose veins that put clear water between Denis and Peter Cleave. Yet it should not have done so. Their respective views on the cause of them were much closer than either was prepared to acknowledge. The clash was more one of personalities and perhaps the pursuit of that age-long quest in both science and the arts, to be the first with an idea and be credited with it. Denis hated the conflict and went out of his way to be conciliatory and to acknowledge Cleave's views in what he wrote. In response, Cleave remained intransigent perhaps seeing the argument as one against his whole theory of the 'saccharine diseases'. To the onlooker of this dispute, in this case Conrad Latto, the real question was: 'Whoever in their right mind could believe that diet had anything to do with varicose veins?' Latto, whom Denis had first met when they were both working at the hospital in Poole in 1939, was now Consultant Urological Surgeon at the Royal Berkshire Hospital in Reading not far from the Burkitt home in Shiplake. Conrad and Denis were lifelong friends sharing the same faith, both were teetotal and had an interest in diet, Conrad being vegetarian. Denis would take Conrad on a trip to West Africa in 1972 [1].

For both Peter Cleave and Denis, any suggestions about the cause of varicose veins must also include haemorrhoids (piles) and deep vein thrombosis (DVT),[1] which trilogy they considered to have a common cause because of their similar distributions across the world. This idea had been in Denis' sights for a number of years, and he would have pursued it sooner had not the lymphoma intervened. On his safaris through Africa, he had noted the almost complete absence of these venous disorders and their rarity as a post-operative complication. In the questionnaire that Denis sent out to the mission hospitals in sub-Saharan Africa, India and Pakistan in 1969, venous disorders received top billing. Of the 23 questions he asked, numbers

---

[1] Deep vein thrombosis. A clot occurring in the leg veins causing pain and swelling. Common after abdominal surgery and can lead to fragments breaking off and passing through the circulation to the lungs causing a pulmonary embolism (PE), which is sometimes fatal.

© The Author(s), under exclusive license to Springer Nature Switzerland AG 2022
J. H. Cummings, *Denis Burkitt*, Springer Biographies,
https://doi.org/10.1007/978-3-030-88563-2_22

2–8 were about them. The responses allowed him to claim that they were very rare, most hospitals seeing fewer than five cases of varicose veins a year, less than 2% of autopsies having a DVT, with pulmonary embolism almost unheard of. Contrast this with the 25–30% prevalence of varicose veins in Western countries and 20–25% incidence of DVT following surgery, even today one of its most frequent complications. As always with Denis, the collection and analysis of these data did not conform to the protocols of contemporary epidemiology; nevertheless, the differences were so great that they attracted little criticism on this count and allowed him to go on and suggest a cause and then a strategy for prevention.

Early in 1972, Denis decided that he had enough evidence about the veins/thrombosis/piles syndrome to question current wisdom about their cause and to write about it suggesting that the real problem was lack of fibre in the diet [2]. But there was a problem. He knew that Cleave had strong views about the cause of varicose veins and had written a substantial review of the subject for *The Lancet* in 1959 [3], followed by a book published the following year [4]. Cleave was a follower of Darwin and subscribed to his theory of the adaptation of species to their natural environment. The transition of mankind to an upright posture millennia ago was believed by surgeons to have led to excess pressure on blood returning to the heart through the leg veins, thus weakening the valves in the veins that ensure blood flow is upwards against gravity. Cleave would have none of this and suggested instead that the real problem arose out of the anatomical relation between the large bowel in the abdomen and the femoral veins carrying blood back from the legs. Cleave argued that

> The evidence reviewed here suggests that the environmental change responsible for varices is the adoption of a diet largely free from vegetable fibre. In many people the effect of such a diet is to slow the passage of food residues through the colon, which in turn overloaded the bowel leading to pressure on the femoral veins where the colon lies on them in the pelvis.

Denis did not disagree with Cleave concerning the importance of fibre but thought his explanation of pressure on the femoral veins due to a loaded colon was in error. Unlike Cleave the physician, Denis was a surgeon and had performed abdominal surgery on both African patients and Europeans, giving him a bird's-eye view of relationships of the abdominal organs and the state of the colon in the two populations. He knew that people consuming a high-fibre diet had a much fuller colon than people with constipation because of the additional residue from the diet and hence also passed more bulky stools. Cleave's femoral vein pressure theory could not be correct. What Denis believed was born out of ongoing conversations with two other surgeons, Neil Painter and Conrad Latto. They noted that varicose veins and diverticular disease not only had similar geographical distributions worldwide but also tended to occur together in individuals. The common factor was raised pressure, not directly on the femoral veins, but inside the bowel and abdomen. Painter had shown this convincingly for diverticular disease, while for varicose veins, pressure within the abdominal cavity was thought to be transmitted to the leg veins when straining at stool due to constipation. Both of these conditions could be attributed to a low-fibre diet.

In his comprehensive and again well-argued paper, Denis wrote about the epidemiological evidence that he had gathered, the accepted wisdom for the cause of these conditions, and about Cleave's explanation involving local pressure on the femoral vein. He then gave a brief summary of his studies of bowel habit, diet and transit time through the gut. His final conclusion is that for varicose veins, DVT and haemorrhoids, 'the fundamental cause of these diseases is faecal arrest which is the result of a low-residue diet'. He goes out of his way to acknowledge the Cleave hypothesis saying 'with the basic principal of which I am in complete agreement' but, perhaps fatally for their relationship, he includes an alternative hypothesis, namely the raised intra-abdominal pressure due to low fibre intakes. For those people who have not spent their life working on fibre and its physiology in the gut, these differences may seem small. For Cleave, they were wholly unacceptable.

In the hope of keeping Cleave onside, Denis sent him, after discussing the matter with Conrad, an early draft of his *BMJ* paper on varicose veins. It was a mistake. Cleave replied with a critical letter which both distressed and upset Denis. Conrad met Denis again a few days later and told him that Cleave was allergic to Denis and advised him to just keep calm. Denis decided to send a letter of apology, presumably for not giving enough prominence to Cleave's ideas in the paper. Whatever was said, it resulted in a telephone call the next day from Helen, Peter Cleave's wife, and Cleave himself thanking Denis and saying the content of his letter 'Totally altered the situation', for which Denis was thankful and relieved. Conrad, a few days later, told Denis that Cleave had accepted changes to the paper and it looked as though the squabble had ended. The final draft was sent to Cleave as a courtesy by Denis but was again returned with many requests for additions by Cleave, who clearly thought that the truce now allowed him to comment freely on the paper. Conrad helped Denis to manage these, and to their surprise, Cleave wrote a complimentary letter on reading the published article, in which his name was mentioned several times, but still found much in it to criticise. Thus, varicose veins, haemorrhoids and deep vein thrombosis became part of the fibre story.

On seeing the published paper, in 1972, Ernst Chain rang Denis to congratulate him but warned him that the Flour Millers and Bakers were issuing a press release, which would not give fibre their backing. They were already worried it would impact the sale of their principal product, white bread. Denis was beginning to get an inkling of the controversy that fibre would engender.

Denis' confidence and determination to press on with his pursuit of the causes of Western diseases was, despite opposition from Cleave and from the millers, buoyed up by the unstoppable tide of invitations to speak about his work, now mostly about fibre. The international recognition he was achieving for his pioneering discovery of the cause of the lymphoma and now bowel cancer with the associated cluster of common Western diseases was an intriguing story, compellingly told. In March, he received the Gold Medal of the Society of Apothecaries in London followed, a week later in Frankfurt with Olive accompanying him, by the Paul Ehrlich and Ludwig Darmstaedter Prize and Gold Medal. Here they were treated as celebrities, met government ministers and were hosted by the British Council, and after the prize-giving ceremony, there was a reception and dinner. Chain, a previous winner, was also at

the ceremony which was shared with Jan Waldenström for his work on the epony-
mous macroglobulinaemia. The expected lecture that is part of events on these occasions was given by Denis the next morning, Wednesday, at 9 am following which he
and Olive headed for the airport for a flight to New York. From there, they journeyed on to Washington arriving late in the evening. Denis was up at 4:45 am the
next day, taken to a meeting of the AFIP[2] where he lectured at 7 am departing immediately to fly to Chicago and on to Wichita for a Cancer Congress where he spoke
on 'Diseases of the West' on Friday morning. It was the start of a 2-week lecture
tour in the USA from which he returned to the UK in early April in time for the
ceremony in London where he was admitted to the Fellowship of the Royal Society.
As the congratulations came in from near and far, he was informed that he was to
receive the Lasker Award later in the year for his work on chemotherapy of the lymphoma. The Lasker is one of the most prestigious awards in medical science,
which he shared with Jo Burchenal and John Zeigler while, unusually in 1972, there
were 13 awards given for similar work in the treatment of other cancers.

Denis was a star, and there was no doubt in his mind as to what he should be
doing. Most important was to bring all the evidence he was gathering into what
would be the first book on the fibre hypothesis [5]. Hugh Trowell managed the project, and he and Denis would write 15 of the 21 chapters with help from Neil Painter,
Ken Heaton and others for subjects such as diverticular disease, peptic ulcer, dental
caries, gallstones and lessons from the animal kingdom. Meeting with the team,
which also included Alec Walker presently writing a paper for *The Lancet* with
Denis and Neil Painter on bowel transit times and their significance, it became clear
to Denis that, although they all talked confidently about fibre, they did not really
understand exactly what it was. Denis, through his many discussions with Cleave,
was convinced this was not a story about sugar. Nor was it simply about bran but
fibre,[3] which he knew had important effects on the bowel, on metabolism in general
and was removed by modern food processing. He had picked up the broader fibre
idea from Walker and Painter but had not yet grasped its essential contribution to a
healthy diet. In his writings about it to date, he had used a number of terms to
describe fibre but was not inhibited in talking about it although not sure about its
nature. As with his work on the lymphoma, he was quite happy to leave what he felt
were technical matters to those more highly skilled in the composition of diet, while
he got on with adding to the story. To Hugh Trowell was given the task of coming
up with a definition of fibre, a job for which he was ideally suited.

The use of the word fibre in nutrition had dated from around 1800 when it was
borrowed from the textile industry to describe the appearance and nutritional properties of animal feeds [6, 7]. Originally known as crude fibre, it was measured by
early animal nutritional biochemists in order to give an indication of the quality of
an animal feed, particularly for cows and other domesticated ruminants. Fibre was
considered non-digestible and to impair digestion overall. The higher the fibre, the
less digestible the feed. Almost an anti-nutrient. This in no way reflected what

---

[2] Armed Forces Institute of Pathology, USA.

[3] Fibre, being plant cell wall components such as cellulose, pectin and hemicelluloses also known
as the non-starch polysaccharides (NSP), is present in all cereals, fruits, vegetables, pulses and
legumes. A much larger concept than bran.

Fig. 22.1 1929 Report by Professor McCance and Dr. Lawrence of the Biochemical Department of King's College Hospital, London, a major part of which discusses fibre, its chemistry, physiology and nutritional value

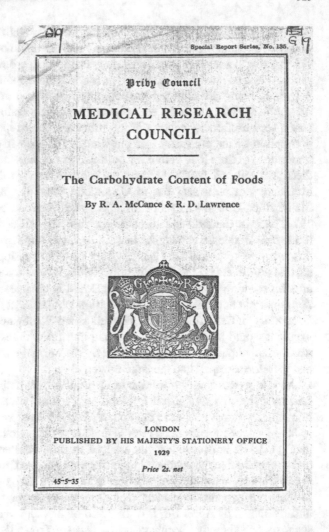

Special Report Series, No. 135.

Privy Council

MEDICAL RESEARCH COUNCIL

The Carbohydrate Content of Foods

By R. A. McCance & R. D. Lawrence

LONDON
PUBLISHED BY HIS MAJESTY'S STATIONERY OFFICE
1929

Price 2s. net

45-5-35

Trowell and Denis now conceived as its importance for human nutrition. But the word fibre did not appear in the indices of the main textbooks of human nutrition or gastroenterology nor was there a listing for it in the Cumulated Index Medicus, the bible for researchers before the digital age. There was no book on the subject nor articles in any of the leading medical journals. Trowell decided that a new definition was needed suitable for human nutrition (Fig. 22.1).

Trowell's African connections provided a vital clue. In 1929, Professor R A McCance [8][4] and Dr. R D Lawrence [9][5] had been asked by the MRC to prepare

---

[4] R A McCance (1898–1993), one of the outstanding nutritionists of the twentieth century. Irish by birth, he became the first Professor of Experimental Medicine in Cambridge.

[5] R D Lawrence (1892–1968), Scotsman from Aberdeen who became a physician at King's College London specialising in diabetes care. Diabetic himself he was one of the earliest recipients of insulin in 1922.

accurate food tables for use when planning diets for diabetics [10]. A fundamental postulate of theirs was that carbohydrate in the diet could be divided into that which was readily available through digestion in the upper bowel, such as sugars and starch, and thus provided glucose in the blood for metabolism, and unavailable carbohydrate being not digested in the upper bowel and not providing glucose for metabolism. This latter is the equivalent of fibre. The available and unavailable idea was a very useful new concept at the time, and their report contains an excellent description of the chemistry and physiology of fibre. It gave a credibility to the importance of fibre long before it would become a characterising feature of healthy diets. Trowell and McCance knew each other through their Africa connections. McCance had been sent out to Uganda in 1966 to take care of the Infantile Malnutrition Unit in Kampala until a permanent director could be appointed. Now, 1970, back in Cambridge and retired, he met Trowell and advised him to use the term 'unavailable carbohydrate' instead of fibre, but Trowell was not happy with this term. He knew that fibre was not inert material passing through the gut but had significant effects on the digestive physiology of the lower bowel or colon. Soon after his meeting with McCance, he met David Southgate who had been a student of McCance and now ran the food composition laboratory at the MRC Dunn Nutrition Laboratory in Cambridge. When Trowell asked Southgate how to define fibre in a way that would have meaning for human nutrition, he responded by saying that it was carbohydrate that was not digested by the normal enzymes in the small bowel and therefore escaped into the large bowel.

And so it was that Trowell came up with a new definition of what he called 'dietary fibre' to distinguish it from the 'crude fibre' used in the animal world. He immediately wrote a letter to the journal *Atherosclerosis* giving his new definition as 'The skeletal remains of plant cells that are resistant to hydrolysis by enzymes of man' [11].[6] He acknowledges the work of Burkitt, Painter, Walker and others who had suggested that this element of the diet was the one that protected against diverticular disease, appendicitis and cancer of the colon. He goes on to add an extension to the fibre hypothesis, namely that fibre might lower blood cholesterol, a fact already shown by Ancel Keys, Walker and others, and thus protect against atherosclerosis and therefore coronary heart disease. Trowell's definition is more of a physiological concept, i.e. not digested in the upper bowel, rather than the precise identification of a dietary component that could be measured. Importantly, as a physiological concept, it did not include perhaps the key function of fibre, namely to be digested by bacteria in the large bowel and act as a primary control of metabolism in this organ. Not even aware at this time of such niceties, Denis was very pleased with Trowell's definition and the way the fibre story was now developing. He began to include in his talks a simple description of fibre as the material of the plant cell wall that passed through the digestive areas of the upper gut unchanged.

---

[6] Note that the term 'dietary fibre' had been used by Hipsley 29 years earlier in an article about the toxaemia of pregnancy [12] but Trowell was probably unaware of this paper. Hipsley defines fibre in his paper as including lignin, cellulose and hemicelluloses, a more practical basis for a definition than the non-digestibility concept of Trowell, Southgate and McCance.

The year 1972 was the year that dietary fibre was launched into the consciousness of the world's scientific, nutritional, medical and public health communities. More than anyone else, Denis must be given credit for this. It was his work on large bowel cancer and fibre that was the key. But he was now more collegiate and would always acknowledge the vital importance of his new colleagues in building the hypothesis and of the early work of animal nutritionists, biochemists and physiologists, especially in the UK, Germany and the USA, who had done pioneering work on fibre in the previous 100 years. The pursuit of an understanding of the nature of fibre, its actions in the body and capacity to prevent disease formed the fibre story and was taken up in laboratories across the world. The first scientific review of 'Dietary Fibre' was published in the journal *Gut* [13] early in 1973 in which the credibility of the Burkitt and Trowell hypothesis was reviewed in the light of 135 papers on fibre published over the preceding 110 years. The first symposium on 'Fibre in Human Nutrition' took place in Edinburgh in April 1973 sponsored by the UK's Nutrition Society and organised by Martin Eastwood. The papers arising from the symposium were published later in the year [14].[7] Media interest, which had always accompanied Denis on his travels, now turned to the fibre story [15], and the food industry saw fibre as having the potential to aid in the marketing of healthy foods. Fibre content started to appear on food packaging. American physicians and scientists now directed attention to fibre, following the pioneering work of Williams, Olmsted and others in the 1930s, and in May 1974, a small meeting was convened at the University of Chicago by Richard Reilly and Jo Kirsner of the Department of Medicine limited to 34 participants and a single day. Talks were given by the usual suspects, Heaton, Eastwood, Painter and Denis, alongside contributions by distinguished medical scientists from the USA including Mark Hegsted, Sid Phillips, James Christensen and Tom Almy. Denis was impressed by the meeting. 'Very high powered and excellent participants. A most useful day'. The first book devoted to fibre was a record of the meeting and was published in 1975 [16] (Table 22.1).

More symposia were convened and further books followed with the topics quickly moving to encompass all aspects of the fibre story. Denis was the star attraction at many of these meetings, supported by Hugh Trowell, and gave his lecture on fibre, but as time went by, he would move more towards the preventive aspects of healthy eating. Other books relating to the fibre story included Cleave's three editions of his treatise on 'the saccharine diseases' and Painter writing about diverticular disease [17], all part of the genre of the time. Soon the diet brigade moved in to exploit the idea that high-fibre diets might help in weight loss, heralded by the popular 'F-Plan Diet' by Audrey Eyton with its tables of the fibre content of foods and menus for F-Plan meals to aid weight loss [18]. By 1990, recommendations for healthy diets started to include recommendations for fibre [19–21], but it would be 20 years from the publication of Denis paper on fibre and bowel cancer in 1971

---

[7]Topics included the chemistry and measurement of fibre, physical properties, serum cholesterol effects, obesity, diabetes, immunity and Denis talking about large bowel diseases and Trowell about diabetes and heart disease.

Table 22.1 The early books on dietary fibre, 1975–1990 (p-number of pages)

Year	Authors/editors	Title. No of pages	Publisher
1975	R W Reilly & J B Kirsner	Fiber deficiency and colonic disorders. 185 p	Plenum Medical Book Company. New York and London
1975	D P Burkitt & H C Trowell	Refined carbohydrate foods and disease. Some implications of dietary fibre. 356 p	Academic Press. London, New York, San Francisco
1976	G A Spiller & R J Amen	Fiber in human nutrition. 278 p	Plenum Press. New York and London
1978	Spiller G A	Topics in dietary fiber research. 223 p	Plenum Press. New York and London
1978	Heaton K W	Dietary fibre: current developments of importance to Health. 158 p	John Libbey and Co. London
1979	D P Burkitt	Don't forget fibre in your diet. 128 p	Martin Dunitz. London
1980	G A Spiller & R M Kay	Medical aspects of dietary fiber. 299 p	Plenum Medical Book Company. New York and London
1980	Royal College of Physicians	Medical aspects of dietary fibre. 175 p	Pitman Medical. Tunbridge Wells
1981	H C Trowell & D P Burkitt	Western diseases: their emergence and prevention. 456 p	Harvard University Press. Cambridge, Massachusetts
1981	I Mclean Baird & M H Ornstein	Dietary fibre; progress towards the future. 99 p	Kellogg Company. Manchester.
1982	G V Vahouny & D Kritchevsky	Dietary fiber in health and disease. 330 p	Plenum Press. New York and London
1983	G Wallace & Bell L	Fibre in human and animal nutrition. 249 p	The Royal Society of New Zealand. Wellington, New Zealand
1985	H Trowell, D P Burkitt, K Heaton	Dietary fibre, fibre-depleted foods and disease. 433 p	Academic Press. London, etc.
1986	G V Vahouny & D Kritchevsky	Dietary fiber. Basic and clinical aspects. 566 p	Plenum Press. New York and London
1986	G A Spiller	CRC handbook of dietary fiber in human nutrition. 504 p (First Edition)	CRC Press Inc. Boca Raton
1990	D Kritchevsky, C Bonfield, J W Anderson	Dietary fiber. Chemistry, physiology and health effects. 499 p	Plenum Press. New York
1990	A R Leeds & V J Burley	Dietary fibre perspectives – reviews and bibliography 2. 299 p	John Libbey. London, Paris

before diets characterised by high fibre would become part of the lexicon of healthy eating and preventive medicine [22].

Alongside the books and meetings, the explosion of interest in fibre led to an exponential growth in the research being undertaken around the world and as a result the number of papers on fibre written for the medical and scientific press [23, 24]. The bibliography prepared by Leeds and Burley in 1989 included 175 pages of fibre-related subjects, lists of authors and thousands of papers from the medical/scientific literature.

Denis decided that a popular book was needed. The fibre story was bigger and more generally applicable than the lymphoma work he had done because of its potential to prevent some of the major diseases in Western societies. How he managed to fit this in amongst all his other writing, which was at a peak in the 1970s, must have required supreme dedication, energy and often less time to spend with his family. During the 6 years from 1972 to 1977, he wrote over 100 articles of which 75 were substantial scientific papers, chapters in books or symposium proceedings and were mainly about fibre and the geography and prevention of Western diseases. For 70% of the main papers, he was the sole author, which would be most unlikely today given the collaborative nature of science and the different skills and expertise required in any project. This output is a reflection of his conviction that here was a story of importance to everyone, which he personally had a duty to propagate having he felt been its originator. This sense of ownership would soon lead to some of his close collaborators in this venture feeling they were not being given the credit due to them (Table 22.2).

Denis was well accustomed to the interest in his work and the publicity given to it by the media. Although he enjoyed being the star attraction at meetings on fibre,

Table 22.2 Burkitt publications 1972–1984

Year	Total	Sole author	Papers	Book chapters, symposium proceedings	Books
1972	10+8[a]	9/10	4	3	
1973	19+8	17/19	8	8	
1974	3+1	1/3	2		
1975	17+6	9/17	8	3	1
1976	12+6	5/12	7	2	
1977	15+5	11/15	8	4	
1978	4+4	4/4	3	1	
1979	8+4	4/8	4	1	1
1980	8+3	7/8	1	6	
1981	8+4	6/8	3	1	1
1982	7+4	6/7	2	3	
1983	6+13	6/6	3	3	
1984	9+7	7/9	4	2	

[a]10+8 First number is papers in scientific journals, chapters in books, letters of substance to journals, editorials etc. The second denotes reports of meetings, shorter pieces for the popular press, or where the full reference has not been confirmed. Main co-authors are Walker, Painter, Trowell, Hutt. Overall total is 199 over 13 years. Details in Appendix 3

he kept reminding himself that in all of this he was trying to fulfil God's will for him and that he must remain humble and show due deference to the scientists and medical academics he met. At night, he would always end the day reading his Bible, praying and in the mornings spend a period of quiet before work began. With retirement now on the horizon, he wanted to finish his experimental studies on bowel transit and gut health, but his options to do this were becoming much more limited as the interval of time following his departure from Africa extended. His final paper containing original data from his fieldwork appeared in *The Lancet* in December 1972 under the title 'Effect of Dietary Fibre on Stools and Transit-Times, and Its Role in the Causation of Disease' [25]. With co-authors Alec Walker and Neil Painter contributing results from their own studies, the paper gives a summary of transit time measurements in over 1000 subjects from about 30 hospitals, schools and research institutes around the world. In all these, volunteers' stool weights were recorded, and the relationship between transit time and stool weight was plotted. The methodology was not unimpeachable, but the message from these studies was clear to Denis and his co-authors. Where people consume diets containing plenty of fibre, then transit times are short and stool amounts are large. Based on the combined experience of the three authors, they conclude that high-fibre diets, through their effect on bowel function, will prevent diverticular disease, appendicitis and large bowel cancer. They add coronary heart disease to the list based on Walker's early work on fibre and blood cholesterol with Trowell now having reported earlier in the year that coronary heart disease could be another of the diseases preventable by fibre. Cleave is not mentioned but cited for his work on bran and constipation. The paper became Denis' fourth most cited, the other three being his colon cancer paper of 1971, that with Painter on diverticular disease the same year and a more general review of dietary fibre and disease with Walker and Painter, published in the *Journal of the American Medical Association* in 1974. With Hugh Trowell, whose expertise lay in a wholly different medical speciality than that of Denis, he published five full papers and was the co-author with him on the three main books on dietary fibre they would publish between 1975 and 1985.

Overseas travel continued to be included in Denis' busy life with four trips during 1972 lasting a total of 11 weeks. Most of the time was spent in North America where the medical and scientific community had taken to the new story about fibre with alacrity but had yet to engage on any great scale in work on it. One of the four trips was to West Africa on which he was accompanied by Conrad Latto, who found the whole experience to be memorable and which provided new insights for him into how Denis thought and worked. The pace of the visit was breath-taking. They arrived in Abidjan on the Ivory Coast where Denis, with Conrad in tow, was immediately taken on a ward round, lectured and then commenced his routine of asking his hosts about their experiences, not now of the lymphoma, but of the fibre-related conditions. All this was very new to Conrad. The next day they flew on to Accra in Ghana for a quick visit, then to Johannesburg, where they visited the new hospital for Africans at Baragwanath. Denis gave the Oettlé Memorial Lecture at the SAIMR, met Alec Walker and Betty his wife with whom they stayed before travelling to Durban. There George Campbell greeted them with his wife Magda, whose relations with George were becoming increasingly difficult requiring Denis' form of psychotherapy: a long talk and a reminder of God's healing hand. From there, they drove to the Transkei for a conference at which Denis spoke before driving on to Pietermaritzburg, Ladysmith, back to Pretoria and then Johannesburg and to stay with the Oettlé family. In Nairobi, they met Philip Howard who was teaching at The Starehe Boys Centre in the Nairobi slums and Judy at a nursery school in the centre of the city.

They were back home a few days later, at the end of what would be one of Denis' last trips to Africa. They had not visited Kampala nor travelled through Uganda. Idi Amin had led a military coup against the president, Milton Obote, in 1971 and declared himself President instigating a purge of Obote supporters which widened to include foreign nationals. Denis would return only briefly in July 1974 to the place where he had spent almost 20 years, made his landmark observations on the childhood lymphoma, brought up his family and served the African community in all sincerity. Now he could give all his energies to convincing the world that he had a solution to what he would come to call 'man-made diseases'.

References

1. http://livesonline.rcseng.ac.uk/biogs/E000570b.htm
2. Burkitt DP. Varicose veins, deep vein thrombosis, and haemorrhoids: epidemiology and suggested aetiology. BMJ. 1972;2:556–61.
3. Cleave TL. Varicose veins. Nature's error or man's? Lancet. 1959;2(7095):172–5.
4. Cleave TL. On the causation of varicose veins. Bristol/Baltimore: Wright/The Williams & Wilkins Co; 1960.
5. Burkitt DP, Trowell HC, editors. Refined carbohydrate foods and disease. Some implications of dietary fibre. London: Academic; 1975.
6. https://doi.org/10.1017/S0954422417000117
7. Van Soest PJ, McQueen RW. The chemistry and estimation of fibre. Proc Nutr Soc. 1973;32:123–30.
8. Widdowson EM. Robert Alexander McCance. Biograph Mem Fellows R Soc. 1995;41:261–80.
9. https://en.wikipedia.org/wiki/Robert_Daniel_Lawrence
10. McCance RA, Lawrence RD. The carbohydrate content of foods. Special Report Series of the Medical Research Council no. 135. London: HM Stationery Office; 1929.
11. Trowell HC. Crude fibre, dietary fibre and atherosclerosis. Atherosclerosis. 1972;16:138–40.
12. Hipsley EH. Dietary 'fibre' and pregnancy toxaemia. BMJ. 1953;2:420–2.
13. Cummings JH. Dietary Fibre Gut. 1973;14:69–81.
14. Proc Nutr Soc. 1973;32:123–204.
15. Sunday Times. 1972;20 August 34.
16. Reilly RW, Kirsner JB, editors. Fiber deficiency and colonic disorders. New York/London: Plenum Medical Book Company; 1975.
17. Painter NS. Diverticular disease of the colon. London: Heinemann Medical Books; 1975.
18. Eyton A. The F-Plan diet. Middlesex: Penguin Books Ltd; 1982.
19. WHO Diet, nutrition and the prevention of chronic diseases. WHO technical report Series 797. Geneva; 1990.
20. Department of Health. Dietary reference values for food energy and nutrients for the United Kingdom. Report on Health and Social Subjects 41. London: HMSO; 1991.

21. US Department of Agriculture and Department of Health and Human Services. Nutrition and your health. Dietary Guidelines for Americans. 1990. https://health.gov/sites/default/files/2019-10/1990thin.pdf
22. WHO Carbohydrate Guideline (In press).
23. Trowell HC. Bran yesterday…bran tomorrow. BMJ. 1984;289:436.
24. Leeds AR, Burley VJ. Dietary fibre perspectives – reviews and bibliography. 2nd ed. London: John Libbey; 1990.
25. Burkitt DP, Walker ARP, Painter NS. Effect of dietary fibre on stools and transit-times, and its role in the causation of disease. Lancet. 1972;2(7792):1408–12.

Chapter 23
'Character Is More Important than Cleverness'

Five years after agreeing with Hugh Trowell that they should write a book on fibre together, *Refined Carbohydrate Foods and Disease* was published in 1975 by Academic Press [1]. When Cleave read it, he was furious and wrote a quite graphic letter to Denis, in his inimitable style incorporating red ink, capitals, underlining and unconventional spacing for emphasis. The book was essentially a series of 21 reviews of topics pertinent to the fibre story written mostly by Trowell and Burkitt with a forward by Richard Doll, now Sir Richard, Regius Professor of Medicine in the University of Oxford. What incensed Cleave was their failure to mention sugar and his saccharine hypothesis, instead ascribing to fibre the means of preventing the now established list of fibre deficiency disorders. Perhaps in anticipation of his reaction, they gave credit to Cleave in the 'Preface', saying that their interest in investigating these conditions was stimulated 'by what we consider to be the perceptive genius in the work of Surgeon Captain T.L. Cleave to whom we wish to pay particular credit and acknowledge a profound debt'. In the 'Concluding Considerations', they again acknowledge the importance of his work writing that 'Stimulated by this hypothesis and also by his far-reaching epidemiological studies in Africa and Asia, we undertook our present enquiry'. But in later years, Denis would gently remind his readers and listeners that some of the mechanisms Cleave had postulated in order to explain disease causation, not unreasonably, had to be modified in the light of subsequent knowledge. Denis had realised this from the first time he met Cleave in September 1967 but had failed to convince him in later years of the importance of fibre.

These placatory remarks did not work, and in reality, the book is devoted entirely to the fibre story. Even the title is qualified by the words 'Some implications of dietary fibre'. Denis, never a sugar hypothesis believer, had recalled to Trowell in a letter his 'Flash of Understanding' moment from 1970 when he realised that fibre was crucial, not sugar, and in 1971 had written, whilst visiting Tehran, to Trowell with 'I have again and again in India and Iran been testing the Cleave Hypothesis. Much is sound, much needs changing'. On page 337 of their new book, they say,

© The Author(s), under exclusive license to Springer Nature Switzerland AG 2022   331
J. H. Cummings, *Denis Burkitt*, Springer Biographies,
https://doi.org/10.1007/978-3-030-88563-2_23

'The Editors started their enquiry considerably stimulated by Cleave's hypothesis of the major role of sugar. As their investigations proceeded, quite independently, they began to lay more emphasis on the beneficial effects of adequate fibre intake and the ill-effects of fibre deficiency'. On page 343, 'This approach, which may be considered as a major modification of Cleave's original hypothesis'.

Cleave's letter to Denis of November 25, 1975, takes him to task for all this and for joining Trowell in dual authorship of these remarks:

Dear Denis
    I consider it very unchristian of the Rev Trowell to undo so completely the value of your dedication at the end of the book by the remarks on page 343, since there is not a shred of evidence that anyone to come will be able to change my conception...
    Yours
    Peter Cleave

In fairness to Cleave, he was an original thinker but worked alone on his ideas all his life only rarely publishing with a co-author. Although he had identified fibre as important in the cause of varicose veins as early as 1959, he never really accepted that removal of fibre and increased consumption of refined carbohydrates such as sugar and starch were really two sides of the same coin. In the first edition of his book about the 'saccharine disease', fibre is mentioned only in the context of his varicose vein theory and is not in the index of the book nor in his 'Food Guide for Health' at the end. A more holistic concept would have left him as one of the more enlightened thinkers in twentieth-century medicine. Denis always gave Cleave credit for introducing him to the concept that where a group of diseases occur together in some populations but not in others and together in the same person, like obesity, diabetes and coronary heart disease, they are likely to have a common cause. He went out of his way to give Cleave credit for this theory, whilst clearly modifying it, in order to minimise the barrage of correspondence and phone calls that ensued when Cleave read one of Denis' papers on the subject of fibre. What was probably fatal to Cleave's sugar theory, despite Yudkin in 1972 pointing the finger at sugar as 'Pure, white and deadly: the problem of sugar' [2], was the evidence that emerged in the ensuing years showing fibre to have important and unique physiological effects on the gut. The most important of these is in directing large bowel function through its role in supplying the principal substrate (food) for the resident bacteria, known as the biomass or more recently the microbiome. This provided a clear explanation that allowed the epidemiology and pathology to be linked and more importantly to point to a clear means of prevention. Sugar, often cited as the demon of diet-related diseases, in fact proved to be linked most clearly only with dental caries (tooth decay) and as one ready source of calories in the development of obesity, especially in children [3] (Fig. 23.1).

Cleave had his supporters, one of the most distinguished of which was Dr. Ken Heaton, Reader in Medicine at the University of Bristol. In writing an appreciation of Cleave after his death in 1983 [4], he says, 'Cleave was certainly one of the most original medical thinkers of the twentieth century. His rare combination of panoramic vision, piercing logic and bulldog tenacity deserves the title of genius. He was a true pioneer who started a major revolution in scientific thought'. Although

Fig. 23.1 'Surgeon Captain Cleave' seated in his garden at home, 1980. (From Lifespan Magazine (health), pages 88–89)

considered for, but not elected to, Fellowship of the Royal Society[1] Cleave did receive two prestigious awards in 1979, the Harben Gold Medal of the Royal Institute of Public Health and Hygiene, previously awarded to Pasteur, Lister and Fleming, amongst others, and the Gilbert Blane Medal for Naval Medicine awarded jointly by the Royal Colleges of Physicians and Surgeon. Denis was invited to this grand occasion at the Royal Naval Hospital in Gosport where the Harben was presented by Vice Admiral Sir John Rawlins, Medical Director of the Royal Navy, and the Gilbert Blane by Sir Douglas Black, President of the Royal College of Physicians. Denis described being present for the awards as giving him enormous joy and went on to write an article entitled 'Ridicule, Replaced by Recognition'. The credit he always gave to Cleave was undoubtedly genuine, but he regretted not being able to convince him that fibre was an essential part of the story of the rise of Western diseases. Cleave published nothing after 1979 and would die of the complications of dementia 4 years later.

By 1976, with his formal retirement from the MRC only a few months away, Denis realised he would need financial support to continue his work. He decided to apply to the MRC for a Project Grant and went to see Richard Doll about it. What advice he received from Doll is not known, but Denis decided to do what he did best and continue his studies of the geography of diseases using his now worldwide contacts. In conjunction with Dr. Ian Prior of Wellington University, New Zealand, he brought together a group of people already engaged in these studies from whom he could get information about what he now unhesitatingly described as diseases

---

[1] Personal communication. Professor Sir Frank Young, 1978.

due to fibre-depleted diets. As a direct result of his extensive travels around the world in the previous 10 years, he had an unrivalled contact list of people in around 50 countries who agreed to collaborate. Questionnaires would be sent out and the results analysed by himself and Ian Prior. The proposal was sent into the MRC in August, but in September, he received somewhat disturbing news. He was told that he would have to vacate his office on Tottenham Court Road in London when he retired. In October, the MRC told him that they would not fund any further studies. 'Naturally disappointed. But commit it with confidence to the Lord'. The question-naires which now included 16 diseases and a line for details of the staple diet in the region were nevertheless sent out, and Denis received replies from over 200 hospitals, but the information was still qualitative with respondents simply asked to record if they saw the conditions or not and to describe diet in a single sentence. The world of epidemiology and medical statistics had moved on. And so would Denis.

Having been in more or less continuous employment since graduating in 1935 with various hospital jobs during the pre-National Health Service days, the Royal Amy Medical Corps from 1941, the Colonial Medical Service from 1946 and finally the MRC from 1964 until retirement in 1976, Denis had an adequate personal pension but needed funds to continue his studies and a base from which to operate. The latter was provided by Michael Hutt, at St Thomas's Hospital, London, a very long-standing and good friend of Denis. Money for travel, secretarial help and his survey work was more problematic. Since the days when he had first met Himsworth of the MRC in 1961 and been offered £150 for his studies, he had been funded to the extent of his needs by the MRC along with grants from the Sloan Kettering in the USA and the UK Cancer Research. He had never had to go out and compete for funding as most scientists did. But Denis' requirements were modest, thanks to his lifelong philosophy of being frugal, and there were organisations with money, particularly in the food industry, which saw benefits in getting him to support their products.

Given that fibre had initially meant bran and wholemeal bread to Denis, the Flour Millers and Bakers were an obvious partner for him. But they had their own research station at Chorleywood and saw Denis more as a threat. On the publication of his paper about varicose veins, Chain had warned Denis that the Millers were going to take defensive action in a press release, in which they 'totally rejected suggestions made by doctors and others to the effect that a shortage of roughage in the diet con-tributed to diseases of the bowel'.[2] Denis had met Hutchison, a senior figure in the National Association of British and Irish Millers (NABIM), and Harold Dodd, a well-known London surgeon who had written extensively about varicose veins, at their headquarters on Arlington Street in London, soon after the publication of Denis' paper. Dodd described NABIM as being very vulnerable, clearly anxious about the threat to sales of white bread. Returning with his team of Alec Walker, Hugh Trowell, Ken Heaton and Neil Painter to meet them again later in 1972, they now found strong opposition to the idea of fibre, which they could not understand in the light of the potential marketing value for them of this new wonder ingredient in

---

[2] Burkitt DP, Autobiography, Chap. 18, p. 1. Wellcome Library WTI/DPB/F/1.

bread. Denis was invited to speak at the annual conference of the Millers and Bakers and further meetings followed in 1973 during which progress must have been made because by the middle of 1974 relations had improved. The change of mind in the leaders of the milling industry, mainly long-established family firms such as Ranks, was brought about largely by discussions with the head of their research establishment at Chorley Wood, Dr. Brian Spencer. He realised the potential importance of fibre and was able to reassure them that no major changes in milling practices were needed. Moreover, there was potentially a cost benefit in wholemeal bread, which could be sold at a higher price than white bread, but with the addition of only the bran, a cheap ingredient. Between them they put the problem of the threat to this national staple food to the UK Department of Health who in response set up an expert panel to look at 'Nutritional Aspects of Bread and Flour' in 1978. Its report [5] in 1981 gave strong support to the Millers and Bakers by emphasising the nutritional value of bread, especially as a source of cereal fibre, and suggested it be promoted to replace some of the fat and sugar in the diet. A few months earlier, in September 1980, a Working Party, set up by the Royal College of Physicians of London to review 'Medical Aspects of Dietary Fibre', reported concluding that 'increasing the proportion of "dietary fibre" in the diet of Western countries would be nutritionally desirable' [6].[3] Fibre had arrived and in the ensuing years more evidence would be gathered to confirm that diets characterised by wholegrain cereals, as well as fresh fruits and vegetables, are beneficial to health.

The Millers never funded Denis' work, but the Kellogg Company did. To Kellogg's the value of fibre was a core part of their business having launched Bran Flakes and All-Bran in 1915/1916 under the guidance of one of the founding brothers, Dr. John Kellogg. They had opened up discussions with Denis in 1971, but his colleagues in the fibre story warned him not to get too close to the food industry lest he be seen as a Kellogg's man and the mouthpiece of the food industry. Denis was aware of this but was reassured when the company put the avuncular Wilf Hamilton, Director of Scientific Affairs for Kellogg UK, in charge of relations with Denis and a mutually beneficial partnership ensued that lasted more than 10 years. After a series of confidence-forming meetings between Denis and senior members of the company, they agreed to sponsor a symposium on fibre in London in 1974, followed by meetings in other parts of the UK and then in Europe and the USA at which Denis would speak in his inimitable style and the audience were able again to say that they had heard the great man.

In the June 1974 Birthday Honours list of the Queen, it was announced that Denis had been appointed a Companion of the most Distinguished Order of Saint Michael and Saint George allowing him to use the letters CMG[4] after his name and to Olive's more immediate pleasure a trip to Buckingham Palace. The award honours people who have given distinguished service overseas, which in Denis' case

---

[3] The Working Group was chaired by Sir Douglas Black, College President, with Dr. Ken Heaton as secretary.

[4] Humorously referred to as Call Me God by those who take delight in such matters.

meant he would have been proposed to the Honours Committee by the Commonwealth Office for his work in Africa. Olive had great pleasure in announcing it to the family before it was published officially, which somewhat worried Denis, always anxious not to cause trouble, but he was pleased with the immediate pleasure it brought everyone at home. The honour was conferred on Denis by the Queen on December 3 but never one for a party Denis went straight from the Palace to lecture in the afternoon to the Christian Union Group at Guy's Hospital before returning home to prepare for an MRC meeting on fibre the next day, then to Finland for a brief visit in which he gave three lectures. It was his 13th overseas trip of the year, the major ones having been to West Africa, South America, the USA and Mexico and India keeping him away from home for a total of 3 months. But the nature of the trips had now changed. Much reduced were the visits to mission hospitals to collect data on disease prevalence; now he was responding to invitations to speak at conferences, collect honours, visit hospitals where he would do ward rounds and talk to staff and meet old friends from his days in Africa. Whilst Burkitt's lymphoma always remained a topic with which he could enthral his audiences, he had moved to lecturing about bowel cancer, then to talks about varicose veins, hiatus hernia, peptic ulcer, gallstones and the importance of fibre. With this topic, he had something that touched everyone's life and ensured he would remain in demand as a speaker.

Denis retired from the MRC on August 31, 1976. Unsurprisingly, he paid little attention to what is a landmark event in the lives of most people. He was on a mission to tell the world how some simple changes in diet could help prevent the major chronic diseases, Western diseases. His studies of their geographical distribution, confirmed in his latest work with Ian Prior, gave him evidence that was similar in nature to that which had allowed him to identify the cause of the African childhood lymphoma, but now the prize was much greater. To his new cause, he devoted all his being, confident that he was doing work to which God had called him and was sustaining him through his word in the Bible, his daily devotions and the Christian community with whom he engaged. Retirement day saw him moving out of his office in Tottenham Court Road, London, and re-establishing it at home in Shiplake whilst retaining a base in London, thanks to Michael Hutt's offer at St Thomas's Hospital. With his research activities coming to an end, he had more time for writing, travel and fulfilling his duties as President of the Christian Medical Fellowship (CMF). All three daughters were now married, so Olive was able to accompany him on some of his trips overseas where she would manage the social occasions that he so hated and collect press cuttings, meetings programmes, recordings of TV and radio interviews and other memorabilia for her rapidly expanding series of scrapbooks[5].

Fibre soon gathered worldwide interest. Following the publication of his paper on bowel cancer, many research groups turned to confirming the findings, looking at the possible mechanisms at a cellular and molecular level and to developing animal models. The IARC in Lyon, the Cancer branch of the WHO, set up a large study

---

[5] Now in the Manuscripts and Archives Research Library at Trinity College Dublin (TCD) MS 11387/2.

in Scandinavia to measure the risk of bowel cancer and relate it to diet, especially fibre and transit time [7]. It concluded that higher intakes of dietary fibre were 'possibly protective', thus giving important credence to Denis' ideas and allowing research funding to flow into its investigation. The National Institutes of Health, the USA's principal organisation for research into health-related issues, convened a conference on fibre in March 1977 attended by Denis and the usual names. Soon after this, the US Senate Select Committee on Nutrition and Human Needs, chaired by Senator George McGovern, as part of its brief to look at 'Diet Related to Killer Diseases' identified fibre as a significant research priority. Both Denis and Hugh Trowell gave evidence to the Committee. Back in England, Denis was called to the Department of Health to meet officials responsible for diet and health. Invitations for Denis to speak came in weekly and were declined only if he had a conflicting engagement. He did not like to disappoint with the result that he would be talking not only at big symposia and international meetings but would be happy to meet groups of medical students and local dieticians/nutritionists. His main destination became the USA with 8 of his 12 overseas trips in 1978 heading in this direction. And gone were the 10–12-week safaris around Africa visiting his collaborating mission hospitals; instead, he would now be away for about 7 days into which time he would fit several locations and give six to ten talks. In the USA, the sister organisation of the CMF was the CMS, Christian Medical Society, who once discovering he was visiting the city's hospital or medical school would ask him to speak to them and would arrange a dinner in his honour. It made for a very full schedule, but if he had an afternoon spare, he would feel his time was wasted. He was usually busiest when in the USA. For example, he records his activities in Phoenix on September 19, 1979, as:

> Up 6.30 Lecture 8. TV 9.20 Flew 10.30 to Los Angeles – Fresno. Immediately after arrival lecture 1.30-2.30 then 5 mins lunch and time for bath & shower. Then TV interview. Then medical dinner with lecture.

Fibre came to dominate his thinking as he gradually realised that here was the means of preventing a group of conditions that included some of the major killer diseases found in the world. He began campaigning for healthier diets and, using a series of clever cartoons drawn by daughter Judy, pointed out that much of our effort in healthcare is devoted to clearing up the mess that we bring upon ourselves by unhealthy lifestyles. He argued that the health services in many countries were complicit by pouring resources into treatment rather than prevention. It almost became a moral issue for him, and here the distinction between his message to his CMS/CMF audiences and those to medical and scientific meetings, who were usually paying his airfares, became less distinct. Preaching to an English congregation one Sunday in Norwich, he no longer started with a text from the Bible but talked about 'Neglected necessities', referring to the importance of diet as well as faith in making a whole person and concluded the day with an illustrated talk entitled 'A Bran for All Seasons'.[6]

---

[6] Diary entry, October 16, 1977.

Although he had never related comfortably to children, he enjoyed talking to medical students and felt it his duty not only to tell stories about his achievements but also to give them some guidance for life. In Durban during his trip to South Africa with Olive in 1977, sponsored by Kellogg's, he spent over a week doing TV and radio interviews, talking to the press and meeting paediatricians, surgeons and other hospital staff, doing ward rounds and renewing his acquaintance with George Campbell and Magda. In the Medical School, he lectured about his lymphoma to the assembled staff, then to the nurses and finally to the medical students.

'Gave my best and main lecture to students. Left them with the message that commitment > ability, character than cleverness and motive than method'. And he left them with a poem by Ella Wheeler Wilcox 'Tis the set of the sail-or-One ship sails east' which ends with the verse:

Like the winds of the sea
    Are the waves of time,
    As we journey along through life,
    'Tis the set of the soul,
    That determines the goal,
    And not the calm or the strife.

An aphorism about ability, character, cleverness and motive, which he left as guidance for the students, may well have been his own, perhaps inspired by the poem. It very much encapsulated his belief, formed during his years at school and Trinity College, that he was not a clever person who had only limited natural ability. Rather, he was someone who had never acquired any laboratory or statistical skills but nevertheless had succeeded beyond his own expectations because commitment, character and motive were key. He was also very modest and worked hard to keep himself humble in the face of the adulation he received on his lecture tours. The maxim that he left with the students on his South African trip in 1977 was one he would develop and use on many occasions in the future, usually leaving it for people who asked for his autograph or writing it into papers about himself and his life [8]. It evolved ultimately into:

Attitudes are more important than abilities.
    Motives are more important than methods.
    Character is more important than cleverness.
    Perseverance is more important than power.
    And the heart takes precedence over the head.

Not everyone was prepared to listen to a talk combining science and spirituality with the result that on one occasion a group of medical students walked out when they realised he was not telling the story of fibre. It was an incident that he was not deterred by at all. But it prompted him to start making notes of those talks he gave that received a standing ovation, which occurred from time to time-fairly unusual for an audience of doctors and medical scientists. And to boost his confidence, if that were needed, honours continued to be bestowed on him. During 1978 and 1979, he was awarded the Gold Medal of the British Medical Association, usually given to someone who had raised the character of the medical profession by scientific work, an Honorary MD by the University of Bristol, the first Ralph Blocksma Award of MAP International in the USA, and he was invited for lunch with Her Majesty the Queen[7]. What perhaps pleased him more than anything, given what he always claimed was his 'unpromising beginning' as a student, was being made an Honorary Fellow of Trinity College Dublin. The letter from the Provost read:

> We regard Honorary Fellowship as the highest award which it is in our power to bestow and we do not do so very often. Our intention is to admit to that company from time to time graduates (and occasionally others) who have reached high eminence in their life's work and who have a deep attachment to the College.
>
> Since you are by common consent our most distinguished living medical graduate it is the fervent hope of your many friends in the college that you will allow me to propose your name.

With the increasing belief that fibre had the potential to benefit the health of everyone living in modern affluent technological communities, Denis decided that he must take his message to a wider audience than the one he engaged with on his lecture tours. He finally found time in 1977 to start writing a book explaining the fibre story including lots of illustrations. With some help from Hugh Trowell and the publisher Martin Dunitz, 'Don't Forget Fibre in Your Diet' was published in 1979 and was a huge success running into four editions and selling well over 200,000 copies in 10 languages [9].[8] It is a good, readable overview of the fibre story up to 1979 starting with the epidemiology, then an explanation of what fibre is, followed by a look at the individual diseases of importance in this context and how fibre in the diet may help to prevent them. He does not shy away from writing about coronary heart disease, blood clotting, diabetes nor obesity despite these being areas of medicine he had, as a surgeon, always left to others to deal with. Hugh Trowell is thanked along with Jim Anderson from the USA[9] whose company Denis always enjoyed, and the book includes recipes for a healthy diet and some food composition data provided by David Southgate. Denis now had greater confidence in his own ability to understand and talk about a range of conditions that lay well outside his training and expertise as a surgeon. Officially retired, he could have opted for a quiet family life, but he was as busy as ever.

The Dunitz book, a paperback, launched Denis on what he described as a 'World Tour' in 1980. Funded in part by Kellogg's, pleased at the popularity of their man, it comprised visits to Australia, New Zealand, Japan and then the USA via Hawaii. Accompanied by Olive as far as San Francisco but flying economy class, probably at Denis' insistence, the itinerary was tightly packed with up to eight events on some

---

[7] February 15, 1978. Present were the Queen and Duke of Edinburgh along with eight guests and two members of the Queen's staff. Denis sat opposite Margaret Drabble the author and alongside Professor Peter Godfrey, Professor of Music at the University of Auckland. Prior to lunch, the Duke of Edinburgh told Denis that he had read Cleave's book. Both Cleave and the Duke were Naval men. Denis concluded 'A memorable occasion' (Wellcome WTI/DPB/F/1/4).

[8] From a letter to Norman Temple Alberta Canada, August 30, 1991. JHC archive.

[9] James W Anderson, diabetologist from the University College of Medicine, Lexington, Kentucky, was one of the pioneers, along with David Jenkins and Jim Mann, of the importance of high-fibre diets for diabetics. Anderson was the first to show the cholesterol-lowering properties of oats.

days. Media coverage was managed by Kellogg's Australia but leaving Denis to choose topics for his lectures to suit the audience. These were very much focussed on fibre although he managed a talk entitled 'Does Moral Fibre Require Faith?' at an evening meeting of doctors. Aware that his independence as a scientist might be called into question, he declined Kellogg's request to use a portrait of him in their advertising. Hospitality was at all times generous especially in Japan where the luxury of the hotels and the price of meals made Denis feel guilty but knowing that Olive was enjoying it all, he refrained from his usual complaint to his hosts that they were wasting their money. Away for over 9 weeks on the tour, Denis remained well apart from an attack of gastroenteritis, to which he had been prone from the earliest days in Africa. Arriving back in the UK, he faced a mountain of post and immediate preparations for two lectures in London a few days later.

The stardom that a popular book clearly engenders made Denis determined to work on his autobiography, which was nearing completion. On learning that he was writing this, Martin Dunitz told him that they would be very interested in publishing it but some months later, after reading the manuscript, turned down this opportunity. Olive, supportive and encouraging as ever, suggested that he separate the biography from the concluding parables, which advice was mirrored by Edward England, publishing director at Hodder & Stoughton heading up their religious division, who was helping Denis to find a publisher. He said it needed to be much more personal, which Denis found somewhat worrying, intent as he was to tell the story of his discoveries rather than his personal life. Hodder & Stoughton also turned it down. Each time it was rejected, he would redouble his efforts, not to dramatise the pivotal moments of discovery in his life but credit God with giving him so much that he did not deserve. In the final chapter, he writes, 'I look back with deep gratitude to God over a life filled with many joys'. When Olive, despite her own strong faith, read it, she condemned much of this chapter clearly realising that such an attribution would not necessarily enhance the story for many readers. The autobiography comprises 19 chapters, about 93,000 words, and would have taken considerable time and effort for Denis to write it. The manuscript lies in the archives of the Wellcome Library in London, still unpublished[10]. It is a valuable reference source for first-hand accounts of his role in the discovery of the childhood lymphoma and his later espousal and development of the fibre story. But his critics were correct in asking for a more personal account of his life and less exegesis. There is little account of his distinguished family except where they became involved with the Church, his school days are barely mentioned and his 3 years as a medical student at Trinity are dismissed in a sentence. Intriguingly, the only diaries that are missing are the ones from those years. They must have existed because the narrative of his life recorded in the years to either side of the missing volumes is continuous with no suggestion that he might be putting down his reflective pen at the end of each day. Efforts to trace them have been unsuccessful. Perhaps they were lost or destroyed. If so, by whom and why? Moreover, his courtship and marriage to Olive is given only a few lines whilst the

---

[10] Wellcome Library, Papers of Denis Burkitt, WTI/DPB/F.

great events in the world that he lived through like the partition of Ireland in 1922, the Second World War and the turmoil in some East and Central African states around 1960 did not lead him to set down his views or comment on these epic events. The autobiography was the least successful of all his writings and sheds little light on the question: who was the real Denis Burkitt?

Lion Publishing came to the rescue. Pat Alexander of Lion read Denis' biographical effort and suggested to him that she find an established author to write the book in a style and with content that would attract readers. The person she found was Brian Kellock who had just finished helping Brian Greenaway, President of Hells Angels, the infamous motorcycling group, to produce his biography, then published by Lion [10]. Kellock had the advantage of access to Denis' unpublished text and in addition much had already been written about him and his work. These included Glemser's *The Long Safari* from 1970, which dealt solely with the lymphoma story, and Galton's 1976 book *The Truth about Fiber in Your Food* which starts with Denis' return to the UK in 1966 and makes only passing reference to the lymphoma. Kellock acknowledges help from Hugh Trowell, Robin Burkitt, Denis' brother, who had a growing archive of the Burkitt family achievements and from Olive who was relieved that someone was taking this on. Denis who was happy to work with him regaled him with stories of meeting graduates of Trinity College working in remotest Africa. The book *The Fibre Man* was published in 1985 and was the most comprehensive biography to date but contained no illustrations to Denis' chagrin given the comprehensive record of his life and work recorded with his camera and made available to Kellock [11].

Denis' 70th birthday early in 1981 seems to have awakened him to his own mortality. In reality, given the exceptional schedule of work he would fulfil especially when abroad, he was quite healthy. His migraine attacks were now much less frequent, and whilst he had developed chest pain and been found to have somewhat raised blood pressure, this was well controlled with tablets. But cutting branches off a tree that were spreading over his workshop at home one day, he slipped and fell through the roof knocking himself out for a short time and injuring his chest. Olive and Rachel who were helping him at the time were quite frightened by the episode, which left him in severe pain, making it difficult for him to turn over in bed, for most of 3 weeks. He had probably cracked a rib but did not seek any medical help.

His writings at this time were punctuated most weeks with reflections on his life as a Christian and his strong belief in life after death. Earlier when in Santa Monica at a conference, he had met Ernst Wynder who worked at the Sloan Kettering Institute in New York and was one of the first people to demonstrate the link between smoking and lung cancer. On hearing him lecture, Denis writes, "He gave an excellent talk and got a standing ovation but inferred that man had no afterlife."

At the back of the 1981 diary, he writes a confessional:

I look back in life to some particularly terrible actions, for which I have long since sought and received forgiveness. I remember during the war cutting the vas deferens[11] during a

---

[11] Tube-like structure that carries sperm from the testis to the seminal vesicle where it is stored prior to ejaculation. There is one for each testis. Both would be cut during a sterilisation operation.

hernia operation and not attempting repair being ashamed to acknowledge that I had done it. In retrospect it would probably have made no difference, but that does not lesson the shameful sinful attitude.

Once…I cut a femoral nerve, mistaking it for a tendon and didn't admit it. Once again repair would probably have been ineffective but I should have admitted my mistake. They reveal terrible flaws in my sinful nature and where would I be but for a God who forgives.

In 1982, he would meet head-on the challenges of working to prevent the common chronic diseases, Western or man-made diseases, by diet.

References

1. Burkitt DP, Trowell HC, editors. Refined carbohydrate foods and disease. Some implications of dietary fibre. London: Academic; 1975.
2. Yudkin J. Pure, white and deadly: the problem of sugar. London: Davis-Poynter Ltd; 1972.
3. World Health Organisation. Guideline: sugars intake for adults and children. Geneva: WHO; 2015.
4. Cleave HK. The founders of modern nutrition. General Editor Geoffrey Cannon. London: The McCarrison Society; Undated.
5. Department of Health and Social Security. Nutritional aspects of bread and flour. Report on Health and Social Subjects, 23. London: HMSO; 1981.
6. Royal College of Physicians of London. Medical aspects of dietary fibre. London: Pitman Medical; 1980.
7. IARC Intestinal Microecology Group. Dietary fibre, transit-time, faecal bacteria, steroids and colon cancer in two Scandinavian populations. Lancet. 1977;2:207–11.
8. Burkitt D. Unpromising beginnings. J Ir Coll Physicians Surg. 1993;22:36–8.
9. Burkitt DP. Don't forget fibre in your diet. 1st ed. London: Martin Dunitz Ltd; 1979.
10. Greenaway B, Kellock B. Hell's Angel. Tring: Lion Publishing; 1982.
11. Kellock B. The fibre man. Tring: Lion Publishing; 1985.

Chapter 24
'Man-Made Diseases'

The fibre story propelled Denis to celebrity status in a way that his discovery of the lymphoma had not done. A rare tumour of African children may not quite have captured the imagination of his new audience so fervently as did dietary fibre. Whilst fibre might have been viewed as just another dietary fad, the surgeon from Africa with no training in nutrition, stimulated and educated by Walker, Painter and Cleave, quickly realised that it was key to the prevention of bowel cancer. Then, with the vision of Trowell, Denis saw a completely new concept emerge, which would explain the huge disparity in the occurrence of heart disease, stroke, diabetes, various cancers and the other non-infective diseases of the West with their rarity in Africa. With the triumph of his lymphoma work to give him credibility, the enormous importance of these killer diseases and Denis' realisation that prevention, in this case by diet, is all important, he had a new gospel to preach. The concept of Western diseases and their prevention is probably the most significant contribution to health made by Hugh Trowell and Denis either jointly or individually, although it is not what either of them are remembered for.

In Trowell's seminal book *Non-infective Disease in Africa* [1], he describes the infrequent occurrence in Uganda and 'indigenous inhabitants of Africa south of the Sahara' of coronary heart disease, diabetes, peptic ulcer, high blood pressure and various bowel and urinary tract disorders. He is intrigued by the rarity of these non-infective conditions in Africa when they were the major killing diseases in Europe, North America, Australia and New Zealand, the so-called Western or industrialised countries. These are diseases normally seen by a physician; therefore, the list does not include more surgical conditions such as bowel cancer, but the concept of non-infective diseases is original and credit for it must go to Trowell. He already had a worldwide reputation for his work on kwashiorkor and now, by bringing these conditions together under a single heading, he was able to suggest there may be a common factor in their cause. There is no significant discussion of diet in the book, or even of fibre, but the idea of non-infective diseases evolved into 'Western diseases'

© The Author(s), under exclusive license to Springer Nature Switzerland AG 2022   343
J. H. Cummings, *Denis Burkitt*, Springer Biographies,
https://doi.org/10.1007/978-3-030-88563-2_24

[2] and is the forerunner of the now essential concept of non-communicable or chronic diseases (NCD) adopted by the WHO [3]. The contribution of Cleave to this progression of ideas deserves mention, although it is highly unlikely that Trowell would have read Cleave's 1956 paper on 'natural principles' and 'The Saccharine Disease' was not published until 1966. However, Cleave regarded diet, particularly sugar, as key to understanding these diseases, whilst Denis and Hugh Trowell came to diet through fibre. Denis and Hugh worked closely together for the next 15 years following their reunion at the Albert Cook Memorial Meeting in Kampala in 1970. Trowell brought to the partnership his skills as a physician, long-established list of contacts in the medical world, his natural ability as an editor and patience to search the literature. Denis brought a proven reputation as a geographical pathologist and the belief that diet, particularly fibre, was the key to preventing Western diseases.

Trowell watched as Denis travelled the world to great acclaim, in demand as a speaker and receiving honours and awards, whilst he was somewhat in the background not getting the credit he felt was due to him for an idea he had first promulgated in 1960. Trowell's world travel was however restricted because his wife Margaret (Peggy)[1] with whom he had a genuinely loving and caring relationship was struggling with dementia. Nor was he naturally a combative person, as was Cleave. He had been trained in the doctrines and liturgy of the Church of England before practising as a parish priest for 10 years in Stratford-sub-Castle, a suburb of Salisbury, so there was never a falling-out between him and Denis. But quietly and privately and not addressed to anyone but himself he set down in a memo dated August 15, 1971, a 'Timetable of all opinions received concerning my views on the different disease patterns in Western man and Africans? Atheroma? Fibre, evolution etc'.[2] In this he gives the dates in 1971 when he discussed fibre with McCance, Doll, Painter, Burkitt and others and his need to gain their support for his ideas on the role of fibre in control of atheroma, serum cholesterol and blood clotting.[3] Denis, however, picked up the idea of Western diseases very quickly and was using the term in his publications from 1973 onwards [4].

Trowell decided to follow up the success of their book *Refined Carbohydrate Foods and Disease* with a more updated version but including a wider perspective than just fibre. Denis, quite sensitive to ownership issues, realised that Trowell had in mind a new version of his 'Non-infective Diseases' and wanted to align the two concepts, of non-infective diseases and fibre, and perhaps bring himself back into the picture. Denis initially said that he would drop out of the shared editorship of the proposed book with Trowell, but Trowell felt somewhat exposed by this suggestion and quickly they agreed that Trowell would be the lead editor and manager of the project. The Trowell/Burkitt book *Western Diseases: Their Emergence and Prevention* published in 1981 became quite influential. The intent was to discuss

---

[1] Margaret Trowell (1904–1985) was a highly talented artist and teacher who had founded the School of Fine Arts at Makerere, which became known as the Margaret Trowell School of Industrial and Fine Arts.

[2] Wellcome Library, Papers of Hugh Trowell, PP/HCT/A.6.

[3] Trowell would have met these people as part of his quest to find a new definition of fibre.

diseases of civilisation, but it is made clear in the 'Preface' that this was an inappropriate term because there was evidence that some of the diseases discussed in the book, such as diabetes, had existed in the ancient civilisations of Egypt, Greece, Rome, India and China. The book takes a much wider look at the emergence of these diseases than did their first book together, very much because of Trowell's contribution. In their introduction, they describe Western diseases 'as essentially man-made diseases', thus paving the way for a broad approach to their eradication.

In 2001, long after Denis and Hugh Trowell had died, the IUNS (International Union of Nutritional Sciences) gave leaders in the field of nutrition and other scholars inside and outside the academic community a task 'to formulate a new definition, new goals and a new conceptual framework for a science fully equipped to meet the challenges and opportunities of the world in which we now live'. At a workshop held in Giessen in April 2005, attended by many of the world's leaders in nutrition, a declaration was agreed unanimously that now was the time to 'practice nutrition as a biological and also a social and environmental science'. A special issue of the *Journal of Public Health Nutrition* was published, and a poll was taken of all participants at the meeting asking which writers had most influenced them and who would they most highly recommend to students of the new nutrition science. Denis Burkitt was one of a small group of highly recommended authors and the outstanding source mentioned most often in the poll was the Trowell/Burkitt book *Western Diseases* [5]. The journal says of them:

> Earning reputations as indefatigable campaigners, they did more than any others to identify chronic diseases of most systems of the body as having environmental causes, and in particular broadly the same types of inappropriate nutrition.
>
> It is part of the foundation and structure of what is now established thinking on food, nutrition and the prevention of chronic diseases. Ahead of its time, some of the proposals remain ahead of our times.

Sotheby's, the fine art auctioneers, chose 'Western diseases' as one of the items to go into a time capsule they set up, sometime between 1981 and 1992, to be opened 50 years later.[4]

From 1980 onwards, Trowell was usually the lead author on anything he and Denis published together including scientific papers, other books, letters to journals and even minor articles for the popular press. On receiving an early copy of *Dietary Fibre, Fibre-Depleted Foods and Disease* in 1985, to which he had contributed four chapters and was a co-editor with Trowell and Heaton, Denis remarked 'Received copies of Trowell's new book'. This very much reflected his new working relationship with Trowell. But now with three jointly edited books, Denis realised that there were more immediate ways of spreading the message about diet, especially fibre, through radio and television, articles in both the scientific and popular press and his schedule of lectures at home and abroad. Neither Denis nor Trowell could have

---

[4] This information about Sotheby's is from a videotape recording of a lecture Denis was due to give to the Fourth Washington Symposium on Dietary Fiber that took place in April 1992. He declined the invitation to attend, because he had recently been in hospital, sending the video instead.

anticipated that 20 years later 'Western diseases' would become so influential and that they had worked together as such an effective and complementary team.

As the fibre story developed, Neil Painter felt increasingly that he was being left out of it, and he was. Whilst Denis constantly credited Peter Cleave with being the founding source of his ideas about diet and health, it was Neil Painter and also Alec Walker who brought fibre to Denis' attention. Painter had done original studies of diverticular disease identifying lack of fibre in the diet as the major cause and starting a revolution in its treatment. Like Denis, he was invited to meetings all over the world and continued to write about the subject for more than 10 years after the publication of his iconic study of the treatment of diverticular disease with bran in 1972. But he was never a contributor to the fibre hypothesis as it developed embracing other bowel diseases, such as cancer and appendicitis, and its widening to include heart disease and diabetes. Painter was a busy surgeon at the Manor House Hospital in London without access to research facilities, and those papers on which he was an author in later years were often written with Denis. He had made his contribution, but in a letter to Helen Cleave in 1987, he writes, 'Joy (Painter's wife) says that I was a fool and so was Peter to fill Burkitt with all our ideas'.[5] In other correspondence, he disagrees profoundly with Trowell's detailing of the history of fibre in which Trowell once again claims the initial idea was in his 1960 book about non-infective diseases. It was frustration rather than a real loss of entitlement. He had received worldwide acknowledgement for his work on diverticular disease, having changed medical thinking about its cause and management radically and had written a book about it published in 1975 [6]. Painter wrote to me around this time asking what research he should do next. I replied saying that he had made his mark and been recognised for it, an achievement not many could look back on. He retired early to pursue his hobby of photography, which he had started when a pilot in the Fleet Air Arm where he had served with distinction from 1941 to 1946.

Alec Walker spent his whole working life in South Africa and so was rather cut off from the rest of the world for a time, but he wrote up the findings of his research, mainly observational and experimental diet studies in adults on a wide range of nutrients, in the scientific press and was discussing fibre long before either Denis or Neil Painter. His work was very far-sighted, carefully done and always published in good journals. Visitors to his house in Johannesburg found him and his wife Betty to be engaging and generous hosts, highly active with their work and in conversation were well informed, enthusiastic, always interested in what you were doing but never claimed anything for their own studies. But one left them always with new ideas and needing to take a more critical look at one's own work.

Perhaps it was Peter Cleave's sad journey into dementia and his death in 1983 that gave Denis a new freedom to pursue the fibre and Western disease hypothesis without having to look over his shoulder for incoming criticism anymore. By 1983, he was travelling the world very much with support from the food and drug industry,

---

[5] Neil Painter (1923–1989). Many files of his correspondence can be found in the Wellcome Library under PP/NSP/C. The sentiments expressed in some of them are quite strong so in fairness to all parties are not quoted here.

particularly Reckitt & Colman and the Kellogg Company with his audiences now more varied including nurses, doctors, nutritionists, medical students, senior school pupils and non-scientific societies anxious to know about this new healthy diet. He was still in demand at international events and recognition of his achievements continued with the highly prestigious Bristol Myers Award for Distinguished Achievement in Cancer Research, which he shared with Tony Epstein as the first recipients outside the USA. The equally notable Charles Mott General Motors Cancer Award, the Lucy Wortham James Clinical Research Award and the Diploma and Gold Medal of the French Académie Nationale de Médecin all in 1982 were followed in 1983 by the Beaumont Bonelli Award for Cancer Research. Denis enjoyed these occasions but found the amounts of money that came with some of them quite embarrassing. Bristol Myers was $50,000 US, General Motors $100,000 US and the Italian prize 2 million lira. A tenth of his and Olive's income went to Christian charities, but Judy and Philip were given £1000 to invest, whilst Rachel and David received £1000 for Edward. The gift for Edward was significant. Rachel had given birth to Edward in January 1982, and it was quickly discovered that he was a Down's baby. Denis and Olive hearing of this whilst having lunch with Judy were in tears, but the family came together, and they were all supported by the prayers of their friends in the church. Later that week on learning that Bristol Myers would also pay for First Class fares for Olive to accompany him to New York, Denis wrote, understandably, 'What do wealth and honours count?'.

Denis was a little ashamed when it came to the event because he and Olive enjoyed what he described as 'ridiculous luxury'. It contrasted in his mind with the early years in Africa when he had travelled in an old station wagon with his two colleagues, camped when they could to avoid hotel costs and bought the cheapest food in the market to sustain them whilst laying the foundations of the lymphoma story. Staying one night on the shores of Lake Nyasa, he recalled 'bat droppings falling on us from a tattered hessian ceiling and the bed linen far from clean'. Having made these contrasts, he and Olive reminded themselves that their happiest memories were of times spent in the simplest of circumstances but in an atmosphere of love and laughter 'for contentment can never be bought and is independent of material possessions'. The occasion brought home to Olive Denis' iconic status. Perhaps some recompense for the many weeks he spent away from home and she was grateful and appreciative of 'the unstinting generosity and helpfulness and courtesy shown to us on every occasion' as they were piloted through the days of ceremonial and publicity that took place on both sides of the Atlantic.[6]

The year 1983 was a defining year for Denis. The resources he now had to hand allowed him to travel, respond to the countless invitations to speak that came in and write about his work. But there was no opportunity to gather further data on the geography of Western diseases nor to add anything more to the fibre story. It had been taken over by scientists, public health bodies, governments, the food industry,

---

[6] In his Autobiography, Chapter 19, p. 4–7, Denis gives a more detailed and emotive description of this Award Ceremony than he does for any other such event.

Fig. 24.1 At the Vahouny Conference in Washington, 1984
Left to right: Sheila Bingham – MRC scientific staff, Dunn Clinical Nutrition Centre, Cambridge; the author; Alec Walker; Isidor Segal – physician at Baragwanath Hospital, Soweto, South Africa; Denis Burkitt and Martin Eastwood

the media and the lay public anxious as ever to find the ideal diet for their health (Fig. 24.1).

He remained very busy as, for example, his domestic programme between June 7 and 23, 1983, shows:[7]

> 7th Train to Northampton via Birmingham. Good evening meeting for Reckitt & Coleman
> 8th Train to London. Picked up at Thomas's & driven to meeting in Hotel on edge of Epping Forest. A good attendance.
> 9th Morning discussing filming with David Prill and film Director (?name) Driven to Hornchurch by Peter Scott. About 50 at meeting chaired by C.M.F. Doctor Burnham.
> 10th Layton(sic) Buzzard via London. Picked up by Standford's secretary. Spoke at Nutrition day at Aylesbury. Landslide victory for Conservatives.[8]
> 11th Lucy's birthday party.
> 12th (Sunday – always a day of rest for Denis)
> 13th Cut lawns and orchard. Wonderful to be able to enjoy doing this.
> 14th Tom Shiplake drove me pm to Penzance. Arrived Old Smugglers Inn overlooking sea at 8.15. (This would be a journey of 270 miles by road).

---

[7] The spellings and punctuation are as near as possible faithful to Denis' handwriting.

[8] General Election in the UK. Mrs. Thatcher returned to power with huge majority following Falklands (Malvinas) War.

15th Spoke at lunch meeting in Penzance hospital. Then driven to Truro. <u>Very</u> good evening meeting in Post Grad Centre. Many good talks.

16th Long 340 mile drive to Haverford West in West Wales. <u>Very</u> good meeting at Post Graduate Centre.

17th Driven home.

18th Lovely day. Very busy with grandchildren.

19th (Sunday) Preached Eastcombe.

20th In London for lecture at L.S.H.& Trop Med.

21st To Sheffield by train. Driven to Bawtry. Video tape made of my talk.

22nd Radio Sheffield interview. Driven to Last Drop Inn outside Bolton. Pub lunch with film crew. Televised talking to Michael O'Donnell. Over 160 at evening lecture.

23rd Driven short distance to luxurious Balfry Hotel at Handford. Good meeting but poor chairman.

11 journeys, 12 talks, one TV appearance, one video, work for one film (movie), 5/17 days at home. Aged 72.

This type of agenda was fairly constant throughout the year and was organised to be possible between his overseas visits, which were 15 in 1983, 7 in 1984 and 10 in 1985 where his schedule would be even more hectic especially when in his favourite destination the USA. In March 1986, whilst in Memphis, Tennessee, he delivered ten lectures all at different sites over 3 days and flew out to Washington at the end of it. There was clearly a special relationship between Denis and the medical and scientific community in the USA. Returning home from one such trip in October, he found a 'Letter from 3 girls written by Judy pointing lovingly that I was away from Mummy too much'. Olive continued to get bouts of depression not helped by Denis' absences. He noted the letter, but he was on a mission to save the world, which Olive understood. She enjoyed travelling with him to some of the more prestigious events to which he was invited, whilst at home she was the matriarch of a large family including nine grandchildren. Denis was taking more interest in the family and their affairs, especially financial needs, as well as noticing political and other events such as the miners' strike in 1984, election results and the Falklands War in 1982 about which he writes that he was very disturbed. This was surprising, given that for 20 years he had seemingly ignored the rewriting of history that went on around him changing the political map of Central and East Africa.

But the world was starting to close around Denis. Isobel, victim of those fateful air raids on the hospital in Plymouth in 1941, died on Saturday, February 19, 1983. 'Learnt that Isobel had been called home – suddenly. A great soldier and sufferer finishing her well-run race. Some tears, but joy at her release'. He, and Olive as well, had known Isobel since those fateful days in 1941 when life changed so drastically. She had lived a restricted life due to her injuries but had remained a family friend over the years keeping in touch with Denis.

It was the end of an era. Although his schedule remained very busy, largely in the hands of Reckitt & Colman and Kellogg's, he no longer accepted all invitations but remained a charismatic speaker who attracted often large audiences and when travelling abroad his visits would usually be given generous TV and newspaper coverage. His favourite talk was now 'Diet and Health', which was always captivating. It would be illustrated by his daughter Judy's cartoons, which illuminated what was now his main message, that we must all turn our efforts towards preventing these increasingly prevalent killer diseases of our industrialised societies. In addition to the mopping up drawing, favourites with his audiences were the wonky shelves and hospital car parks. At the same time, he could more than hold his own in the highest levels of scientific debate about his work. Newly appointed to a WHO Sub-Committee on Medical Research, he gave, at the first one he attended in 1985, a 12-minute commentary on the proposals under discussion. There was a standing ovation from the Committee members, a rare event within the hallowed walls of the WHO in Geneva – so rare that one of the officials told him at the end of the day that it was the first time such a response had been made to any speaker at the Committee in the 6 years since it had started (Figs. 24.2, 24.3 and 24.4).

After 12 years in Shiplake at the Knoll, they moved to The Old House, Bussage, to be near their daughters, but Olive was not happy there. They were fortunate in 1986 that Hartwell Cottage, adjacent to Paulmead in Bisley where Judy and family lived, came up for sale and they were able to purchase it in December 1986 for £155,000. It was a very old house and in need of substantial upgrading, but Denis saw this as a challenge and used his considerable practical skills to good effect. But for the more structural work on the kitchen area, he had to employ builders, which cost him £16,000, but he was able to pay for this with the $25,000 US initial

Fig. 24.2 The 'Floor moppers' cartoon. Drawn by Judy Howard (Burkitt). With it, Denis would make the point that floor moppers are failing to obviate the problem of water on the floor by turning off the tap on the overflowing basin. The floor moppers represented those who treated diseases, whilst the tap turners are working on prevention through diet

a

b

Fig. 24.3 'Wonky shelf' cartoon by Judy Howard (Burkitt) used to good effect by Denis in his lectures on the prevention of man-made diseases. A. Shows a man repairing broken pots that have fallen off a wonky shelf and is being congratulated for testing a new wonder glue, which has commercial potential. B. A man is saying, 'I have a new idea. Why not straighten the shelf?'

Fig. 24.4 'Hospital car park' cartoon. On the left are the cars of the doctors who treat diseases, 'Floor moppers', and on the right of those who work on their prevention, 'Tap turners'. Always explained in a humorous manner by Denis. Cartoon by Judy Howard (Burkitt)

payment for work on the film about fibre during the early part of 1987. His lifelong interest in carpentry, inherited with the tools from his father James, meant that he had become quite skilled, having also taken practical courses in his early years. He found doing these essentially domestic jobs very therapeutic and fulfilling alongside his impossibly busy and now much more combative lecture schedules. When talking about his lymphoma or fibre and colorectal cancer and the broader dimension of Western diseases, he was listened to by audiences who were eager to hear him and receptive to his account of his work, which was proving transformative. But now, more than ever convinced that prevention was more important than treatment for the medical ills of Western society, Denis was challenging the very structure and ethos of healthcare the world over. Putting the roof on his newly built garden shed, with the help of Philip, or tiling the bathroom gave him a chance to reflect.

Then there was news of Peggy, dearest Peggy, elder sister, companion and playmate throughout the early years of family life in Enniskillen. Following an attack of chickenpox in 1935, when aged 28, she had developed a severe mental health problem and had been taken by James and Denis to a care home in Dublin eventually being moved to a long-term psychiatric unit in Omagh at the inception of the NHS in 1948. Judy, who had been on holiday with her family in Fermanagh in 1985, had met Dick Smith and Lizzie Corrigan, the latter a former maid of Gwen, Denis' mother. In conversation they asked if Judy knew whether Peggy was still alive. Judy, who had never known Peggy, rang Denis and discovered some of the history from him and that when he last saw her, she was in a psychiatric unit in the grounds of the main hospital in Omagh. Judy recalls driving to Omagh in the pouring rain and enquiring at the main hospital enquiry desk whether they had a patient known as Gwendoline (Peggy) Zelma Burkitt. They found her name and that she was in the psychiatric unit. Judy was directed to the communal sitting room where she recognised her aunt, the image of Denis' mother, whom Judy had known well. Peggy was unable to communicate, so Judy simply sat and held her hand. Apparently in earlier years, when James and Gwen would visit, they had been told that such visits were upsetting her, so they had subsequently kept away whilst all the time ensuring that her needs for clothing and money were met.

On return to Bisley, Judy reported the visit to Denis who passed the information on to Robin. Nothing happened, and when Judy spoke to their local Methodist minister, he told her that this matter was not up to her and to 'let it go'. But she did not and once again put pen to paper writing to Denis and Robin, making it clear that she felt Peggy was being neglected by the family. Denis, whose respect for Judy and her views was born of the faith that they shared perhaps in a deeper way than with his other daughters, was not sure what to do. Robin totally rejected the idea that anything would be gained by visiting Peggy and he did not. A year later, whilst on a visit to Belfast on April 6, 1987, Denis 'Drove to Omagh. Saw Peggy for the first time in over 40 years. Quite unrecognisable'.

Robin rang Denis on Tuesday, July 24, 1990, to tell him that Peggy had died. Denis wrote, 'Robin rang about 8 to say that Peggy had died suddenly Monday. A release and relief. She had really gone years ago. Contacted hospital about cremation etc.' This took place two weeks later in Belfast, attended by Robin following which Denis, Judy and Olive collected the ashes taking them to Trory church the next morning where there was a 'beautiful graveside service to deposit Peggy's remains in our parents grave'.

Neil Painter's death a year earlier gets a brief although somewhat enigmatic mention by Denis. 'Learnt of Neil Painter's fatal coronary. A man I was never near to & failed'. Never near to because he was not a Christian in the Burkitt mould and failed because Denis never managed to convert him. But they had worked together for more than 10 years and had published five quite respectable joint papers in good journals, and possibly in a more life-changing way, Painter had introduced Denis to the world of fibre at his lecture to the British Society of Gastroenterology in 1968. But they were opposites. Painter was a clubbable man fond of telling stories over a drink and with an impish sense of humour. His slides of 'a typical Trades Union Movement' always raised a laugh from his audience, laying its emphasis on what would be typical in a diverticular disease patient. Although troubled with a wife and son, both of whom had depressive illnesses, he was always good company and loyal to those who were prepared to acknowledge his contribution to the early days of the fibre story.

Hugh Trowell died a few weeks before Neil Painter. He was a man of faith, brought up in a household where his father was a congregational minister and eventually becoming ordained into the Church of England himself. He shared very much the same beliefs as Denis and had spent almost his whole career with the Colonial Medical Service in Africa. They worked together in Africa for 12 years, Trowell showing Denis his first case of the childhood lymphoma. Later, back in the UK, Trowell brought a new and enlightening dimension to the fibre story with his revelation to Denis that it should include atherosclerosis and heart disease. In later years, the Trowells had been regular visitors to the Burkitt household. Trowell had been ill for some time with prostate cancer, and Denis notes on July 24 1989 'Hugh Trowell died last night. Really a timely release'.

Denis wrote obituaries of Trowell for *The Times* [7], *British Medical Journal* and a longer personal memoir for the McCarrison Society 'Founders of Modern Nutrition' series,[9] in which he paid tribute to his achievements. He gave a eulogy, alongside the clergy, at his funeral, and a few days later after a memorial service at Marylebone Church in London, there was a celebration of Trowell's life at the Royal College of Physicians. Here Denis described his friend as 'one of the greatest, most original and influential biological scientists of our time'[10] highlighting his work on kwashiorkor and fibre and introducing the concept of Western diseases. For Denis' part, the detail in these accounts shows that he fully understood the contribution that Trowell had made to the fibre story and was familiar with his life and work both in medicine and the church. The McCarrison Society paper is a remarkable first-hand record of medical insights in Africa that changed the world and a measured farewell to a lifelong colleague, who shared the same faith and broadened so crucially Denis' vision. Of their relationship, Hugh Trowell had written in earlier years to Denis and Olive, on the occasion of their 40th wedding anniversary describing the joy, anxiety and sorrow that family life can bring and saying, 'Our professional friendship (between Denis and Trowell) has been blessed with much help for one for the other, never marred by strife, dissension, friction or rivalry'.

Throughout all this, Denis remained hugely focussed on changing the world by moving the emphasis in medicine to prevention rather than treatment, buoyed up by his firm belief that God was working through him. His output of papers in the scientific press was now much less than when he was actively involved in research and the subject matter had changed completely. No longer were there papers on the lymphoma, its clinical features, geography and treatment nor on bowel cancer, varicose veins and appendicitis. The subject matter now included the broader dimension of fibre with titles such as 'Putting the Wrong Fuel in the Tank', an argument that our modern lifestyles were the main cause of Western diseases, 'Potential Prevention of Cancer', 'Lessons from Our Palaeolithic Ancestors' and 'The War on Cancer— Failure of Therapy and Research' which he published with Norman Temple from Athabasca University in Canada with whom, latterly, he developed a valuable working relationship. The world had moved on from the early and exciting days of his

---

[9] Burkitt D (probably 1989 or 1990), Dr. Hugh Carey Trowell. Wellcome Library GC/198/B/2/6.
[10] Elizabeth Bray, Hugh Trowell: Pioneer Nutritionist, 1904–1989. Wellcome Library, Papers of Hugh Trowell, PP/HCT/A.5.

lymphoma studies. About a meeting at Wadham College Oxford in April 1989 to celebrate the 25th anniversary of the discovery of the Epstein-Barr virus, at which he spoke, he wrote in a letter to Norman Temple, 'It was so complicated I didn't attend a single lecture as it was entirely in a language which I do not understand'. He continued to travel frequently, the USA and Canada being his main destinations, giving 50–60 talks a year. The standing ovations kept coming as did the honours. The County of Fermanagh in Northern Ireland wanted to give him the Freedom of the City of Enniskillen, but 'Council changes preclude these honours', and so he was invited to dinner and presented with a picture of Devenish Island and the Fermanagh Coat of Arms.

His 80th birthday in 1991 was suitably celebrated by the family but not marked by any change in pace for Denis. However, he began to notice that his ability to take on the impossible schedules of travel and lectures that he had now done for almost 20 years was becoming noticeably harder. In June, during a 23-day trip to North America, he gave ten lectures over a period of 8 days and remarked afterwards that it was the 'End of a heavy schedule'. On return home he, unusually, suffered from jet lag. Sleeping badly and then developing a cough and some breathlessness when out walking he felt unwell and thought he might have contracted infectious hepatitis. Olive noticed this and arranged for him to see the family doctor resulting in Denis' admission to hospital for tests and his being found to have high blood pressure and mild heart failure. He was put on the appropriate treatment, and by the end of October, he was in London for the first time in 4 months to give a lecture.

He was not idle during this period spending considerable time working with two people, Dr. Norman Temple and Dr. Ethel Nelson. Temple[11], a nutritionist with a serious interest in what he felt was the inflated role of the drug companies and medical profession in focussing on the treatment of disease to the exclusion of its prevention. An important philosophy he shared with Denis. Early in his career, he had started to question current nutritional wisdom in papers such as 'Refined Carbohydrates – A Cause for Suboptimal Nutrient Intake' [8] and was aware of Denis and his work having read *Western Diseases*. Temple wrote to him in 1989 asking if he would contribute the 'Preface' for a book he was co-editing [9] on nutritional strategies for disease prevention, but Denis declined on the grounds that he could not give the needed time to it but agreed to look at the manuscript and send some comments back. For all Denis' recent ill-health, age and now once again his busy schedule he sent back to Dr. Temple, within a few days of receipt of the manuscript, a magisterial analysis of the text leaving its author in no doubt that here was someone who was no elderly eccentric but highly intelligent, well informed, a skilled editor, determined but reasonable. This started a correspondence between the two men who quickly realised they had much in common in their outlook. They were both convinced of the need to try and get the world to understand that however early you diagnose a disease, how well you treat it and despite the enormous

---

[11] In 1991, Norman Temple, a nutritionist by profession with a PhD in biochemistry, was a member of the Animal Sciences Division at the Alberta Environmental Centre, Vegreville, Canada. He would later move to Athabasca as Professor of Nutrition.

budgets promised for research, no progress can be made until the cause is found. Following on from this comes prevention, the subject of their first paper together [10] 'The War on Cancer-Failure of Therapy and Research' which, after publication in the *Journal of the Royal Society of Medicine* in February, produced an immediate analysis of it in *The Daily Telegraph*, a leading daily newspaper in the UK. Denis, now with Dr. Temple, was still making waves.

The following year (1990), Denis decided to modify the schedule for his next trip to the USA and divert to Canada in order to make contact with Dr. Temple whom he had never met. They were able to spend time together in Edmonton on June 5 and within a few weeks had agreed to write a book together intended as a sequel to *Western Diseases*. Denis felt that the original book on this theme, ostensibly co-edited by himself and Hugh Trowell, had really been almost entirely managed by Trowell. It had been a means for Trowell to update his earlier work on *Non-infective Disease in Africa* and secure his claim to the concept of diet as a principal factor in the cause of the man-made diseases of Western cultures. Now Denis had the chance to put more of his own views into print working with a new partner who needed no persuasion about the validity of these ideas. And so *Western Diseases: Their Dietary Prevention and Reversibility* was written, mainly by Temple and Denis with contributions from Alec Walker, Ken Heaton and others. The book [11] was not published until after Denis' death in 1993, but together he and Temple continued to collaborate pursuing the theme that a new system of health was needed to meet the challenge of Western diseases [12] and remarking in their continuing correspondence that the WHO had now taken up their ideas in their most recent major publication on diet and health [13]. In reality, the WHO never used the term 'Western diseases', given that the conditions in question were not geographically confined to the 'West' and were starting to emerge in most countries of the world. Instead, they preferred the term 'chronic non-infectious diseases' in their 797 Report although readily acknowledging that the adverse health effects 'of the affluent diet prevailing in the developed industrialized countries – characterised by an excess of energy-dense foods rich in fat and free sugars, but a deficiency of complex carbohydrate foods (the main source of dietary fibre) – have only become apparent over recent decades'. The WHO would stick to the term 'Chronic diseases' in their subsequent Report 'Diet, Nutrition and the Prevention of Chronic Diseases' [3] but by then the principal was established and for this Denis could claim to have played a part. He had progressed from the lymphoma through dietary fibre to take on, undaunted, the health problems of the world.

Ethel Nelson, whose biography of Denis [14] published in 1998, was a medical missionary in Thailand and a member of staff at the New England Memorial Hospital in Stoneham, Massachusetts.[12] As a pathologist, she was aware of Denis' work on the lymphoma and in 1971 had heard him give his landmark first paper on fibre and colon cancer at the American Cancer Society meeting in San Diego. She was intrigued by his suggestion that fibre was involved and why he had become

---

[12] A Seventh-Day Adventist facility latterly known as the Boston Regional Medical Center before closing in 1999.

interested in nutrition. Three years later, she and her surgeon husband invited Denis to speak at a 1-day conference about advances in nutrition on which occasion he stayed with them in their home. Not only were the Nelsons delighted by his visit but when Denis told them he been invited to talk to the CMS that evening their ties were strengthened by a common faith. This understanding probably contributed greatly to Denis agreeing, many years later, to her writing his biography. They met again in 1990 when Denis was in Ottawa where he told her that he was happy for her to write about him despite the prior Kellock biography because he wanted the inspiration that his family had given him to be emphasised, credit to be given to his medical friends and colleagues and 'Most of all, I want to glorify God for His providential leading in my life'. They started by discussing his life in general terms, and on his return to the UK, Denis sent her a copy of his unpublished autobiography and started going through his diaries making tape recordings of the essential information he wanted her to include. Nelson never read the diaries herself, but over the course of a year, they exchanged much correspondence such that by May 1991 she and her sister Francis visited Bisley and gave Denis a first draft of the biography for him to comment on.

Olive, always guardian of the great man's reputation, made time to read the manuscript and was not pleased with it. How much was changed we shall never know but there was little need for Olive to protect Denis' reputation.

References

1. Trowell HC. Non-infective disease in Africa. London: Edward Arnold; 1960.
2. Trowell HC, Burkitt DP, editors. Western diseases: their emergence and prevention. Cambridge, MA: Harvard University Press; 1981.
3. World Health Organisation. Diet, nutrition and the prevention of chronic diseases, WHO Technical Report Series 916. Geneva; 2003.
4. Burkitt DP. Some diseases characteristic of modern Western civilization. Br Med J. 1973;1(5848):274–8.
5. Public Health Nutrition. 2005; 8(6a):800–4.
6. Painter NS. Diverticular disease of the colon – a deficiency disease of Western Civilisation. London: William Heinemann Medical Books Ltd; 1975.
7. The Times, 28th July 1989.
8. Temple NJ. Refined carbohydrates – a cause for suboptimal nutrient intake. Med Hypotheses. 1983;10:411–24.
9. Wilson T, Temple NJ, editors. Nutritional health: strategies for disease prevention. New Jersey: Humana Press; 2001.
10. Temple NJ, Burkitt DP. The war on cancer-failure of therapy and research. J Roy Soc Med. 1991;84:95–8.
11. Temple NJ, Burkitt DB, editors. Western diseases: their dietary prevention and reversibility. New Jersey: Humana Press; 1994.
12. Temple NJ, Burkitt DB. Towards a new system of health: the challenge of western disease. J Comm Health. 1993;18:37–47.
13. WHO Study Group. Diet, nutrition and chronic diseases, WHO technical report Series 797. Geneva; 1990.
14. Nelson ER. Burkitt, cancer, fiber. New York: Teach Services Inc; 1998.

Chapter 25
Preparing for Departure

The final 15 months of Denis' life were more gentle. The hospital admission for heart failure had given him notice that he was not immortal and he was now reminded of this daily as his tolerance for exercise was much reduced. His very long-standing friend, Dr. Ted Williams, who had accompanied him on the long safari, climbing Mount Kilimanjaro and with his wife and brother, played host to Denis on his travels, died in October 1992. Denis realised that he might be 'called' at any time. Writing in his diary on October 3rd he says 'Learnt that Ted crossed the River and entered the Gates of the City last night. "Well done thou good and faithful servant"', raising hope in Denis that he might be welcomed in the next life as he was sure his soulmate and at times mentor in faith had been. He set about putting his affairs in order in a responsible way. Overseas trips were now very much curtailed. He declined an invitation to attend the 4th Vahouny Conference on fibre in Washington in April but was persuaded to write an introductory chapter for the book [1]. This, like his talks and other writing, was now more historical in its perspective and auto-biographical in content paying tribute to those who had influenced his thinking about fibre and Western diseases.[1] He started gathering what he called memorabilia together to send to Trinity College Dublin and the Wellcome Tropical Institute, both of which now have substantial archives of Burkitt papers. He arranged for money to be passed over to the three families of his children and must have written valedictory letters to key people from his past because one provoked a reply that must have given him a sense of fulfilment.

Sir Ian McAdam, Head of the Department of Surgery at Makerere where Denis had worked for 18 crucial years, wrote:

> I find difficulty in replying to your very generous letter…Denis you know only too well that progress in any field of investigation is a team effort, that an original observation requires

---

[1] Denis mentions Walker, Cleave, Trowell, Eastwood, Heaton, Painter, Southgate, Cummings, Van Soest, Anderson and Jenkins.

© The Author(s), under exclusive license to Springer Nature Switzerland AG 2022   359
J. H. Cummings, *Denis Burkitt*, Springer Biographies,
https://doi.org/10.1007/978-3-030-88563-2_25

the spark which ignites the imagination of people thousands of miles away. I must remind you of the six pathologists waiting outside the theatres of the old Mulago Hospital for a piece of the Burkitt tumour. That single episode ensured that every continent on earth had the ability to contribute to the work you started.

The story of arriving at the conclusion that this lymphoma was indeed a "new" tumour, and that this was recognised by you as the end of that particular road – that others with a very different background should now take over with their highly sophisticated and special-ised medicine.

It would have been a tragedy had you continued to concern yourself with your tumour, because the medical community was ready to be taken aboard the next idea which was perhaps more important than lymphomas. The idea of geographical medicine was not new but it required an enthusiast with ideas and a type of leadership which would appeal to the medical officer in remote areas and make him proud to be a contributor. That you achieved this not with a multi-million-dollar research grant but with a camera, a tent and an old land-rover, speaks volumes for your approach.

Denis you earned the recognition you have been given: the Surgical department at Mulago, the Makerere Medical Faculty and the whole medical set-up are proud of your contributions – and no-one can minimise your achievement. Well done.

Love to all the family –

Ian

Other organisations around the world continued to acknowledge his achievements. Always close to his heart, Trinity College Dublin invited him and Olive to a sympo-sium at the Medical School in connection with the 400th anniversary celebrations of the founding of the University. He was one of eight former Trinity graduates from around the world chosen to speak for 15 minutes on their work. Denis was first and surprised his audience of 4–500 medical scientists by once again talking about the values and attributes other than intellect that shape a man's life. He and Olive were accommodated in refurbished rooms that were the same ones where he had met in 1930 with his Christian Union friends, who had made such a dramatic impact on him. He told the assembled scientists that although entering University as probably the most unpromising student of the year [2], through his friends in the CU, he had found a sense of identity, purpose and direction that had counted far more than intel-lect in shaping his future. His listeners were warm and sympathetic in accepting his ideas, and at the dinner in the evening, he was acclaimed as the most famous of living Trinity Medical Graduates. His comment on this was 'Rubbish!'

The Medical University of South Carolina wanted to give him an Honorary DSc, but he felt that he was not yet ready to start travelling to the USA and asked if it could be awarded in absentia. They agreed and sent him a fine desk chair with a suitably engraved plaque for his study. By September, he was feeling well enough to visit Canada where he gave two lectures in Cornwall at the annual CHIP meeting, then on to Ottawa to be given the Honorary Fellowship of the Royal College of Physicians and Surgeons of Canada. Further lectures were given in Toronto, after which he returned home having been accompanied throughout by Olive, whose concern for his health was growing. The Buchanan Medal of the Royal Society fol-lowed in November, but prior to that, he had been told by the Franklin Institute that he was to receive one of the USA's most prestigious honours in the world of science, the Bower Award and Prize for Achievement in Science, at a ceremony to take place in Philadelphia in January 1993. A television crew was dispatched immediately to make a video about him for the day of the presentation, and a few days later, the award was announced in the press. But Denis was worried by all this because he had been told that with the award came a cheque for $373,000 USD, the largest amount given for a prize in any field of science including the physical sciences in the USA at that time.[2] He did not know what he should do with such a large sum of money. Both he and Olive were horrified when the *Daily Mail* reported on the Award a few days later [3] emphasising the money aspects. He decided that apart from ensuring Olive had enough money for the future he would give it all away.

Leaving Gatwick on January 8, 1993, they flew to Atlanta where they were able to meet Ruth and Ralph Blocksma, their very long-standing friends, who were to accompany them to the award ceremony. Arriving in Philadelphia a couple of days later, Denis was thrown into a busy round of press interviews, dinner parties and tour of the Franklin Institute, a schedule he was very well used to. At the Awards ceremony, the prize for Business Leadership, Advancing Science through Business, went to Dr. Arnold Beckman[3] who had developed the first pH meter and later founded Beckman Instruments. Next came the award to Denis, who was introduced to the audience by Dr. C Everett Koop,[4] with the words:

> Dr Denis Burkitt, a surgeon with no formal training in research techniques, no access to lab facilities and with research grants of only $75 became in 1957 the first person ever to identify a link between viruses and human cancer. More recently using the same instincts for discovery, Dr Burkitt is examining why people around the world suffer from distinct diseases and how important diet may be in explaining the causes of disease.

The Citation for the Award read:

> For innovative and creative research, under extremely difficult circumstances in a tropical developing region, leading to the establishment of a virus-cancer linkage in the widespread childhood disease that has become known as Burkitt's lymphoma and leading to the redirection of cancer research and treatment throughout the world. For his inventive and methodical documentation of factors that explain the geographic distribution of disease among world populations. For his advocacy of the hypothesis implicating a deficiency of dietary fiber as a fundamental cause of health afflictions in the industrialised world and for his humanitarian devotion to the health of mankind.

The award was then presented, together with the large cheque that was troubling Denis, following which there was a reception at which he met many old friends from all over North America, then dinner. The next day, he gave his Bower Award Lecture. The audience must have been intrigued as well as being expectant because

---

[2] The Nobel Prize for Physiology and Medicine in 1992 was worth about $1,100,000.

[3] Dr. Arnold Beckman (1900–2004) who had developed the first pH meter and a spectrophotometer that could measure light across a wide range including UV. Later he went on to develop diagnostic instruments and pioneer the use of laser technology.

[4] Dr. C Everett Koop (1916–2013) was a US Paediatric Surgeon and Surgeon in Chief at the Children's Hospital in Philadelphia. He was conservative in outlook but had a distinguished record of public service becoming the 13th US Surgeon General working for the rights of children and especially the handicapped. He had first met Denis at Mulago a few months before the start of the long safari in 1961.

the title of his talk was 'Western Diseases: Emergence, Prevention, and Regression'. Those who were expecting a first-hand account of the discovery of the lymphoma and its subsequent impact on the world of cancer were about to be disappointed. He started with:

> If I had been talking to you towards the end of the last century, I would have opened my remarks by saying that the major health discovery in the last century was the fact that infectious disease, then the commonest cause of death throughout the world, was due to factors in the environment which could be controlled, namely microorganisms. And it was by controlling microorganisms, clean water, clean milk, adequate sewage disposal and so on that infectious disease was largely conquered. Treatment had nothing whatever to do with it.
>
> Now I am talking to you in the last decades of the twentieth century and the same applies to the chronic non-infective diseases which are the major health problems in this country. They too are due to factors in the environment which can be controlled. That is lifestyle, and once again treatment is going to play no significant role at all despite that we put so much emphasis on it.[5]

There followed, delivered with his usual panache and good humour, a 40-minute sermon without any data, any statistics, any graphs or bullet points, an exhortation to his audience to change their way of thinking about health and start putting their efforts into prevention. Commencing with his list of chronic non-infectious diseases he was quickly into the importance of diet and of large stools versus the small stools characteristic of Western communities, the failure to respect nature and evolution, thanking people who had inspired him on his journey to this enlightenment and finishing with his three favourite slides, the moppers-up, the wonky shelf and the hospital car park. It was clear by the end that the audience were with him and he received a standing ovation, of course.

Back at home, he was occupied mainly with editing chapters for the book he was working on with Norman Temple, then remarkably paid a quick visit to Boston to speak at an AAAC[6] Convention at the invitation of Harry Goldsmith. His 82nd birthday on February 28 was duly celebrated by all members of the family at a lunch in the house of Judy and Philip. The next day, what he called a catastrophe happened in which the contents of the village sewage system emptied themselves into their house. It was clearly a major disaster for him and Olive, requiring them immediately to move out to stay with family. Leaving Olive, once more, this time in charge of negotiations with the Water Board whilst he travelled to Stourbridge by train via Birmingham to give a lecture at the Corbett Hospital. A report of this by the meeting Chairman records: 'His talk then (on disease prevention and diet) was as ever forthright, clear, and punctuated by vigorous slide illustrations drawn by his daughters'. There he reminded his audience that 'life is not a sprint but a marathon' and that if young doctors are to make anything of their professional lives they must be 'prepared to work hard for long periods'.[7] It would prove to be his last. Home was

---

[5] Tape recording made at the meeting, now in the family archives.

[6] Possibly the AOAC, which in 1993 would have been the Association of Official Agricultural Chemists.

[7] Dr. Adria Hamlyn. Notes written for the first Burkitt memorial lecture at Stourbridge Hospital; 14.11.94.

uninhabitable and a dispute arose with the Water Board as to whose insurance policy would cover the costs of making good the damage done to the fabric of their
house, garden, furniture, carpets, etc. The Water Board relented and sent Olive a
bouquet of flowers, but stress levels were high for Denis. He decided to pull out of
a lecture tour of South America later in the year and began to feel unwell. On
Monday, March 15, he wrote in his diary, his last entry after 71 years of diary
keeping:

> Still feeling unwell. Spent day writing. Judy asked to be Assistant Diocesan Youth Training
> Officer with a teacher's salary. She is just the person for this and it is a real accomplishment.
> Wrote cheques for family members from Bower Award.[8]

On the 16th, he suffered a stroke and was admitted to Gloucester Royal Hospital
where he died on Tuesday, March 23, 1993. His funeral took place shortly afterwards at the Parish Church in Bisley. It was attended by many people and officiated
by Rev Richard (Dick) Drown. Denis was buried in the village graveyard at Bisley.
A Memorial Service followed on May 12 at All Souls Langham Place, under the
auspices of the Christian Medical Fellowship. It was attended by the High
Commissioner for Uganda. Prebendary Richard Bewes officiated, assisted by Dick
Drown. Judy read two passages from the Bible, and tributes were paid by Dr. Murray
Baker, Dr. Keith Sanders, Dr. Jack Darling and Dr. David Maurice (Rachel's husband). It was attended by Olive, other family members and representatives of the
wide range of organisations who had honoured Denis, including people with whom
he had worked both in Africa, North America and the UK and by representatives
from industry (Fig. 25.1).

To conclude, here are some of Denis Burkitt's remarks about his life, mostly not
mentioned elsewhere in the text:

> What do you possess that was not given to you? If then you really received it as a gift why
> take the credit to yourself' (I Cor 4:7. New English Bible). Framed and above the desk in
> Denis' study

> How superficial is most dinner talk'. Mulago 1948

> One of Olive's friends was due to be presented and the two stood close to each other as the
> Queen Mother approached. Denis was behind Olive but felt that as Her Majesty chatted to
> Olive's friend that 'she might almost have been chatting to me'. (A true royalist) Queen
> Mother's visit to Makerere in 1959.

> Discussing Judy's wedding. I mustn't be miserly as I would tend to be without Olive's
> help'. December 11, 1970

> Olive wonderful at Xmas but it is always a trial for me'. December 24, 1970

> How useless are arguments in life'. 20 02 1972

---

[8] Denis' estate was valued at £187,936, indicating that he had been very effective at giving away
his money.

Fig. 25.1 Denis in
Uganda. Wellcome images

The waste of Christmas hit me. The wine and food. Not happy'. December 25, 1974

Lonely in wealthy materialistic society'. December 7, 1970, away from home

Upset Peter Cleave badly by sending him references on thrombosis. He put down the phone in the middle of the conversation. I hate upsetting people'. December 8, 1971

Lord keep me humble when applauded'. After first talk on bowel cancer at San Diego conference, January 8, 1971

Appalled at the expense and the endless efforts to prolong life even in the aged'. June 5, 1973, when at Massachusetts General Hospital

All concerned with prolonging life – no thought for the hereafter'. September 14, 1979. At Human Functioning Conference.

Brights took us out to dinner. Always to me an unnecessary and sinful luxury'. September 25, 1973

I have always wondered why the Anglican services of morning and evening prayer conclude the psalms and canticles with the words, "As it was in the beginning is now and ever shall be, world without end". Whereas everything that is living is moving and changing, and unchangeable foundation truths have to be re-interpreted for different cultures and successive generations'. (Autobiography 13:3)

This morning knowing Jim was a keen Catholic, I went to mass with him. I had never been to an RC service. The sermon was quite good & could have fitted an Anglican Church but all the worship was totally meaningless. Mumbled in Latin with ringing of bells and various religious exercises. Only some of the people present took the Sacrament'. Comment on going to mass in an RC church in Dar es Salaam. Diary 1964

Given a sauna and massage. Enjoyed neither. Probably expensive rubbish'. August 19, 1978

Very expensive hotel and terribly extravagant dinner with others in team. Only I (Denis) no drinks and no starter. £15+ a head in a starving world'. July 16, 1986, Manchester

Better to build a fence at the top of the cliff than to park ambulances at the bottom'. From a poem by Joseph Malins, modified by Denis

In later years, he would write, 'It is certainly disputable to what extent we are justified in seeking God's guidance through isolated texts of scripture.

I have been repelled by luxury and extravagance throughout my life, and consequently have too often denied Olive the occasional treat she deserves'.

Cure' means power, money, dependency; prevention gives equality and empowerment.

DPB after Kilimanjaro. 'Surely one of the things that count most in life is relationships' (Fig. 25.2).

Remarks made about Denis Burkitt.

Those attributable to people who are no longer alive are given with their names. For those from people still alive they are given anonymously, unless permission has been sought to identify their contribution, but include remarks from Professors Norman J Temple, JJ (George) Misiewicz, Stuart Truswell, Keith McAdam, Owen Smith, Claus Leitzmann, and Meg Barham, Joan Church and others.

Extracts of references about Denis after some of his junior hospital jobs.

He is a very gentlemanly and likeable man, and I always found him cheerful and ready for more work at the end of a heavy day'. H.L. Warren Woodroffe, M.D. Chester, where Denis did his first job as House Surgeon starting on July 7, 1935

…he leaves the Preston Infirmary with the good wishes of every member of the staff. Personally I very much regret his departure and wish him a very prosperous professional career. The fact that he has an extensive knowledge of medicine and has the best interest of his patients at heart will I feel go a long way to ensure this for him'. S.S. Sumner F.R.C.S. Preston where Denis was a House Surgeon in the ENT department, March 1937

He has proved himself most hard-working and reliable in every way. He is a good organiser and keeps his knowledge up-to-date. He is a willing colleague and I am very pleased to speak highly of his services which have been of great value to us'. S. Gordon Luker, FRCS Cornelia & East Dorset Hospital

We are all very sorry to lose him'. F.R. Forrest. F.R.C.S. Cornelia & East Dorset Hospital

The fact that he was re-appointed for a further six months is evidence of the good opinion that the staff as a whole had of is capabilities'. W.F. Adeney, F.R.C.S. Cornelia & East Dorset Hospital

He is a pleasant and loyal colleague, a good surgeon and a trustworthy man'. C.M. Kennedy, F.R.C.S. after Denis completed a year at Plymouth, June 1940

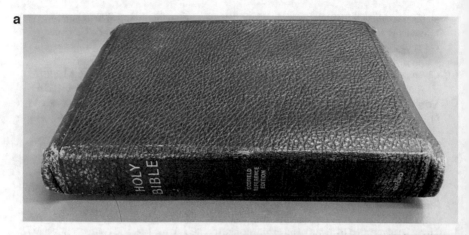

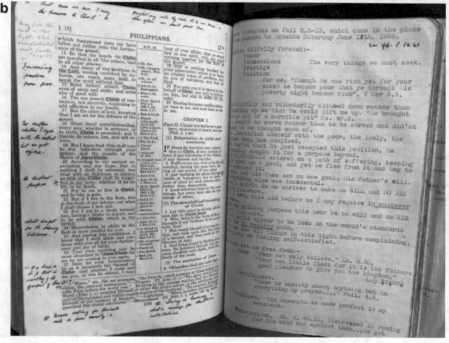

Fig. 25.2 (**a** and **b**) One of Denis' Bibles, heavily annotated and interleaved with comments. (Trinity College Dublin Archives and Manuscripts. TCD MS 11387/3)

John Church recounting the 1964 safari to Rwanda with John Billinghurst, Brian Rogers and Denis (see Chap. 17)

> Denis also had an uncanny way of handling difficult situations. When we finally reached the Rwanda/Uganda border, after our highly successful safari, we found a dismally long line of vehicles waiting to process their papers for the border crossing. Denis could not afford the obvious delay if we joined the queue. So he drove us off the road to a shady tree, reasonably

near to the Customs shed, and we set up a tea party. It was mid-afternoon, and very hot. We rapidly produced our picnic kit, but embellished the table with a clean cloth and flowers (picked from a nearby fragrant franjipani). We pulled out all the stops, from the remains of our 'tuckbox', and set the table for our afternoon tea. Then Denis strode into the Customs office, and invited the Customs officer to join us for a 'quick cuppa'. This he did with alacrity. The rest was easy. He signed our papers, we were through and away in double quick time, and returned rejoicing to Kampala'. (Recollections sent to JHC dated January 28, 2021)

About the Kampala Medical School:

You are quite right about the Kampala Medical School. In its hey day, in the 1950s and 60s, it attracted an amazing array of outstanding people… Denis, with his gentle and unassuming nature, yet tremendous drive, was a'supremo', and a very special friend and colleague. (Sent to JHC 2021)

Arriving in Amudat, Kenya, March 1970

I shall never forget his arrival. The Volkswagon Beetle pulled up outside our house, he leapt out, said "Hello", and before we had even reached the front door he had launched into his theme. "You see it's not the diseases that you get that I'm really interested in. It's the diseases that you don't get!" In his Irish accent the words tumbled over one another. He was one bundle of enthusiasm, energy, excitement. He couldn't wait to get started. And yet he was also one of the humblest, most self-effacing people I have ever known'. David Webster; The Shimmering Heat, p. 96. See also Chap. 20.

In spite of his remarkable work Denis remains humble, simple, natural and a real man of God, and gives all the credit for any success to those with whom he works'. New Day, Thursday, February 3, 1966 (Ugandan newspaper) Dorah Billington

We have known him as a great Christian and we are pleased to know him also as a distinguished scientist'. Shiplake Paris Church Journal, October 1968, on award of Arnott Gold Medal

I met him in Edinburgh when there was the Vahouny Conference on dietary fibre organized by Martin Eastwood. Martin took me to the canteen for the lunch, found an old gentleman, and asked me "Do you know Denis?". At first I was not able to react, because it is not usual to address such a respected person as Denis Burkitt with the first name in Japan. Then, my brain slowly started to work and recognized that the gentleman in front of me was the Denis Burkitt the fibre giant in real. Takashi Sakata, circa 1980

When we arrived there was a crowd of students, residents and faculty filling every seat and the steps between. The mood did not match the attendance…his work on fiber was considered nearer to quackery than valid medical research. He left to a standing ovation, one like I've never seen there before or since'. Markham Berry letter to Olive after hearing of Denis' death. He reminisces about Denis' visit to Grady Hospital Atlanta and giving a talk at Grand Rounds. Letter dated January 30, 1994

Brilliant. Did things, lymphoma and fibre, which most people would be pleased to have done'.

A child of his times'

Passion, curiosity, optimism, potential, saw the best in people, a tree with deep roots.

He talked non-stop and obsessively about high fibre diet....a highly driven man, with appreciable dose of obsessionalism and impatience with those who chose to differ. He kind of knew he was right, a real prophet therefore'. (Circa 1970)

As a colleague he was the most loyal and conscientious person I have ever known. He is a practising Christian who thinks action more important than words. He left behind in Africa a store of goodwill, and I doubt whether there has ever been any doctor working in Arica who has either been better known or more loved'. Letter from Professor Ian McAdam to Bernard Glemser, 1970

A few years ago when I spoke at a medical meeting in Durham my hosts told me that their last guest had been Denis Burkitt. So persuasive had he been that by 11 o'clock the following morning Durham had sold out of bran.

And 'Burkitt meets my criteria for greatness. He thinks clearly and imaginatively and his work has had a profound effect on other people's thoughts and behaviour'. –Michael O'Donnell BMJ. 1983;287:368

Boasted about the costs of his research being low.

Uganda, the place where he fully developed his exceptional observational and analytical skills that allowed him to turn simple observation into major scientific discovery. The most famous Trinity Medical Alumnus of the twentieth century, the man who had made a truly massive impact on the field of oncology. Typical of the man he played this down attributing much of what he had achieved to the work of others. Denis Burkitt had natural gifts and drive, but he also had curiosity in abundance and was willing to take risks (chemotherapy administration to children when he knew little or nothing about it) and forge new paths (what you eat is what you are – fibre and disease) – this made him the remarkable clinician-scientist he was to become.

Denis did not invent or discover dietary fibre nor was he the first person to suggest that it might be important for health. Yet his name is inscribed at the head of the leader board of scientists who developed the fibre story and for a time he carried the flag behind which he rallied a growing battalion of people with interests in everything from the minutiae of microbial metabolism in the human large bowel to a general wish to find a healthier diet. With Hugh Trowell he went on to show that diets characterised by high fibre were important in the prevention of some of the most important killer diseases, which they called Western, or Man Made, Diseases. Denis achieved this status because of a unique combination of skills, knowledge, enthusiasm, attitude and being unafraid to take on major problems of health in the world.

Beliefs gave him an attitude to death that left him without fear of it. He saw it as passing over to a new life closer to God. At times his judgement of people who were not of a similar belief was rather harsh.

His contributions by identifying dietary fibres as a key element in human nutrition is a milestone in nutrition and medicine in the second half of the 20th century.

Dr Burkitt is one of that very select few – the truly great and the truly good. The story of his remarkable life and achievements in medical research is an inspiration. Audrey Eyton, author of 'The F-Plan Diet.

This ability to see the importance of facts available to, but ignored by, less perspicacious mortals, was certainly shown by Denis in his work on the tumour which now bears his name'. Jack Darling in note to Ethel Nelson 1991

They were a warm, loving and outgoing family, and it was a privilege to share in their happy domesticity'. (JD)

True, Denis had never been a notable technical surgeon.
Denis described himself as a "simple hack surgeon" for whom God had found a place in Uganda.
Truly a humble man of God. Cliff Nelson in letter to Ethel Nelson (not related), March 23, 1991

Though very intelligent Denis did not have an overwhelmingly superior intellect. What he did have was a penetrating insight into problems'. Cliff Nelson 1994

I owe Denis a lot'.

Olive was not allowed to sail with Denis to his new posting with the Colonial Medical Service in 1947 in Uganda because she was pregnant. She asked Denis to postpone his departure by a few months but

He wouldn't contemplate such a move. His one idea was to get on and establish himself in the country to which he was certain God had called him.

I suppose I felt jealous of God because I had to take second place in Denis' affections.

…there has been absolutely no sacrifice on my part in giving priority to my role as wife and mother'. 'However, in spite of finding fulfilment in my home and family and in local activities I found it quite difficult when my husband having returned from an extremely interesting overseas trip was not able to show enthusiasm for my small world'. A remark made in the papers of Olive describing the period of separation they had after their first baby was born. Olive Burkitt 1991 Biographical memoire by Olive pages 74–85. Probably 1980s

…but he did have an innate curiosity which allowed him to express himself and made him into the great Dr. he was, but not necessarily an ideal husband.

It is impossible to grasp the number of lives that have been improved or saved and will continue to be improved as a result of Dr Burkitt's epidemiological acumen and his missionary zeal when promoting health'. From Nelson and Temple 1994 (see Appendix 4)

He was an unusual lecturer, who held people's attention for an hour without the slightest trouble, and had them laughing in the aisles, while persuading them firmly against the views current at the time'. Daily Telegraph Obituary 1993

Great talker. Raconteur. Lots of confidence.' 2016

A man of God and bowels…a man of his time'. 2021

Cliff Nelson 25 03 1991 commenting on the occasion when Denis was shown the plaster busts of lymphoma patients by Prof Prates and pointing out that the various tumours were all part of the same syndrome. 'It was then that I began to realise that I stood in the presence of a medical genius'.

One of the most exciting figures I ever met-very humble, very pious, very strong-minded and extremely opinionated'. Crawford DH, Rickson A. Cancer virus. The story of the Epstein-Barr virus. Oxford, England. Oxford University Press, 2014. Quoting Volker Diehl

Denis Burkitt was a remarkable man and truly deserved the many honours, prizes, awards and accolades given him. Meeting him in his later years there was never any discussion of his past successes, no anecdotes about his earlier life just total commitment to the cause of saving the world from what he saw as its self-inflicted diseases. His passion and fervour, together with a unique style of oratory, were born out of his early wish to 'win souls for Christ', which together with his experiences in Africa led to his remarkable contributions to cancer medicine, healthy eating and public health policy.

Finally,

Letter from Sir Harold Himsworth, retired Secretary of the Medical Research Council, UK, to Denis in response to a letter from him.

June 6, 1983

My dear Denis,

Many thanks for your very kind letter of the 24th May which has now reached me (via) Northwick Park.

I greatly appreciated your writing as you did, for the real reward of doing a job like that of Secretary of the MRC is to know that one has managed to help somebody like you. So the indebtedness is by no means one sided. I shall never forget the thrill I felt when you first showed me the map of Africa showing the distribution of your tumour. It pulled back the shutters and let in a flood of light.

You select for particular mention the fact that the MRC gave you a free hand. This wasn't strange. It has – always been their policy – not tacit but avowed – once they have found their man. Found him, back him and don't breathe down his neck.

And once I had found you I knew that I had met somebody who was a natural explorer of the unknown. That was why at our very first meeting I had no hesitation in giving you a grant to (underwrite) that journey – to Mozambique wasn't it?

But there was one thing I was particularly glad about. After you had tied up your (ends) about your tumour you didn't, as so many do, go on simply elaborating details about it. You looked for new fields to conquer – and found it in the correlation between dietary fibre and the incidence of certain diseases.

In the latter connection, I think would be correct in suspecting that you thought I was rather lukewarm. I was about some of the correlations Cleave was claiming but not about your main ones. So I was relieved to know from your letter that you had strayed from some of his conclusions!

The upshot of all this is, of course, that if there is any indebtedness between us it is reciprocal. For certainly you've repaid me many fold by the sheer excitement of your discoveries.

With all best wishes

Yours truly,

Harold Himsworth.

References

1. Kritchevsky D, Bonfield C, editors. Dietary fiber in health and disease. St Paul: Eagan Press; 1995.
2. Burkitt DP. Unpromising beginnings. J Ir Coll Physicians Surg. 1993;22:36–8.
3. Daily Mail. Friday October 16, 1992.

Appendices

Appendix 1: Timeline: The Life of Denis Parsons Burkitt[1]

1870 August 20 Father, James Parsons Burkitt born at Killybegs, Donegal, Ireland

1877 April 3 Mother, Gwendoline Hill born at Audley House, Cork, Ireland

1903 February 10 James Burkitt marries Gwendoline Hill

1907 June 24 Sister, Gwendolyn Zelma (Peggy) born at 3, Alexandra Terrace, Enniskillen, Fermanagh, Ireland

1911 February 28 Denis Parsons Burkitt born at 3, Alexandra Terrace

1911 April 5 Family move to a rented house named Lawnakilla, 2.25 miles north of Enniskillen

1912 September 28 Brother, Robert (Robin) Townsend Burkitt born at Lawnakilla

1916 (or thereabouts) Educated at home by Governess

1918 Tonsils removed by Uncle Roland during a visit to Lawnakilla

1919 To school with Peggy and Robin in Enniskillen

1920 April 3 Olive Elsie Lettice Mary Rogers born near Reading, Berkshire, England

1920 To Portora Royal School, Enniskillen

1922 June 15 Injures right eye in playground at school and has it removed on July 24 in Belfast

1922 Denis and Robin sent to small preparatory school at Trearddur Bay, Anglesey

1925 September 22 attends Dean Close School Cheltenham

1926 March Family move from Lawnakilla to Laragh, Ballinamallard, 2.75 miles further north. Purchased for £3350

1927 March 25 Denis and Robin confirmed as members of the Church of England

1929 October 5 To Trinity College Dublin to study engineering

1929 October 25 First meeting with the No. 40 group at Trinity

---

[1]This relates mainly to personal and domestic matters. The landmarks in his career are given in Appendix 3, Curriculum Vitae.

© The Author(s), under exclusive license to Springer Nature Switzerland AG 2022

J. H. Cummings, *Denis Burkitt*, Springer Biographies,
https://doi.org/10.1007/978-3-030-88563-2

1930 February 14 Makes a commitment to 'do my best to follow Christ'.

1930 September 2 Tells parents he wishes to change from engineering to medicine. October 1 returns to Trinity to study medicine

1933 June Graduates BA from Trinity College Dublin

1935 May Exams for Hudson Scholarship. Came second to JNP Moore who had played rugby for Trinity and Ireland. Awarded plaque and silver medal

1935 June Graduates MB, ChB and BAO from Trinity College Dublin

1935 July 7 Arrives in Chester to start first job as House Surgeon

1936 January–June House Surgeon, Adelaide Hospital, Dublin

1936 July 10–18 or 19 Adventure in Spain with Allan

1936 August Locum in General Practice in Blackwood, Wales

1936 September–1937 March Ear, Nose and Throat (ENT) and Eye, House Surgeon, Royal Infirmary, Preston

1937 April 17–1938 March 31 Resident Surgical Officer (RSO), Cornelia and East Dorset Hospital, Poole, Dorset

1938 April 20–June 19 Edinburgh FRCS course

1938 Nov 12–1939 March 2 Voyage on M.V. Glenshiel as ship's surgeon

1939 March 16 Locum in General Practice, Chacewater, Cornwall

1939 April 7–14 Locum in General Practice in Blackwood, South Wales

1939 June 20–1940 July Resident Surgical Officer, RSO Prince of Wales Hospital, Plymouth

1939 September 1 Hitler invades Poland and 2 days later Britain and France declare war on Germany

1940 April Olive Rogers enrols as a trainee nurse at the Prince of Wales Hospital, Plymouth

1940 July 15–1941 May RSO Becket Hospital, Barnsley, Yorkshire

1941 January Plymouth Hospital bombed. Olive Rogers an eyewitness to events there. Correspondence between Denis and Olive begins

1941 February 21 Rejected for work with Colonial Medical Service allegedly on account of his being blind in one eye

1941 May 19 Accepted for service in the Royal Army Medical Corp (RAMC). July 27 started as Medical Officer to the 342 S/L Battery, Royal Artillery at Herstmonceux, Norfolk

1942 January Posted to Somerleyton, Lowestoft, Suffolk, as part of 219 Field Ambulance

1942 Promoted Captain

1943 January 14 to February 7 Surgical Locum at Addenbrooke's Hospital Cambridge

1943 February 20 Becomes engaged to Olive Rogers

1943 July 28 2.30 pm Marries Olive Rogers at Seaford Parish Church, Sussex. Captain Walter Thompson best man

1943 October 22 Departs for army service overseas, initially in Kenya

1944 March 26–June 14 Posted to Somaliland

1944 June Promoted Major

1944 June 25 Posted to Gilgil in Kenya

1944 November 16–December 7 First army leave in Africa. Journeys from Gilgil in Kenya through Uganda and Belgian Congo (DRC). Types 13,000-word record of his journey

1945 January returns to Gilgil. January 30 applies to the Colonial Office to join the Uganda Medical Service. May 11 Accepted for work in Uganda

1945 March 16 Arrives in Colombo Ceylon. Posted to 48 IGH SEAC April–June

1945 December 12 Leaves Colombo for the UK arriving back on February 7, 1946

1946 March 12 Demobilised from the army

1946 May 11 Accepted for post as Surgeon and then Senior Surgeon to the Uganda Ministry of Health, 1946–1964

1946 July 2 MD Trinity College Dublin 'Spontaneous rupture of abdominal viscera'

1946 September 5 Departs for Africa, Colonial Medical Service

1946 October 3 Arrives in Lira as Medical Officer

1946 December 24 First daughter Judith (Judy) born at Enniskillen Hospital

1947 March 9 Meets Hugh Trowell for the first time on train to Nairobi

1947 March 14 Olive and Judy arrive in Lira having travelled on RMS Winchester Castle

1947 August 2 First paper on hydrocele of the testis published in the *British Medical Journal*

1947 September 6 Reads paper on hydrocele at a meeting of the British Medical Association in Kampala

1948 January 4 Leaves Lira arriving at Mulago Hospital, Kampala. January 7 appointed as Government Surgeon and Lecturer in Surgery

1949 January 30 Second daughter Carolyn (Cas) born at Mengo Hospital

1949 July 30–1950 April 31. Home leave spent in England and Ireland.

October 1, 1949, to March 17, 1950. Training course at Royal National Orthopaedic Hospital in London to improve his skills in bone and joint surgery

1950 August 21 Car accident returning from Fort Portal to Kampala. Family unscathed but car badly damaged

1951 Denis starts private practice at the European Hospital in Kampala

1951 July 8 Tooth out with 'hammer and chisel'

1952 March 21 Third daughter Rachel born at Mengo Hospital. Olive develops post-natal depression

1952 Promoted to Specialist Surgeon

1953 April 18 To England for home leave. Away 7 months including:

August 28–September 18 'European Tour' to celebrate 10th wedding anniversary. October 23 aboard SS Uganda heading for home. November 11 back in Kampala

1954 April 29 Queen Elizabeth II with the Duke of Edinburgh open Owen Falls Dam

1955 Letter about typical week. (Family archives)

1955 March 22–June 14 Home leave

1956 April in Karamoja. Mentions jaw tumours

1956 Coastal holiday trip to Tanga, July 25 to August 22

1956 November 22–1957 February 11 Home leave

1957 Between February 11 and early March. Probable date when Hugh Trowell showed him first case of childhood lymphoma. March 14 a 'few weeks' after February 11, was in Jinja where he saw the second case of the lymphoma

1957 July 31 Kalde visit

1957 November Northern Province visit

1958 August 2 Western Province visit

1958 January Climbs Kilimanjaro Tuesday 7 to Sunday 12

1958 First paper on the lymphoma published in *British Journal of Surgery*

1959 February describes visit of Queen Mother

1959 March 30 10:10 am Father, James Parsons Burkitt dies at Laragh, Ballinamallard, and is subsequently buried in Trory Churchyard

1959 March Visits Northern Province

1959 April 5 Embarks on Tripolitania from Mombasa with family en route for Europe

1959 June 14 Cycle accident whilst in Ireland. Breaks right arm

1959 July Karamoja trip for 10 days

1960 Northern Province February–March

1960 Northern Province July

1960 August 23 Mother Gwendoline Burkitt dies at Laragh

1961 March 22 5:15 pm Lecture at Courtauld Institute, Middlesex Hospital London, about the childhood lymphoma and first meeting with Dr. MA (Tony) Epstein

1961 May 25 Lecture at Royal College of Surgeons of London on the lymphoma attended by only 12 people

1961 October 7–December 15 Long safari

1962 October 9 Uganda becomes independent

1962 May 15–June 7 West Africa Safari

1963 February 11–15. At UICC conference to discuss 'Tumours of the Lympho-Reticular System in Africa' the name 'Burkitt tumour' was given to Denis' childhood cancer, soon after to be known by common consent as 'Burkitt's lymphoma'

1963 February 16–March 1 First trip to the USA

1963 November 22 John F Kennedy assassinated. Denis and Olive in New York

1964 February 1. Epstein and Barr paper in *The Lancet* describing culture of Burkitt lymphoma cells. Eight weeks later a further paper in *The Lancet* from Epstein, Achong and Barr reporting the finding of a new virus in the tumour cells

1964 April 1 Resigns from Senior Surgical post with Uganda Ministry of Health and joins UK Medical Research Council Clinical Scientific Staff – member until 1976

1964 Safaris:

> January 15–February 1. Northern Uganda with Bob Harris
> April 18–25. Kenya, Tanzania, Dar es Salaam
> May 20–30. South-West Uganda with Olive
> June 19–July 18. Tanzania, Zanzibar
> September 2–18. Kenya coastal region
> October 27–November 3. Kenya Rift Valley and north of Tanzania
> November 28–December 7. Rwanda with John Church, John Billinghurst and Bryan Rogers

December 12–19. Sudan

1965 Safaris:

> May 18–June 3. Kenya with Olive
> July 7–25. Southern Tanzania with Bob Harris
> October 8–13. North-West Uganda with Olive and others
> November 7–18. Sudan

1966 January 25 Leaves Kampala for final journey out of Africa. February 14 arrives in London

1966 February Member of MRC external staff. Office in Tottenham Court Road, London

1966 June 4. Has an offer of £14,500 accepted for The Knoll, Shiplake August 16 moves into The Knoll

1966 October 16–December 5. Visits Singapore, Hong Kong, Tokyo, Dunedin and Christchurch New Zealand, Sydney, Australia, Port Moresby and Goroka in New Guinea, Brisbane, Townsville, Melbourne, Adelaide, Bangkok, Karachi and Baghdad

1967 July–August East African safari; Zanzibar, Pemba, Tanzania, Burundi, Rwanda, Kampala

1967 September 14 Meets Surgeon Captain Cleave for the first time

1968 January 7 West African safari commencing in Accra

1968 June 7–23 USA, New York, Washington, Chicago, Houston, Los Angeles, Loma Linda

1968 August 10–September 6 Uganda, Kenya. Visits about 40 hospitals

1968 November 7 Lectures to British Society of Gastroenterology and hears Neil Painter speak about fibre and diverticular disease

1968 November 28–December 23 Safari to Kinshasa (DRC), Johannesburg, Fort Portal with Michael Hutt, West Nile with Ted Williams, Gulu, Lira, Makerere, then December 16 WHO Conference in Nairobi where he talks about viruses and cancer

1969 February 28. Meets George Misiewicz at the Central Middlesex Hospital in London to discuss bowel transit studies

1969 August 1–30. Kenya, Uganda, Tanzania and South Africa to organise transit studies. Meets George Campbell for the first time

1969 September 4–18. Brazil and USA, September16 receives Judd Award in New York

1970 March 6 Sir Albert Cook Centenary Meeting in Kampala. With Hugh Trowell. Denis gives two talks, one about the lymphoma but the second on the topic of the geography of other diseases, 'Related Disease-Related Cause'.

1970 April 1–9 Holiday with Olive in Mombasa. 'Flash of inspiration' moment

1971 July 24 Eldest daughter Judith marries Philip Gravely Howard

1971 January 25 Idi Amin seizes power in Uganda

1971 October 21–November 13 Safari to Afghanistan, Pakistan and India

1972 April 13 Elected to Fellowship of Royal Society

1972 August 10 Olive's father dies

1972 March 14 Paul Ehrlich and Ludwig Darmstaedter Prize and Gold Medal

1972 November 16 Lasker Award

1972 Cas marries Dr. Andrew Boddam-Whetham. They would work in Kenya together at Chogoria Hospital for 8 years

1974 June Awarded CMG in Queen's birthday honours

1976 July 10. Rachel marries Dr. David Maurice, the sixth generation of Maurice's in General Practice in Marlborough: a continuous line of father to son since 1792

1976 August 31 Retires from MRC post

1976 Hon Research Fellow at St Thomas's Hospital, London

1978 February 15 Lunch with the Queen at Buckingham Palace

1979 October 3 'Don't forget fibre in your diet' (Martin Dunitz) published

1979 February Honorary Fellowship of Trinity College Dublin

1980 February 13–April 18 'World tour' to promote his book

1981 July Admitted to Roll of Fellows of the British Medical Association

1981 'Western Diseases" published. (Trowell HC, Burkitt DP (1981) Western Diseases: their emergence and prevention. Harvard University Press, Cambridge, Massachusetts.)

1982 April 7 Bristol Myers Award for Distinguished Achievement in Cancer Research with Tony Epstein

1982 June 16 Charles Mott General Motors Cancer Award.

1983 February 19 Isobel Brown dies

1987 Moves to Hartwell Cottage, Bisley, Gloucestershire

1987 April 6 Visits Peggy in Omagh Hospital

1990 July 24 Peggy dies

1991 October Admitted to hospital for investigation of his heart

1992 January Declines invitation from Charlie Bonfield to Vahouny Conference because of heart trouble

1992 June 5 Buchanan Medal, Royal Society

1993 January 14 Bower Award of the Franklin Institute, Philadelphia. Feb 12–18, Boston

1993 March 17 Stroke whilst at home. March 23 died Gloucestershire Royal Hospital

1993 March Funeral at Bisley Parish Church and burial in the village graveyard

1993 May 12 2 pm Service of Thanksgiving at All Souls Langham Place, London

1995 February 8 De Villiers Medal (Posthumously) Leukaemia Society of America

Estate at death £187,936

Appendix 2: Curriculum Vitae: Denis Parsons Burkitt[2]

Schools:
1914[3] Governess at home
1917 Local school
1920 Portora Royal School, Enniskillen
1922–1925 Trearddur Bay School, Holyhead, Anglesey, Wales
1925–1929 Dean Close School, Cheltenham

University:
1929–1930 Trinity College Dublin to read engineering
1930–1935 Trinity College Dublin, reading medicine

Qualifications, all from Trinity College Dublin
1933 BA
1935 MB, ChB, BAO[4]
1946 MD 'Spontaneous rupture of abdominal viscera'

Postgraduate training:
1935 July–December House Surgeon, Royal Infirmary, Chester
1936 January–June House Surgeon, Adelaide Hospital, Dublin
1936 September–1937 March Ear, Nose and Throat (ENT) and Eye, House Surgeon, Royal Infirmary, Preston
1937 March–1938 March Resident Surgical Officer (RSO), Cornelia and East Dorset Hospital, Poole, Dorset
1938 March–September Studying for Fellowship of Royal College of Surgeons, Edinburgh
1939 June–1940 July RSO Prince of Wales Hospital, Plymouth
1940 July–1941 May RSO Becket Hospital, Barnsley, Yorkshire
1949 October–1950 March Postgraduate Course in Orthopaedic Surgery at Royal National Orthopaedic Hospital, London

Posts and appointments:
1938 October–1939 March Ship's Surgeon aboard The Glenshiel, a merchant vessel
1941 May–1946 March Military Service, Royal Army Medical Corp (RAMC)

   1941–1944 Graded Surgeon
   1944–1946 Surgical Specialist (Major)

1946 Appointed to HM Colonial Medical Service, Uganda
1946 October 3–1948 January 4 Posted to Lira, Northern Region

---

[2] This information is derived from several quasi-independent sources and where conflicting dates are given has been checked as far as possible.

[3] The 1914 and 1917 dates are best guesses from available evidence.

[4] "Batchelor in Medicine, in Surgery, and in Obstetric Science" BAO, Batchelor of Obstetrics, is a degree peculiar to Irish Universities.

1948 January 7 to Mulago Hospital as Special Grade Medical Officer-Surgeon and part-time Lecturer in Surgery, Makerere College Medical School
1952 Specialist Surgeon
1961 Senior Specialist Surgeon
1964 External Scientific Staff, Medical Research Council, Makerere
1966–1976 External Scientific Staff, Medical Research Council, London
1976 August 31 Retires from MRC
1976–1988 Honorary Senior Research Fellow, St Thomas's Hospital Medical School, London
1976–1988 Vice President, International Christian Medical and Dental Association
1967 President, Christian Medical Fellowship (CMS)

Vice President of the International Medical and Dental Association

Lay Reader in the Church of England

Honours
1974 Companion of the Order of St Michael and St George (CMG)

Fellowships, by examination
1938 October 13 FRCS (Edinburgh)

Fellowships by appointment or honorary
1972 FRS Royal Society
1973 FRCS(I) Royal College of Surgeons of Ireland
1976 FRCP(I) Royal College of Physicians of Ireland
1979 Honorary Fellowship of Trinity College Dublin
1981 Fellow of the British Medical Association
1989 Foreign Member, Académie des Sciences, Paris
1992 Fellowship of the Royal College of Physicians and Surgeons of Canada

Honorary Degrees
1970 DSc University of East Africa[5]
1979 MD Bristol University
1982 DSc Leeds University
1984 DSc University of London
1985 DSc University of East Virginia
1989 DSc University of Ulster, Northern Ireland
1992 DSc Med. Medical University of South Carolina

Awards and Prizes

UK
1966–1967 Harrison Prize in Laryngology, Royal Society of Medicine

---

[5] Only two DSc degrees were conferred by this university which later was divided into three institutions in Kenya, Tanganyika and Uganda. The other recipient was Professor J.N.P. Davies.

1966 Stuart Prize, British Medical Association for his work on the geographical distribution of malignant lymphoma
1971 Walker Prize, Royal College of Surgeons
1972 Gold Medal, Society of Apothecaries, London
1978 Gold Medal, British Medical Association (Received 27 06 1979)
1992 Buchanan Medal, Royal Society

Ireland
1935 Hudson Silver Medal, Trinity College Dublin
1968 Arnott Gold Medal, Irish Hospitals and Medical Schools Association

USA
1969 Katherine Berkan Judd Award, Sloan Kettering Institute, New York
1969 (or 1970) Robert de Villiers Award, The American Leukaemia Society
1972 Albert Lasker Award
1978 Ralph Blocksma Award, M.A.P. International[6]
1982 Lucy Wortham James Clinical Research Award
1982 Bristol Myers Award for Distinguished Achievement in Cancer Research with Mark Epstein
1982 Charles Mott General Motors Cancer Award
1990 Lucy Wortham James Award, Society of Surgical Oncology
1992 Bower Award and Prize for Achievement in Science, Benjamin Franklin Institute, Philadelphia

Canada
1973 Gairdner Foundation International Award

Germany
1972 Paul Ehrlich and Ludwig Darmstaedter Prize and Gold Medal

France
1982 Diploma and Gold Medal of the Académie Nationale de Médicine
1987 Le Prix Mondiale Cino del Duca

Italy
1983 Beaumont Bonelli Award for Cancer Research

Honorary Fellowships/Membership of Professional Societies

    1976 East African Association of Surgeons
    Brazilian Society of Surgeons
    Sudan Association of Surgeons
    International Medical Club, Washington, DC
    1980 'In recognition of your contribution to Modern Industry and World Peace you have been elected a Knight of Mark Twain'[7]

---

[6] Medical Assistance Programme International
[7] Cyril Clemens, Editor, *Mark Twain Journal*

Appendix 3: Principal Publications of Denis P Burkitt[8]

Part 1. Books written or edited by DPB: n = 8

Burchenal JH, Burkitt DP, editors. Treatment of Burkitt's Tumour. UICC Monograph Series 8. Berlin, Heidelberg, New York: Springer-Verlag; 1967.

Burkitt DP, Wright DH, editors. Burkitt's lymphoma. Edinburgh: Livingstone; 1970.

Burkitt DP, Trowell HC, editors. Refined carbohydrate foods and disease: Some implications of dietary fibre. London: Academic Press; 1975.

Burkitt DP. Don't forget fibre in your diet London: Martin Dunitz; 1979.

Trowell HC, Burkitt DP, editors. Western diseases: their emergence and prevention. Cambridge, Massachusetts: Harvard University Press; 1981.

Trowell HC, Burkitt DP, Heaton KW, editors. Dietary fibre, fibre-depleted foods and disease. London: Academic Press; 1985.

Hutt MSR, Burkitt DP. The geography of non-infectious disease. Oxford: Oxford University Press; 1986.

Temple NJ, Burkitt DP. Western diseases: their dietary prevention and reversibility. New Jersey: Humana Press; 1994. (Paperback edition by Springer) Please leave space of one line before next heading

Part 2. Original papers in scientific journals, chapters in books, symposium proceedings and letters of substance: n = 272

Burkitt DP. Saccular aneurysm of left common iliac artery. BMJ. 1938; (4036) 1:1051.

Burkitt DP. Biliary peritonitis without demonstrable perforation. BMJ. 1946;2:155–7.

Burkitt DP, Hamburger HS. Case of bilateral congenital hydro-ureter and hydronephrosis. J R Army Med Corps. 1946;137:34-7.

Burkitt DP, Kununka BN. Enormous intra-abdominal hydrocele. BMJ. 1947;2(4517):175.

Burkitt DP. Spontaneous rupture of the spleen. East Afr Med J. 1948;25:167–74.

Burkitt DP, McAdam IWJ, Ladkin RG. The surgery of gonococcal stricture of the male urethra. East Afr Med J. 1949;26:370–77.

Burkitt DP. Primary hydrocele and its treatment. Review of 200 cases. Lancet. 1951;1(6669):1341–43.

Burkitt DP. Acute abdomens – British and Baganda compared. East Afr Med J. 1952;29:189–94.

Burkitt DP. A simple serviceable artificial leg. East Afr Med J. 1953;30:177–91.

---

[8]This list is as close to being complete and comprehensive as is possible given the nature of the original 'List of Publications' compiled by Denis where he does not include his own name as an author on any of the papers and very often dates are missing. His list is also incomplete. Attempts have been made to track down all his writings, exclude duplicates and include those found from other sources such as the Web of Science, Ovid and archives in London, Bisley and Dublin. Letters to journals are included but abstracts and short published summaries of papers presented at meetings are not.

Burkitt DP, Fairbank HA. An unusual lipoid reticulosis of bone. J Bone Joint Surg. 1954;36B:109–13.

Burkitt DP. A sarcoma involving the jaws in African children. Brit J Surg. 1958;46:218–23.

Burkitt DP. A boot and caliper bank. East Afr Med J. 1960;37:109–12.

Davies JNP, Burkitt DP. Lymphoma syndrome in Uganda and Tropical Africa. Med Press. 1961;245:367–69.

Burkitt DP. Transposition of the scrotum and penis. Brit J Surg. 1961;48:460

Jelliffe DB, Burkitt DP, O'Conor GT, Beaver PC. Subcutaneous phycomycosis in an East African child. J Pediatr. 1961;59:124–27.

Burkitt DP. A simple crutch. BMJ. 1961;1:1031–32

Burkitt DP, O'Conor GT. Malignant lymphoma in African children. I. A clinical syndrome. Cancer. 1961;14:258–69.

Burkitt DP. Observations on the geography of malignant lymphoma. East Afr Med J. 1961;38:511–14.

Burkitt DP. A tumour syndrome affecting children in tropical Africa. Postgrad Med J. 1962;38:71–9.

Burkitt DP. A children's cancer dependent on climatic factors. Nature. 1962;194:232–34.

Burkitt DP. Un cancer infantile relacionada con el clinica. Acta Oncologica. 1962;11:3–12.

Burkitt DP. A lymphoma syndrome affecting African children. Ann Roy Coll Surg Eng. 1962;30:211–19.

Burkitt D. Determining the climatic limitations of a children's cancer common in Africa. BMJ. 1962;2:1019–23.

Burkitt DP. A 'tumour safari' in East and Central Africa. Brit J Cancer. 1962;16:379–86.

Shaper AG, Burkitt DP. Acute pancreatitis in childhood. Postgrad Med J. 1962;38:704–6

Jelliffe DB, Wilson AMM, Burkitt DP. Subcutaneous phycomycosis responding to oral iodine therapy. Trop Paed. 1962;61:448–51.

Burkitt DP, Wilson AM, Jelliffe DB. Subcutaneous phycomycosis. East Afr Med J. 1963;40:34–8.

Burkitt DP. A climatic dependent children's cancer. Ethiopian Med J. 1963;1:254–8.

Burkitt DP. A children's cancer related to climate. New Scientist. 1963;17:174–6.

Burkitt DP. A children's cancer with geographical limitations. In Cancer Progress. Edited by Reaven RW. London: Butterworth; 1963. 102–113.

Burkitt DP, Nelson CL, Williams EH. Some geographical variations in disease pattern in East and Central Africa. East Afr Med J. 1963;40:1–6.

Burkitt DP, Oettgen HF, Clifford P. Malignant lymphoma involving the jaw in African children. Treatment with alkylating agents and actinomycin D. Cancer Chemoth Rep. 1963;28:25–34

Burkitt DP. A children's cancer dependent on environment. In: Cumley RW et al, editors. Viruses, Nucleic Acids and Cancer, A collection of papers presented at the 17th Annual Symposium on Fundamental Cancer Research Editors, Published for the University of Texas, M.D. Anderson Hospital and Tumour Institute. Baltimore: The Williams and Wilkins Co; 1963. P. 615–29.

Burkitt DP, Wright DH. A lymphoma syndrome in tropical Africa, with a note on histology, cytology and histochemistry. In: Richter GW, Epstein MA, editors. Int Rev of Exp Pathol. New York, London: Academic Press; 1963;2:67–96.

Oettgen HF, Burkitt DP, Burchenal JH. Malignant lymphoma involving the jaw in African children. Treatment with methotrexate. Cancer. 1963;16:616–23.

Davies JNP, Dodge OG, Burkitt DP. Salivary gland tumors in Uganda. Cancer. 1964;17:1310–22.

Burkitt DP, Wilson AM, Jelliffe DB. Subcutaneous phycomycosis: A review of 31 cases seen in Uganda. BMJ. 1964;1(5399):1669–72.

Burkitt DP. A lymphoma syndrome dependent on environment. Part 1. Clinical aspects. In: Roulet FC, editor. Symposium on Lymphoma-reticular tumours in Africa. Basel, New York: Karger; 1964; p. 80–93.

Burkitt DP. A lymphoma syndrome dependent on environment. Part 2. Epidemiological aspects. In: Roulet FC, editor. Symposium on Lymphoma-reticular tumours in Africa. Basel, New York: Karger; 1964; p. 120–36.

Burkitt DP, Williams EH. Abdomino-scrotal hydrocele. Brit J Surg. 1964;51:154–55.

Burkitt DP. Chemotherapy of jaw lymphomata. East Afr Med J. 1965;42:244–48.

Burkitt DP. Relics of tradition in medicine. East Afr Med J. 1965;42:305–12.

Burkitt DP. A propos due frequent des cancers de l'enfant en milieu tropical Afrique. Medicale. 1965; January 11–14.

Burkitt DP. Observational Research. Makerere Med J. 1965; May p. 21.

Hutt MSR, Burkitt DP. Geographical distribution of cancer in East Africa – a new clinicopathological approach. BMJ. 1965;(5464) 2:719–22.

Burkitt DP. Malignant lymphomata involving the jaws in Africa. J Laryingol Otol. 1965;79:929–39.

Burkitt DP, Hutt MSR, Wright DH. The African lymphoma: preliminary observations on response to therapy. Cancer. 1965;18:399–410.

Burkitt DP. A whole man. Makerere Med J. 1965; November, 19.

Burkitt DP, Wright DH. Geographical and tribal distribution of the African lymphoma in Uganda. BMJ. 1966;(5487) 569–73.

Burkitt DP, Hutt MSR. An approach to geographic pathology in developing countries. Int J Path. 1966;7:1–6.

Burkitt DP. Malignant lymphoma of the jaws. J Dent Res. 1966;45:554–59.

Burkitt DP. African lymphoma Observations and response to vincristine sulphate therapy. Cancer. 1966;19:1131–37.

Burkitt DP. Surgical pathology in the course of the Nile. Ann Roy Coll Surg Engl. 1966;39:236–47.

Burkitt DP. The African lymphoma: Clinical features, response to therapy; and epidemiological aspects. J Roy Coll Surg Edin. 1966;11:170–84.

Burkitt DP. A great pathological frontier. Postgrad Med J. 1966;42:543–47.

Booth K, Burkitt DP, Bassett DJ, Cooke RA, Biddulph J. Burkitt lymphoma in Papua, New Guinea. Brit J Cancer. 1967;21:657–64.

Hutt MS, Burkitt DP, Shepherd JJ, Wright B, Mati JKG, Auma S. Malignant tumours of the gastrointestinal tract in Ugandans. In Malignant tumours of the alimentary tract in Africans NCI Monograph (Bethesda, Maryland). 1967;25:41–47

Burkitt DP, Kyalwazi SK. Spontaneous remission of African lymphoma. Brit J Cancer. 1967;21:14–16.

Burkitt DP. A malignant syndrome in tropical Africa. In: Davey WW editor. Companion to Surgery in Africa. Edinburgh: Livingstone; 1967. p. 7–16.

Burkitt DP. Long-term remissions following one and two-dose chemotherapy for African lymphoma. Cancer. 1967;20:756–59.

Burkitt DP. Natural history in relation to chemotherapy. Some clinical features. In: Burchenal JH, Burkitt DP, editors. UICC Monograph Series. Berlin, New York: Springer-Verlag; 1967. Vol 8; p. 1–6.

Burkitt DP (1967) Epidemiology and etiology in relation to chemotherapy. Recent developments in geographical distribution. In: Burchenal JH, Burkitt DP, editors. UICC Monograph Series. Berlin, New York: Springer-Verlag; 1967. Vol 8. p. 36–41.

Burkitt DP. Chemotherapy of jaw tumours. In: Burchenal JH, Burkitt DP, editors. UICC Monograph Series. Berlin, New York: Springer-Verlag; 1967. Vol 8. p. 94–101.

Burkitt DP. Clinical demonstration of long-term survivors. In: Burchenal JH, Burkitt DP, editors. UICC Monograph Series. Berlin, New York: Springer-Verlag; 1967. Vol 8. p. 102–4.

Burkitt DP. Clinical evidence suggesting the development of an immunological response against African Lymphoma. In: Burchenal JH, Burkitt DP, editors. UICC Monograph Series. Berlin, New York: Springer-Verlag; 1967. Vol 8. p. 197–203.

Burkitt DP. African lymphoma: epidemiological evidence suggesting a viral aetiology. In: Shivas AA, editor. Racial and Geographical Factors in Tumour Incidence. Edinburgh: The University Press;1967. p. 19–23

Burkitt DP. Geographical distribution of cancer in East Africa. In: Shivas AA, editor. Racial and Geographical Factors in Tumour Incidence. Edinburgh: The University Press; 1967. p. 147–51.

Burkitt DP. Possible relationships between the African Lymphoma and acute leukaemia. Leukaemia Research Fund Third Guest Lecture Belfast: W & G Baird Ltd; 1967.

Burkitt DP. Clinical and epidemiological aspects of the Burkitt-Tumour. Medizinische Klinik. 1967; 17:653–656 (Kongresbericht uber die III Tagen der Deutschen Tropenmedizischen Gesellschaft e.V. April 22nd Hamburg p 210–218)

Burkitt DP. Burkitt's Lymphoma. A study in medical detection. Abbottempo. 1967;4:26–33.

Burkitt DP. Burkitt's lymphoma outside the known endemic areas of Africa and New Guinea. Int J Cancer. 1967;2:562–65.

Burkitt DP. Chemotherapy of African (Burkitt) lymphoma – clinical evidence suggesting an immunological response. Brit J Surg. 1967;54:817–19.

Burkitt DP. Burkitt's lymphoma. Clinical, therapeutic and epidemiological features. Stockholm: Nordiska Bokhandelus Forlag Thule International Symposia, Cancer and Ageing. 1967; 167–77.

Kisia IA, Burkitt DP. Cancer patterns at Kaimosi Hospital in Western Kenya. East Afr Med J. 1968;45:706–12.

Burkitt DP. The epidemiology of Burkitt's lymphoma. West Afr Med J & Nigerian Practitioner. 1968;17:258–9.

Burkitt DP. [The Burkitt's tumor from the clinical and epidemiological viewpoint.] [German] Der Burkitt-tumor in klinischer und epidemiologischer Sicht. Medizinische Klinik. 1968;63:653–56.

Burkitt DP. Surgery and pathology on the Nile. JAMA. 1968;205:604–6.

Burkitt DP, Hutt MS, Slavin G. Clinico-pathological studies of cancer distribution in Africa. Brit J Cancer. 1968;22:1–6.

Burkitt DP & Slavin G. Patterns of cancer distribution in Tanzania. In: Cancer in Africa: Nairobi East African Publishing House; 1968. p. 321–30.

Burkitt DP. The African lymphoma – epidemiology and therapeutic aspects. (Possibly chapter in a book but source unknown.) 1968.

Burkitt DP. Readjusting the balance in medicine. I.S.M. 1968;55:1–5.

Burkitt DP. The African lymphoma – epidemiological and therapeutic aspects. In: Zarafonetis C, editor. Proc Int Conf on Leukemia-Lymphoma Philadelphia. Lea and Febiger. 1968. p. 321–30.

David J, Burkitt DP. Burkitt's lymphoma: remissions following seemingly non-specific therapy. BMJ. 1968;4:288–91.

Burkitt DP. Cancer epidemiology in tropical Africa. Brit J Hosp Med. 1968;1:214–18.

Burkitt DP. Concepts of geographical pathology in the study of cancer. J Pakistan Med Assoc. 1968;18:15–18.

Burkitt DP, Bundschuh M, Dahlin K, Dahlin L, Neale R. Some cancer patterns in Western Kenya and North West Tanzania. East Afr Med J. 1969;46:188–93.

Burkitt DP, Williams EH, Eshleman L. The contribution of the voluntary agency hospital to cancer epidemiology. Brit J Cancer. 1969;23:269–74.

Burkitt DP, Kyalwazi SK. African (Burkitt's) lymphoma: characteristic features of response to therapy. In: Neoplasia in Childhood. Chicago Year Book of Medicine. 1969 p. 281–?

Burkitt DP. Etiology of Burkitt's lymphoma – an alternative hypothesis to a vectored virus. JNCI. 1969;42:19–28.

Burkitt DP. A study of cancer patterns in Africa. Scientific Basis of Medicine Annual Reviews. University of London: Athlone Press; 1969. p. 82–94.

Burkitt DP. The challenge of geographical pathology. Pakistan Med Forum. 1969;4:13–18.

Burkitt DP. Cyclone over cyclamates. BMJ. 1969;4:495.

Burkitt DP. Burkitt's lymphoma. In: Taylor S, editor. Recent Advances in Surgery. London: J and A Churchill Ltd: 1969. p. 26–47.

Burkitt DP. Related disease-related cause? Lancet. 1969;2(7632):1229–31.

Burkitt DP, Stanfield JP, Church JCT. A medical research safari: fruits and frustrations. Cent Afr J Med. 1969;16:197–201.

Burkitt DP. Disease and death from avoidable causes. St Barts Hosp J. 1970;77:202–5.

Burkitt DP. Host defence mechanisms in Burkitt's lymphoma and Kaposi's sarcoma: the clinical evidence. BMJ. 1970;4(5732):424–26.

Kafuko GW, Burkitt DP. Burkitt's lymphoma and malaria. Int J Cancer. 1970;6:1–9.

Burkitt DP. An epidemiological approach to gastrointestinal cancer. CA: A Cancer Journal for Clinicians. 1970;20:146–49.

Burkitt DP. Are our commonest killing diseases preventable? Rev Eur d'Etudes Clin Biol. 1970;15:253–54.

Cook P, Burkitt DP. An epidemiological study of seven malignant tumours in East Africa. Medical Research Council Report (?unpublished – about 30 pages) January 1970.

Burkitt DP. Relationship as a guide to the etiology of disease. Int Path. 1970;11:3–6.

Burkitt DP. Disease and death from avoidable causes. Pakistan Med Forum. 1970;5:39–43.

Burkitt DP. Relationship as a clue to causation. Lancet. 1970;2(7685):1237–40.

Burkitt DP. Clinical evidence suggesting host defence mechanisms in Burkitt's lymphoma and Kaposi's sarcoma. Brit J Radiol. 1971;44:904.

Burkitt DP. Some neglected leads to cancer causation. JNCI. 1971;47:913–9.

Burkitt DP. Possible relationships between bowel cancer and dietary habits. Proc Roy Soc Med. 1971;64:964–65.

Burkitt DP. Epidemiology of Burkitt's lymphoma. Proc Roy Soc Med. 1971;64:909–10.

Burkitt DP. The aetiology of appendicitis. Brit J Surg. 1971;58:695–99.

Painter NS, Burkitt DP. Diverticular disease of the colon: a deficiency disease of Western civilization. BMJ. 1971;2(5759):450–54.

Cook PJ, Burkitt DP. Cancer in Africa. Brit Med Bull. 1971;27:14–20.

Burkitt DP. Cancer and the way we live. Third David Kissen Memorial Lecture In Cancer Priorities, British Cancer Council, 3rd Symposium, Edinburgh: 1971. p. 124–137.

Burkitt DP. Cancer and environment. Int J Environ Stud. 1971;1:275–79.

Burkitt DP. Epidemiology of cancer of the colon and rectum. Cancer. 1971;28:3–13.

Burkitt DP. Cancer of the colon and rectum. Epidemiology and possible causative factors. Minn Med. 1971;55:779–83.

Burkitt DP. Burkitt's lymphoma. Summary of Lasker Award. JAMA. 1972;222:1164.

Burkitt DP. Classics in oncology. A sarcoma involving the jaws in African children. CA: A Cancer Journal for Clinicians. 1972.22:345–55.

Burkitt DP, Walker AR, Painter NS. Effect of dietary fibre on stools and the transit times, and its role in the causation of disease. Lancet. 1972;2(7792):1408–12.

Burkitt DP. Varicose veins, deep vein thrombosis, and haemorrhoids: epidemiology and suggested aetiology. BMJ. 1972;2(5813):556–561.

Burkitt DP. Aetiology of varicosity. BMJ. 1972;4(5834):231.

Burkitt DP. Distribution of cancer in Africa. Proc Roy Soc Med. 1972;66:312–18.

Burkitt DP. The importance of fibre in food. Bull Brit Nut Foundation. 1972; May 7th 29–35.

Burkitt DP. Burkitt's lymphoma. In: Godden JO, editor. Cancer in Childhood. The Ontario Cancer Treatment and Research Foundation; Proceedings of the 17th Clinical Conference; 1972. p. 209–13.

Burkitt DP. The trail to a virus – A review. In Oncogenesis and herpesviruses. Biggs PM, Thé G de, Payne IN, editors. IARC Scientific Publications. 1972;2:345–48.

Burkitt DP. Geographical pathology related to diet. In: The Medical Annual. Bristol: John Wright and Sons Ltd; 1972. p. 5–15.

Burkitt DP. Epidemiology of large bowel disease: the role of fibre. Proc Nut Soc. 1973;32:145–49.

Burkitt DP. Some diseases characteristic of modern western civilization. A possible common causative factor. Clin Radiol. 1973;24:271–80.

Burkitt DP. Diseases of the alimentary tract and western diets. Path Microbiol. 1973;39:177–86.

Burkitt DP, James PA. Low-residue diets and hiatus hernia. Lancet. 1973;2(7821):128–30.

Burkitt DP. Distribution of cancer in Africa. Proc Roy Soc Med. 1973;66:312–14.

Burkitt DP. Varicose veins, deep vein thrombosis, and hemorrhoids. Am Heart J. 1973;85:572–73.

Hutt MS, Burkitt DP. Aetiology of Burkitt's lymphoma. Lancet. 1973;1(7800):439.

Burkitt DP. Some diseases characteristic of modern Western civilization. BMJ. 1973;1(5848):274–78.

Burkitt DP. Epidemiology of large bowel diseases. In: Taylor S, editor. Recent Advances in Surgery. London: Churchill Livingstone; p. 257–73.

Burkitt DP. Burkitt's lymphoma. Proc. 17th Clin Conf of the Ontario Cancer Treatment and Research Foundation; 1973. p. 209–213.

Burkitt DP. Cancer and other non-infective diseases of the bowel. Epidemiology and possible causative factors. Rendic Gastroenterol. 1973;5:33–9.

Burkitt DP. Carcinoma of the colon and rectum. In: Raven RW, editor. Modern Trends in Oncology. London: Butterworth; 1973. p. 227–41.

Burkitt DP. The role of refined carbohydrates on large bowel behaviour and disease. Plant Foods for Man. 1973;1:5–8.

Burkitt DP. Diseases of modern economic development. In: Howe GM, Loraine JA, editors. Environmental Medicine. London: Heinemann Medical Books Ltd; 1973. p. 140–44.

Burkitt DP. Cancer of the colon and rectum: epidemiology and suggested causative factors. Walker Prize Lecture. In: Cancer at the crossroads and the challenge for the future. London: Wm Heinemann Medical books Ltd; 1973. p. 103–15.

Burkitt DP. Diverticular disease of the colon. Epidemiological evidence relating it to fibre-depleted diets. Trans Med Soc London. 1973;89:81–4.

Burkitt DP. Cancer and infectious diseases. Trans Med Soc London. 1973;89:279–82.

Burkitt DP. Carcinoma of the colon and rectum. In: Raven RW, editor. Modern Trends in Oncology Part 1 Research in Progress. London: Butterworth; 1973. p. 227–41.

Burkitt DP. Varicose veins among the Masai. Lancet. 1973;1(7808):890.

Burkitt DP. An epidemiological approach to cancer of the large intestine: the significance of disease relationships. Dis Colon Rectum. 1974;17:456–61.

Burkitt DP, Walker AR, Painter NS. Dietary fiber and disease. JAMA. 1974;229:1068–74.

Trowell HC, Painter NS, Burkitt DP. Aspects of the epidemiology of diverticular disease and ischemic heart disease. Am J Dig Dis. 1974.19:864–73.

Burkitt DP, Graham-Stewart CW. Haemorrhoids-postulated pathogenesis and proposed prevention. Postgrad Med J. 1975;51:631–6.

Burkitt DP, Jansen HK, Mategaonker DW, Phillips C, Phuntsog YP, Sukhnandan R. Varicose veins in India. Lancet. 1975;2(7938):765.

Burkitt DP, Tunstall M. Common geography as a clue to causation. Trop Geogr Med. 1975;27:117–24.

Burkitt DP, Tunstall M. Gallstones: geographical and chronological features. J Trop Med Hyg. 1975;78:140–4.

Burkitt DP. Hemorrhoids, varicose veins and deep vein thrombosis: epidemiologic features and suggested causative factors. Can J Surg. 1975;8:483–8.

Goldsmith HS, Burkitt DP. Stool characteristics of Black and White Americans. Lancet. 1975;2(7931):407.

Cranston D, Burkitt DP. Diet, bowel behaviour, and disease. Lancet. 1975;2(7923):37.

Burkitt DP. Dietary fibre and 'pressure diseases'. J Roy Coll Phys London. 1975;9:138–46.

Burkitt DP. Fibre-depleted carbohydrates and disease. Comm Health. 1975;6:190–4.

Burkitt DP. Cancer of the GI tract: Colon, rectum, anus. Epidemiology and etiology. JAMA. 1975;231:517–8

Burkitt DP. Large-bowel cancer: an epidemiologic jigsaw puzzle (editorial). JNCI. 1975;4:3–6.

Painter NS, Burkitt DP. Diverticular disease of the colon, a 20th century problem. Clin Gastroenterol. 1975;4:3–21.

Burkitt DP. Diet and disease, a plea for potatoes. J Indian Coll Phys Surg. 1975;4:141–45.

Burkitt DP. Varicose veins in developing countries. Lancet. 1975;2(7983):472.

Burkitt DP. Fiber deficiency and colonic tumors In: Reilly RW, Kirsner JB, editors. Fiber Deficiency and Colonic Disorders. New York and London: Plenum Medical Book Company; 1975. p. 139–47.

Burkitt DP. Some diseases related to fibre-depleted diets. In: The Man/Food Equation. Proceedings of a Symposium held at The Royal Institution, September 1973. London: Academic Press; 1975. p. 247–256.

Trowell HC, Burkitt DP. Faecal fibre fortunes. BMJ. 1975;3 (5978):305.

Walker AR, Burkitt DP. Colonic cancer--hypotheses of causation, dietary prophylaxis, and future research. Am J Dig Dis. 1976;21:910–7.

Burkitt DP. Varicose veins: facts and fantasy. Arch Surg. 1976;111:1327–32.

Walker AR, Burkitt DP. Colon cancer: epidemiology. Seminars in Oncology. 1976;3:341–50.

Burkitt DP, Walker AR. Saint's triad: confirmation and explanation. S Afr Med J. 1976;50:2136–8.

Burkitt DP, Townsend AJ, Patel K, Skaug K. Varicose veins in developing countries. Lancet. 1976;2(7978):202–3 and (7983):472.

Burkitt DP. A deficiency of dietary fiber may be one cause of certain colonic and venous disorders. Am J Dig Dis. 1976;21:104–8.

Phillips C, Burkitt DP. Varicose veins in developing countries. BMJ. 1976;1(6018):1148.

Balasegaram M, Burkitt DP Stool characteristics and western diseases. Lancet. 1976;1(7951):152.

Burkitt DP. Two blind spots in medical knowledge. Nurs Times. 1976;72:24–7.

Burkitt DP. The role of fibre in human diets. Finska Lakaresallskapets Handlingar Arg. 1976;136:123–27.

Burkitt DP. Epidemiology of some human cancers In: Symington T, Carter RL, editors. Scientific Foundations of Oncology. London: William Heinemann Medical Books; 1976. P. 232–37.

Burkitt DP. The etiological significance of related diseases. Can Fam Physician. 1976;22:63–71.

Burkitt DP. Are our commonest diseases preventable? Prev Med. 1977;6:556–9.

Flynn JF, O'Beirn SF, Burkitt DP. The potato as a source of fibre in the diet. Ir J Med Sci. 1977;146:285–8.

Burkitt DP, Trowell HC. Dietary fibre and western diseases. Ir Med J. 1977;70:272–7.

Burkitt DP, Latto C, Janvrin SB, Mayou B. Pelvic phleboliths - epidemiology and postulated etiology. New Engl J Med. 1977;296:1387–9.

Burkitt DP. Relationships between diseases and their etiological significance. Am J Clin Nutr. 1977;30:262–7.

Burkitt DP. Appendicitis and diabetes. BMJ. 1977;1:1413.

Burkitt DP. Fibre hypotheses. Lancet. 1977;1(8023):1214–15.

Burkitt DP. Burkitt hypothesis – reply. Am J Dig Dis. 1977;22:75.

Burkitt DP. Diet and diseases of affluence. Qualitas Plantarum – Plant foods for man. 1977;27:227–38.

Burkitt DP. Economic development – Not all a bonus. Nutr Today. 1977;76:21–4.

Trowell HC, Burkitt DP. Dietary fiber and cardiovascular disease. Artery. 1977;3:107–19.

Burkitt DP. Epidemiology. In: Reaven RW, editor. Principles of Surgical Oncology. New York and London: Plenum Medical Books; 1977. p. 205–26.

Burkitt DP, Hutt MSR. Epidemiology of cancer. In: Baron DN, Compston N, Dawson AM, editors. Recent Advances in Medicine. London, Edinburgh and New York: Churchill Livingstone; 1977. 17:1–22.

Burkitt DP. Some diseases characteristic of modern western civilisation. In: Logan MH, Hunt EE Jr, editors. Health and the Human Condition: perspectives on medical anthropology. Massachusetts: Duxbury Press; 1977. p. 137–47.

Burkitt DP. Diseases of affluence and diet. In Speaking of Science, Proc Roy Inst. 1977;50:103–13. (?1978)

Burkitt DP. Mechanical effects of fibre with reference to appendicitis, hiatus hernia, haemorrhoids and varicose veins. J Plant Foods. 1978;3:35–40.

Burkitt DP. Dietary fiber, a maintainer of health. Int J Environ Stud. 1978;12:5–8.

Burkitt DP. Colonic-rectal cancer: fiber and other dietary factors. Am J Clin Nutr. 1978;31 Suppl 10:S58–S64.

Burkitt DP. A discarded protection: dietary fibre. Ir Med J. 1978;71:244–7.

Burkitt DP. Fiber and cancer – summary and recommendations. Am J Clin Nutr. 1978;31:S213–15.

Burkitt DP, Trowell HC. Nutritional intake, adiposity, and diabetes. BMJ. 1979;1(6170):1083–4.

Trowell HC, Burkitt DP. Diverticular disease in urban Kenyans. BMJ. 1979;1(6180):1795.

Burkitt DP. Epidemiological features of gastrointestinal cancer. Front Gast Res. 1979;4:86–95.

Burkitt DP. The protective value of plant fibre against many modern western diseases. Qualitas Plantarum, Plant Foods in Human Nutrition. 1979;29:39–48.

Burkitt DP, Moolgaokar AS, Tovey FI. Aetiology of appendicitis. BMJ. 1979;1(6163):620.

Burkitt DP, Meisner P. How to manage constipation with high-fiber diet. Geriatrics. 1979;34:33–5, 38–40.

Burkitt DP. Burkitt's lymphoma. In: Fry J, Gambill D, Smith R, editors. Scientific Foundations of Family Medicine. London: Heinemann; 1979. p. 12–20.

Burkitt DP. Gastrointestinal cancer. In: Frontiers of Gastrointestinal Research 86:1–10 and/or Epidemiological features of gastrointestinal cancer. 1979;4:86–95.

Trowell HC, Burkitt DP. Diverticular disease in urban Kenyans. BMJ. 1979;1(6180)1795.

Burkitt DP. Dietary fibre, a protective from disease. Univ Toronto Med J. 1980;57:34–5.

Burkitt DP. The importance of high fiber: update 1979. In: Proceedings of the third international conference on human functioning. Wichita, Kansas: Biomedical Synergics Institute; 1980. p. 25–37.

Burkitt DP. Fiber in the etiology of colo-rectal cancer. In: Winawer SJ, Schottenfeld D, P Sherlock P, editors. Progress in Cancer Research and Therapy 13, Colorectal Cancer. New York: Raven Press; 1980. p. 13–18.

Burkitt DP. Diseases of affluence. In: Howe GN, Lorraine SA, editors. Environmental Medicine. 2nd ed. London: Heinemann Medical; 1980. p. 143–147

Burkitt D. Diet and its relation to haemorrhoids. Coloproctology. 1980;2:315–16.

Burkitt D, Morley D, Walker A. Dietary fibre in under- and overnutrition in childhood. Arch Dis Child. 1980;55:803–7.

Burkitt DP. The background to the Epstein-Barr virus, In: Giraldo G, Beth E, editors. The Role of Viruses in Human Cancer. New York, Amsterdam: Elsevier, North Holland; 1980. p. 1–6.

Burkitt DP. Colon cancer: the emergence of a concept. In: Spiller GA, Kay EM, editors. Medical Aspects of Dietary Fiber. New York, London: Plenum Medical Book Company; 1980. p. 75–81.

Burkitt DP. The protective properties of dietary fiber. N C Med J. 1981;42:467–71.

Burkitt DP. Hiatus hernia: is it preventable? Am J Clin Nutr. 1981;34:428–31.

Trowell HC, Burkitt DP. Diseases of modern civilization. BMJ. 1981;283 (6301):1266.

Trowell HC, Burkitt DP. Blood-pressure rise with age – a Western disease. Lancet. 1981;2(8248):693–94.

Burkitt DP. Diet and its relation to haemorrhoids. Coloproctol. 1981;80:315–16.

Burkitt DP. No relation of sigmoid volvulus to fiber content of African diet. N Eng J Med. 1981;304:914.

Burkitt DP. Geography of disease. Purpose and possibilities from geographical medicine. In: Rothschild H, Chapman C, editors. Biocultural Aspects of Disease. New York, London: Academic Press; 1981. p. 133–51.

Burkitt DP. Diet and disease. Ir Med J. 1981;74:36–8.

Burkitt DP. Western diseases and their emergence related to diet. S Afr Med J. 1982;61:1013–5.

Burkitt DP. Procedures of unproved value. JAMA. 1982;247:1278.

Walker ARP, Burkitt DP. Plant fiber intake in the pediatric diet. Pediatr. 1982;69:130–1.

Burkitt DP. Dietary fiber: is it really helpful? Geriatrics. 1982;37:119–26.

Burkitt DP. Diet and diseases of affluence. In: Rees AD, Purcell HT, editors. Disease and Environment. London: John Wiley and Sons; 1982. p. 167–73.

Burkitt DP. Dietary fibre a protective against disease. In: Jeliffe EFP, Jelliffe DB, editors. Disease in Adverse Effect of Foods. London, New York: Plenum Press; 1982. p. 483–95.

Burkitt DP. Diseases of affluence. In: Rose J, editor. Nutrition and Killer Diseases. The Effects of Dietary Factors on Fatal Chronic Diseases. Park Ridge, New Jersey: Noyes Publications; 1983. p. 1–7.

Burkitt DP. The discovery of Burkitt's lymphoma. Cancer. 1983;51:1777–86.

Burkitt DP. Eradicating sources or removing results. Postgrad Med J. 1983;59(690):232–5.

Burkitt DP. The development of the dietary fibre hypothesis. In: Birch GG, Parker KJ, editors. Dietary Fibre 2. London: Applied Science Publishers; 1983. p. 21–27.

Burkitt DP. Dietary fiber. In: Bland J, editor. Medical Applications of Clinical Nutrition. Connecticut: Keats Publishing; 1983. p. 269–86.

Burkitt DP. Nutrition, Health and Food supplies. In: Academy of Independent Scientists. Food and Climate Review 1982–83. 970 Aurora St, Boulder, Colorado 80309; 1983. p. 23–6.

Burkitt DP. The discovery of Burkitt's Lymphoma. Charles Mott award. Cancer. 1983;51:1777–86.

Burkitt DP. Non-infective disease of the large bowel. Brit Med Bull. 1984;40:387–9.

Burkitt DP. Western diseases and their emergence related to diet. In: Symposium of the Hans Snyckers Institute: University of Pretoria; 1984. p. 12–18.

Burkitt DP. Changing patterns of disease associated with increased affluence. Rev Environ Health. 1984;4:121–31.

Burkitt DP. Etiology and prevention of colorectal cancer. Hospital Practice (Office Edition). 1984;19:67–77.

Trowell HC, Burkitt DP. Bran yesterday…bran tomorrow. BMJ. 1984;289 (6442):436.

Burkitt DP. Why isolate cancer? In: Butterworth CE, Hutchison ML, editors. Nutritional Factors in the Induction and Maintenance of Malignancy. New York, London: Academic Press; 1984. p. 1–9.

Burkitt DP. Possibilities and priorities of Cancer prevention. In: Giraldo G, Beth E, editors. The Role of Viruses in Human Cancer. Amsterdam: Elsevier Science Publishers; 1984. Vol 2, p. 1–9.

Burkitt DP. Fiber as protective against gastrointestinal diseases. Am J Gastro. 1984;79:249–52.

Burkitt DP. Religious fundamentalism in medical school. BMJ. 1984;(6457) 289:1624.

Burkitt DP. The beginnings of the Burkitt's lymphoma story. IARC Scientific Publications. 1985;60:11–15.

Burkitt DP, Clements JL Jr, Eaton SB. Prevalence of diverticular disease, hiatus hernia, and pelvic phleboliths in black and white Americans. Lancet. 1985;2(8460):880–1.

Jones BA, Demetriades D, Segal I & Burkitt DP. The prevalence of appendiceal fecaliths in patients with and without appendicitis. A comparative study from Canada and South Africa. Ann Surg. 1985;202:80–2.

Burkitt DP. Changing patterns of disease associated with increased affluence. In: James GV, editor. Review of Environmental Health. Israel: Freund Publishing House; 1985. Vol 1V No. 2.

Burkitt DP. Varicose veins, haemorrhoids, deep vein thrombosis and pelvic phlebo-liths. In: Trowell H, Burkitt D, Heaton K, editors. Dietary fibre, fibre depleted food and disease. London: Academic Press; 1985. p. 317–30.

Burkitt DP. The pros and cons of economic development. In: Kaplan RH, Criqui M, editors. Behavioural epidemiology and disease prevention. Nato ASI Series. New York: Plenum Press; 1985.

Trowell HC, Burkitt DP. Physiological role of dietary fiber- A 10-year review. J Dent Child. 1986;53:444–47 and Boletin-Asociacion Medica de Puerto Rico 78:541–44.

Burkitt DP. Forward. In: Vahouny G, Kritchevsky D, editors. Dietary Fiber. Basic and clinical aspects. New York: Plenum Press; 1986. p. ix–xii.

Burkitt DP. Fibre – its contribution to surgical problems. Postgrad Doctor. 1986;9:254–57.

Trowell HC, Burkitt DP. The development of the concept of dietary fibre. Mol Aspects Med. 1987;9:7–15

Burkitt DP. Dietary fibre – historical aspects. Scand J Gastroent. 1987;22:S129:10–13.

Burkitt DP. Dietary fiber and cancer J Nutr. 1988;118:531–3.

Burkitt DP. Concepts of Western diseases. Geriatric Med. 1988; Suppl. p. 5–8.

Burkitt DP, Eaton B. Putting the wrong fuel in the tank. Nutrition. 1989; 5:189–91.

Burkitt DP. Relationships between malignant and non-malignant disease of Western culture. In: Levine AS, editor. A Cancer Growth and Progression Series. Etiology of Cancer in Man. Boston, London: Kluwer Academic Publishers; 1989.

Burkitt DP. Hugh Trowell. In: Leeds AR, Burley VJ, editors. Dietary Fibre Perspectives 1 reviews and bibliography. London, Paris: John Libby; 1990.

Burkitt DP. In: Pioneers in Paediatric Oncology. Taylor G, editor. Houston, Texas: The University of Texas M.D. Anderson Cancer Center, 77030; 1990.

Geelhoed GW, Burkitt DP. Varicose veins: a reappraisal from a global perspective. South Med J. 1991;84:1131–4.

Temple NJ, Burkitt DP. The war on cancer--failure of therapy and research: discussion paper. J Roy Soc Med. 1991;84:95–8.

Burkitt DP. An approach to the reduction of the most common western cancers. The failure of therapy to reduce disease. Arch Surg. 1991;126:345–47.

Burkitt DP. Potential prevention of cancer. Singapore Med J. 991; July 47–49.

Burkitt DP. Lessons for health from our palaeolithic ancestors. Cardiovasc Risk Fact. 1991;1:353–57.

Geelhoed GW, Burkitt DP. First order prevention and second order treatment. Health care in the developed world is too long on the latter. South Med J. 1992;85:119–20.

Burkitt D. Epidemiology of cancer of the colon and rectum. Dis Colon Rectum. 1993;36:1071–82.

Temple NJ, Burkitt DP. Towards a new system of health: the challenge of Western disease. J Comm Health. 1993;18:37–47.

Burkitt D. Unpromising beginnings. J Ir Coll Physicians Surg. 1993;22:36–38

Burkitt DP. Dietary fiber – forgotten fraction to significant dietary component in thirty years. Bibliography and Reviews, An International Newsletter on Carbohydrates, Fiber and Health. 1993;1:1.

Burkitt D. Historical Aspects. In Dietary fiber in health and disease. In: Kritchevsky D, Bonfield C, editors. St Paul Minnesota: Eagan Press; 1995. p. 3–10.

Cook-Mozaffari P, Newton R, Beral V, Burkitt DP. The geographical distribution of Kaposi's sarcoma and of lymphomas in Africa before the AIDS epidemic. Brit J Cancer. 1998;78(11):1521–8. Please leave space of one line before next heading.

Part 3. Other writing: Including unverified, untraceable or incomplete references. DP Burkitt author of all items. Where there are additional authors, their names are given but order cannot be guaranteed. n = 117

Removal of plaster casts. BMJ. 1940; 2

Spot diagnosis. Medicine Illustrated. 1951. p. 471.

Cancer syndrome with possible viral aetiology. Panorama, 1963, September, 4

Can tumours be transmitted by insects? Triangle The Sandoz journal of Medical Science. 1964;6:222–26.

Are we too conservative? Nursing Mirror. 1965; July 369–71.

Burkitt's Lymphoma. A study in medical detection. Abbottempo. 1967;4:26–33. (In colour)

The importance of geographical pathology. Iraqi Cancer Bull. 1967;1:37–43.

Possible relationships between the African lymphoma and acute leukaemia: third guest lecture delivered by invitation at the Hospital for Sick Children, Great

Ormond Street, London, WC1; 1967. Published for the Leukaemia Research Fund by Queen Anne Press. 28 pages. BNB GB 6706941.

The African lymphoma – epidemiology and therapeutic aspects. Possibly chapter in a book but source unknown. 1968; 321–30.

Epidemiological studies of cancer in Africa. Medical News Magazine. 1968; November 10–12.

The most curable children's cancer. Mother and Child. 1968;39:7.

The proper study of mankind in "cancer research" In: People and Cancer and the problem of relevance in cancer research. 1969;83–9.

Some cancer patterns in Africa. Medicine Today. 1969;3:15–9.

Diet and non-infective disease of the large bowel. The Uganda Practitioner. 1970;3:136.

Limitations of scientific medicine. World Medicine. 1970;6:32–3.

Attitude: a neglected priority. J Christian Med Assoc India. 1970; XLV: 192–94.

Recognition and treatment of Burkitt's lymphoma The Uganda Practitioner. 1971;4:12–14.

Some lessons from cancer patterns in Africa. Medical Practitioner. 1971;20.2.71:52–8 and 6.3.71:70–4.

Diet and disease. World Medicine. 1971; February: 71–9.

Medical missions adapting to changing circumstances. Saving Health. 1972;11:61–4.

The disease that goes with prosperity. World Medicine. 1972;8:43–9.

The importance of fibre in food. Middlesex Hosp J. 1972;72:32–5.

Research opportunities in up-country hospitals. Nairobi J Med. 1972;5:7–9.

Some diseases associated with economic development. Cambridge University Med J. 1972;6:10–6.

Epidemiology of cancer of the large bowel. Liverpool Medical Institution, Transaction and Report. 1972;23–6.

Second Maurice Grimes Memorial Lecture, Canadian Cancer Society News Letter. 1972;8–10.

Health and environment. World Medicine. 1972. February: 90–1.

Some diseases related to deficiency of dietary fibre. Update. 1973; April: 1213–20.

Some thoughts on tackling disease at source. Assoc Old Adelaide Students. 1973; 17(October):9–11.

Humble pie. World Medicine. 1973;8:26–7.

Fibre – is it a dietary requirement? Nutritional Problems in a Changing World. London: Applied Science Publications; 1973.

Western civilisation, diet and disease. Drug Therapy. 1973; (January) 51–62.

Bran, Nature's problem solver. Life and Health. 1973;89:25–6.

Are some of Scotland's commonest diseases potentially preventable? Surgo. 1973;41:8–10.

Dead wood. World Medicine. 1973;9:70–71.

Some diseases characteristic of modern western civilisation related to diet. Medikon/ Int III. 1974;8:6–10.

1975 Neglected necessities. ISM. 1975;76:1–3.

M.D. blames low fibre diet for bowel diseases in North Americans. Can Fam Physician. 1975; May 21.

Health Hazards of economic development. Spectrum. 1975;13:43–6.

Food fiber and disease prevention. Compr Ther. 1975;1:19–22.

Questionable dogma. World Medicine. 1975; November 5th :29–31.

Diet and disease, dietary fibre (a) Diet determined pressure phenomena. J Roy Soc Health. 1975; August.

Dietary fibre and disease. Chemtech. 1976; March: 188–9.

Not by bread alone. St Mary's Hospital Gazette. 1976;5:18–9.

Surgical disease and fibre-depleted diets. Kellogg Nutritional Symposium Toronto: 1976; 32.

Diet and diseases of affluence. Rural Life. 1976;21:3–9.

The role of refined carbohydrates on large bowel behaviour and disease. McCarrison Society, London. 1976.

G.I. disease and fibre-depleted diets. Dimensions in Health Service. 1976;53:27–9.

Economic development – not all Bonus. Nutr Today. 1976;11:6–13.

Food fibre: Benefits from a surgeon's perspective. Cereal Foods World. 1977; 22: No. 1.

If your diet lacks fibre… Life and Health. 1977;92:21–4.

Link between low-fiber diets and disease. Human Nature. 1978;1:34–41.

Dietary fibre. Some mechanical effects of fibre-depleted diets. In: Proceedings of the Niles Symposium. Dalhousie University: The Nutrition Society of Canada; 1977. p. 5–12.

Les phlebolithes pelviens. Cah Med. 1977; T3 No. 17:18–19 (?)

Trowell HC, Burkitt. Fibre and disease. On Call. 1977; October 27.

The link between low-fibre diets and disease. Human Nature. 1978; Dec 1:34–41.

Burkitt's lymphoma. Scientific Foundations of Family Medicine. 1978;17–20.

Fibre in our diet. British Society of Baking 24th Annual Meeting, London, 1978; October 30th.

Fibre: don't wait for the proof. General Practitioner. 1979; Jan 12th p. 26.

A dietary protection against disease. St Thomas's Hospital Gazette. 1979; Spring: 9–12.

Medical priorities and the nature of man. Ministry. 1979; April: 25–7.

Epidemiological aspects of haemorrhoids, In: Haemorrhoids – Current concepts in management and causation. Symposium published by Royal Society of Medicine. 1979. (?)

Appendicitis. The present state of knowledge. Norgine. 1980; No. 9.

Dietary fiber – essential or disposable. Osteopathic Medicine. 1980; May 35–4.

Prevention a priority over cure. The Bulletin. Business and Estate Planning Consultants. 1980; November, 27–29.

Dietary fiber – a protection against disease. In: Beecher GR, editor. Human Nutrition Research (BARC Symposium No. 4). Totowa: Allanheld. 1981; p. 43–8.

Research. Proc Assoc Surgeons E Africa. 1981;121–23.

Disease prevention. Science and Public Policy. 1981; December 422–43.

Medical progress. The Veins. GP. 1981; May 15 p. 35.

Eradicating causes of disease. Primary Care. 1982;13, 4–5.

The fiber story today. Update. 1982; January 1ˢᵗ, 37–42.

Where are you going? London: Christian Medical Fellowship; 1982. ISBN 0906747104 (Unbound and only 11 pages.)

The development of the fibre hypothesis. Fibre Forum. 1982; Spring, 1–3.

How poor nations can contribute to the welfare of the rich. International Health. 1982; Oxfam, February.

Reaction in crisis. Saving Health. 1983;22:15–6.

Dietary fibre – A way of preventing Western Diseases? Fibre Forum. 1983 or 1984; Symposium Issue 1:1–6.

Mixed cultures influence disease. General Practitioner. 1983; March 18, 47–9.

Dietary fiber and cancers of the Western World. Health and Healing. 1983; Spring, 6–9.

Diseases of affluence. Irish Times. 1983; March 28–30, three articles.

A sarcoma of the jaws in African children. Citation Classics, Current Contents. 1983; 21 :20

Alimentation et maladies dans les societies Occidentales. Medicine et Nutrition. 1983; 19:83–5.

Looking for reasons. Medical Missions of Mary. 1983; Spring 23–5.

Diet and disease in affluent societies. Healthline. 1983;11:5–6.

Kost fiber ett skydd mat ajukdom. Tike Information. 1983;5:1–4.

Your health in their hands. School Science Review. 1983; December, 264–69.

Carton and content. Medical Missions of Mary. 1983; September, 26–8.

La constipation. I.M.S. 1983;23:8–10.

Etiology and prevention of colorectal cancer. Hospital Practice. 1984; February, 67–77.

Where should our priorities be? In Forum I.R.C.S. Med Sci 1984;12:99–100.

Where are you going? Saturday Evening Post 1984; April 66–8.

Burkitt's route to avoid constipation. Patient Care. 1984; July 15, 252–53.

Lack of funds for preventing cancer. Letter to the Times (London). 1984; October 13ᵗʰ.

Are breast, bowel and lung cancers preventable? Primary Care and Cancer. 1984; December 14–6.

Headed in the wrong direction. Lancet. 1984;2(8417 or 8418):1475.

Man, carton or content. J Christian Med Soc. 1985; Winter p. 16.

Tackling problems at source. In: Serv Med. 1985;31:15–7.

What are you aiming at? Guidelines No. 89 CMF. 1985.

Forum – Where should our health priorities lie? Barts Hospital Postgraduate Printout. 1985.

The protective role of dietary fibre. Reckitt & Colman. 1985.

The development of interest in dietary fibre. In: Williams DL, editor. The status of dietary fibre. Physiology and current practice. Reckitt and Colman Foundation. 1985.

The case for fibre. Pulse. 1985; November 2;31–6.

Burkitt DP, Hutt MSR. Maladies of the Christian church. I.S.M. 1986;32:21–4.

Nutritional factors in diseases of Western civilisation. Proc Brit Students Health
    Assoc. 1986;27–31.
Trowell HC, Burkitt DP. Physiological role of dietary fibre. A ten year review.
    Contemporary Nutrition, General Mills Nutrition Department. 1986;11:7.
Burkitt DP, Browne S. Striking a balance in medicine. Saving Health. 1986; 25:2–4.
Examining the maladies of the church. CMSJ. 1987. Fall 1986;13–4.
Browne S, Burkitt DP. What's the difference? CMSJ. 1987; winter 22–5.
First turn off the tap. Evangelicals Now. 1987; August: p. 5.
Lessons for health from our distant ancestors. Scientific J Georgia Baptist Med
    Centre. 1988;8:16–8.
Returning – the only way forward. CMFJ. 1988; Oct: 26–7.
Dietary fibre as protective against large bowel cancer and diabetes. Atti del primo
    Convengo Internazionale di Patologies Ambietale, Napoli. 1988. p. 11–23.
Don't just mop the floor; turn the taps off. McCarrison Soc Newsletter. 1989.
Biography of Hugh Trowell. McCarrison Society. 1989.
Man…Carton or Content. Pamphlet of 12 pages reviewing the importance of pre-
    vention and includes a biblical perspective. Lifestyle Medicine Institute, Loma
    Linda, California USA. 1990.
Flexibility of Interpretation. (Letter) CMFJ. 1990; April; 16–7.
True understanding depends on interpretation. Faith and Thought. 1990; October 8th.
Are our commonest diseases preventable? The Pheros of Alpha Omega Alpha.
    1991;51:19–21.
Direction determines destination. Assoc Old Adelaide Students J. 1994;9–11.

Appendix 4: Books and Papers About Denis Parsons Burkitt and His Work, Biographical Memoires, Tributes, Principal Obituaries

Part 1. Books

Glemser B. The Long Safari. London: The Bodley Head; 1970.
Published in the United States as Mr Burkitt and Cancer.
Kellock B. The Fibre Man. Tring, Herts: Lion Publishing; 1985.
Nelson ER. Burkitt, Cancer, Fiber. New York: Teach Services Inc; 1998.Please
    leave space of one line before next heading.

Part 2. Papers in medical and scientific journals/chapters in books

Lee TC. Seeing the wood for the trees – the early papers of Denis Burkitt. J Ir Coll
    Physicians Surg. 1970; 25:126–130.
The American Cancer Society. Denis Parsons Burkitt (1911–) Classics in Oncology.
    1972:22;345–348. https://acsjournals.onlinelibrary.wiley.com/doi/epdf/10.3322/
    canjclin.22.6.345
Galton L. The truth about fiber in your food. New York: Crown Publishers Inc; 1976.

Bendiner E. A missionary, heart and soul. Hosp Pract. 1970; July 15th: 166–186.

Holmberg SB. Kartlade malignt lymfon hos afrikanska barn-föregångare även inom kostfiberforskningen. In the series The man behind the syndrome. Lakartdningen, Stockholm 1991:88;4333–34.

Corman ML. Denis Parsons Burkitt 1911–1993. Dis Colon Rectum. 1993:36:1071–1082. Combined with a reprinting of his Classic Article, Burkitt DP (1971) Epidemiology of cancer of the colon and rectum. Cancer. 1971:28;3–13.

Story JA, Kritchevsky D. Denis Parsons Burkitt (1911–1993) J Nutr. 1994: 124:1551–54. https://doi.org/10.1093/jn/124.9.1551

Jay V. Extraordinary epidemiological quest of Dr Burkitt. Pediatr Devel Pathol. 1998:1:562–564.

Coakley D. Denis Burkitt and his contribution to haematology/oncology. Brit J Haem. 2006:135:17–25. doi:10.1111/j.1365-2141.2006.06267. See also Professor Davis Coakley. Trinity Monday Memorial Lecture; Denis Burkitt; An Irish Scientist and Clinician working in Africa. 2011:1–22.

Smith O. Denis Parsons Burkitt CMG, MD, DSc, FRS, FRCS, FTCD (1911–1993) Irish by Birth, Trinity by the grace of God. Brit J Haem. 2012:156:770–76.

Denis Burkitt. In Pioneers in Paediatric Oncology, Ed Taylor G. University of Texas M.D. Anderson Cancer Center, Houston (1990) 39–44.

Cady B. Denis Parsons Burkitt, an overlooked surgical oncologist. Ann Surg Oncol. 2018:25:1112–15. https://doi.org/10.1245/s10434-018-6386-9

Cummings JH, Engineer A. Denis Burkitt and the origins of the dietary fibre hypothesis. Nutr Res Rev. 2018:31(1);1–15. https://doi.org/10.1017/S0954422417000117

Esau D. Denis Burkitt: A legacy of global health. J Med Biog. 2019: 27::4–8 First Published September 28, 2016.

O'Keefe SJ. The association between dietary fibre deficiency and high-income lifestyle-associated diseases: Burkitt's hypothesis revisited. Lancet Gastroenterol Hepatol. 2019: 4:984–96.Please leave space of one line before next heading.

Part 3. Biographical memoires

Epstein A, Eastwood MA. Denis Parsons Burkitt. Biographical Memoirs of Fellows of the Royal Society. London, The Royal Society; 1995:41:89–102.

Epstein A. Denis Parsons Burkitt (1911–1993) Oxford Dictionary of National Biography. 2004. https://doi.org/10.1093/ref:odnb/57333

Andrews H. Burkitt, Denis Parsons (1911–1993). Dictionary of Irish Biography. 2009: 2:77–78.Please leave space of one line before next heading.

Part 4. Tributes and other articles

Nelson CL, Temple NJ. 'Denis, as he was known to his friends, was undoubtedly one of the great physicians of the twentieth century'. Tribute to Denis Burkitt. 1994. J Med Biog; 2:180–183

Coakley D. Unique achievement of a Deeply Humble Man. Trinity Medical News. 1995;12:3.

O'Farrall F. A report on Dr Denis Burkitt OD for TMH-S. 2002. Trinity College Library, Manuscripts & Archives Research Library 11387/6/41

Nicole Blazek Hem/Onc Today. July 25, 2008 'His approach to medicine led to the discovery of Burkitt's lymphoma and established the importance of a fiber-rich diet.

https://www.healio.com/hematology-oncology/news/print/hemonc-today/%7B9c3f27de-b67e-4dc1-93cd-d9566f6e1692%7D/denis-parsons-burkitt-19111993

http://www.whonamedit.com/doctor.cfm/2199.html Denis Parsons Burkitt

Scott Harrison – Mercy (2005) "he scratch the eye" – a study of Burkitt's lymphoma.   http://www.onamercyship.com/2005/04/he-scratch-eye-study-of-Burkitts_06.html

Part 5. Principal obituaries Please leave space of one line before this heading.

The Times March 27, 1993. Denis Burkitt. '…making him one of the few men with two "citation classics" in unrelated fields of medicine'.

The Daily Telegraph March 27, 1993. Denis Burkitt 'the medical scientist,…discovered the childhood cancer now known around the world as "Burkitt's Lymphoma"'.

The Guardian March 27, 1993. McColl I, Fibre in our diet. '…that rare breed of clinicians to become a Fellow of the Royal Society and a household name'.

The Independent April 3, 1993. Heaton K, Denis Burkitt. 'the man who forced the world to take dietary fibre seriously'.

British Medical Journal April 10, 1993. DP Burkitt 'He was a superb lecturer, and his Irishness showed in his love of an audience'.

The Lancet April 10, 1993 341:951–952 Heaton K, Denis Burkitt. 'thanks largely to Burkitt, the science of nutrition was galvanised into new life and people's eating habits all over the western world changed drastically'.

New York Times April 16, 1993. Altman LK, Dr Denis Burkitt is dead at 82; Thesis changed diets of millions. 'His major medical contributions came from a passion for plotting diseases on maps'.

Irish Times March 1993 Denis Burkitt. 'widely acknowledged as one of the greatest Irish doctors of the twentieth century'.

Lifeline March-April 1993 8:1 Denis P. Burkitt 1911–1993 '…undoubtedly one of the most distinguished physicians and health educators of the 20th century'.

Medical Monitor 14 04 1993 Michael O'Donnell Credits Burkitt with fibre.

Lectures to medical societies attracted a record turnout followed by 'a local sell-out of bran'

Le Monde, Mardi 11 Mai 1993 Lenoir G La mort de Denis Burkitt 'Un grand précurseur de la cancérologie modern'

Medical Update May 1993 Lymphoma and fibre.

Dietary Fiber Bibliography and Reviews 1993 1:2 Hamilton WDB The fibre of life. In appreciation of the late Denis P Burkitt CMG, MD, FRCS Ed, DSc (Hon), FRCS I (Hon), FRS. '…inspiring young and old alike to reappraise their dietary habits and their lifestyles in the interests of their whole well-being'.

MD Magazine A journal for Physicians and Family Practice. June 1993. The Burkitt
   Legacy, Bendiner E. 'A tireless surgeon, scientist, nutritionist, and evangelist'
Bulletin of Tropical Medicine and International Health. 1993 Hutt MH, Vol 1 No. 2.
   Denis Parsons Burkitt FRCS, CMG, FRS, 1911–1993. In Uganda 'He showed
   that isolated doctors could contribute to research and in so doing raised the status
   and effectiveness of medical care at District level'.
The Association of Old Adelaide Students 1993 Denis Burkitt 'He remained all his
   life a devout Christian, and the tenets of his faith were the guiding principles of
   his lifetime. "I shall not look upon his like again". J.N.P.M.'.[9]
Medical Update May 1993 Denis Parsons Burkitt – Surgeon, humanitarian, and
   friend. 'It is unlikely that any member of the medical profession had had a more
   direct effect on the health of most, if not all, of our readers than Dr Denis Burkitt'.
Medical Missionary Association June 1993 Peter Bewes. 'The medical profession
   has lost one of its most illustrious Christian practitioners ...We don't need larger
   hospitals but larger stools!'
The Saturday Evening Post (Palm Coast, Florida, USA) 1995 March/April 54 et sec.
   Cory Servaas, Dr Denis Burkitt: A passion for preventing disease. '...gained
   worldwide recognition for his ingenious discovery of what is now known as
   Burkitt's lymphoma' and '...how fiber in the diet could prevent many common
   diseases'.
From Denis' diary 1985, at the back:

> 'The only thing of lasting importance in a man's obituary is whether he was a
> child of God and has consequently gone to his Father. To die young and go to
> one's Father is no disaster. To die old and go into darkness is tragedy'.

---

[9] JNP Moore qualified at the same time as Denis in 1935 and was the academic star of the year
being awarded the Hudson Scholarship and Gold Medal. See Chap. 4, first page.

Name Index

© The Author(s), under exclusive license to Springer Nature Switzerland AG 2022   401
J. H. Cummings, *Denis Burkitt*, Springer Biographies,
https://doi.org/10.1007/978-3-030-88563-2

General Index

© The Author(s), under exclusive license to Springer Nature Switzerland AG 2022
J. H. Cummings, *Denis Burkitt*, Springer Biographies,
https://doi.org/10.1007/978-3-030-88563-2

Subject Index

© The Author(s), under exclusive license to Springer Nature Switzerland AG 2022
J. H. Cummings, *Denis Burkitt*, Springer Biographies,
https://doi.org/10.1007/978-3-030-88563-2